MATHLETICS

MATHLETICS

HOW GAMBLERS, MANAGERS,
AND FANS USE
MATHEMATICS IN SPORTS

2ND EDITION

WAYNE L. WINSTON, SCOTT NESTLER, AND KONSTANTINOS PELECHRINIS

PRINCETON UNIVERSITY PRESS • PRINCETON & OXFORD

Published by Princeton University Press
41 William Street, Princeton, New Jersey 08540
6 Oxford Street, Woodstock, Oxfordshire OX20 1TR

press.princeton.edu

All Rights Reserved
ISBN 9780691177625
ISBN (e-book) 9780691189291

Library of Congress Control Number: 2021949227

British Library Cataloging-in-Publication Data is available

Editorial: Susannah Shoemaker, Diana Gillooly, and Kristen Hop
Production Editorial: Nathan Carr
Text Design and Cover: Lauren Smith
Production: Jacquie Poirier
Copyeditor: Bhisham Bherwani

This book has been composed in Adobe Text Pro, Bee Four, and Courier New
Printed on acid-free paper. ∞
Printed in the United States of America

10 9 8 7 6 5 4 3 2

"In God we trust; all others must bring data."

—W. Edwards Deming

CONTENTS

PART VI METHODS AND MISCELLANEOUS

PREFACE

When WW (Wayne Winston) began writing the first edition of *Mathletics* in 2005, many North American sports teams did not have analytics departments and the first annual famed M.I.T. Sloan School Sports Analytics Conference had not taken place. Very few schools taught sports analytics courses. Now every major (baseball, football, basketball, and hockey) North American sports team has an analytics department, tickets to the Sloan School Conference quickly sell out, and many schools teach sports analytics courses. Google searches for the term "sports analytics" have quadrupled since 2007, and surely the number of scholarly publications involving sports analytics has grown exponentially. New data based on cameras in venues and fitness devices worn by athletes have opened up new areas of analysis. The joke is that some teams (like the Houston Rockets) have more analytics personnel than players!

The first edition of *Mathletics* was well received, but with the rapid development of sports analytics, the time seems ripe for a second edition. I am fortunate to have two great co-authors: Scott Nestler (SN) of Notre Dame and Kostas Pelechrinis (KP) of the University of Pittsburgh.

If you want a self-contained introduction to the use of math in sports, we feel this is the book for you. No prior knowledge is assumed, and even a bright high school student should be able to absorb most of the material. Even if your career does not involve sports, we are sure that anyone using analytics in his or her career will benefit from learning how analytics lead to better outcomes for sports teams. We also think that you will watch your favorite sporting events with a new perspective.

WHAT'S NEW

We have added 17 new chapters and made substantial changes to all the chapters in the first edition. A summary of the major changes follows:

- We have updated player references. (Mike Trout, Clayton Kershaw, and James Harden play key roles!)
- We have added extensive discussion of the camera data used in baseball, basketball, and football.
- We walk the reader through how to compute WAR (Wins Above Replacement) in baseball and basketball.
- Newer basketball player evaluation metrics such as Real Plus Minus and RAPTOR are included.
- Sanjurjo and Miller's non-intuitive new work on the hot hand in sports is discussed.
- New chapters have been added on soccer, hockey, volleyball, golf, and even e-sports.
- The gambling chapters have been updated with more recent data, and a chapter on Calcuttas is included.
- The criteria used for selection to the NCAA basketball tournament and the College Football Playoff are discussed.
- Daily fantasy sports are discussed.
- Newer chapters cover the newest methods utilized in sports analytics, including collecting and visualization of data, Bayesian analysis, matrix factorization, and network analysis in sports.

BOOK RESOURCES

We would also like to make an important note about the software tools used throughout the book. While we have presented the analysis through calculations in Excel, this was a choice we made after long deliberation. Given that the focus of the book is on the analytical approaches used to gain insights and make decisions, the specific tool used to implement these is of secondary focus. Furthermore, using Excel allows for democratizing accessibility of the material to people not proficient in programming, which can include, for example, high school students and non-STEM graduates. Nevertheless, understanding the importance of programming skills in today's world, we have also provided a companion website and code repository, where we have provided the implementation in Python of the most challenging analytical tasks covered in the book. While these are not meant to teach from scratch how one can program in Python, we hope that they will be an additional valuable resource for people with minimal programming background. Whenever possible, the code repository content will also be dynamic, and periodically updated with more recent data.

The website for this book is http://www.mathleticsbook.com. You will be able to find a variety of resources there, including errata, datasets, Excel files, Python scripts, and R scripts (forthcoming). Code will also be available in a GitHub repository at https://github .com/mathletics-book/.

CONTACT US

We hope you enjoy this book as much we enjoyed writing it. Feel free to contact us via email.

- Wayne Winston (winston@indiana.edu)
- Kostas Pelechrinis (kpele@pitt.edu)
- Scott Nestler (scott@nestler.com)

ACKNOWLEDGMENTS

Our thanks go to our former editor Vickie Kearn, whose unwavering support of both editions has been truly gratifying. We are eternally grateful to her successor, Susannah Shoemaker, whose patience and support were invaluable throughout the writing process.

Wayne would like to thank his seventh grade social studies teacher from Livingston, New Jersey, the late Thomas Spitz, who taught him the art of outlining. Most of all, he would like to recognize his best friend and world-class sports handicapper Jeff Sagarin. His discussions with Jeff about sports and math have always been stimulating. If he had not known Jeff, this book would not exist.

Kostas would like to thank Mike Lopez, for providing permission to use and present results from the first Big Data Bowl's NFL tracking data. He would also like to thank Kirk Goldsberry for the interesting discussions on basketball that many times led to interesting research problems, some of which are covered in the book. Kostas would also like to thank his Ph.D. mentor Srikanth Krishnamurthy for instilling in him genuine scientific curiosity, as well as Michalis Faloutsos for useful advice and support throughout the years. Finally, Kostas wants to thank his department chairs through the years, David Tipper, Martin Weiss, and Prashant Krishnamurthy, for not putting any obstacles before him to venture to research areas beyond the norm.

Scott would like to thank a number of students, both undergraduate and graduate, who contributed to revisions to existing chapters, including: Mark Giannini, Chase Thompson, Dan Verzuh, Charlie

Puntillo, Katie McCullough, and Sebastian Amato. Their assistance in collecting data, revising existing models, and updating for younger readers was invaluable. Comments and suggestions from numerous students in his Sports Analytics courses (for extra credit) helped identify and correct errors in early drafts of chapters. He would also like to thank his co-authors for their patience with him during the writing process when he fell behind and wasn't the best collaborator. Finally, Scott thanks his family (Kristin, Anna, and Sophia) for the time and attention that this project took away from them in the form of unbalanced carpool responsibilities, missed dance competitions, etc.

ABBREVIATIONS

2B:	Double
3B:	Triple
3PT:	Three point
AB:	At Bat
ACC:	Atlantic Coast Conference
API:	Application Programming Interface
AST:	Assists
AV:	Approximate Value
BARKS:	Bar-Eli, Azar, Ritov, Keidar-Levin, and Schein
BB:	Walk
BCS:	Bowl Championship Series
BLK:	Blocks
BSB:	Berri, Schmidt, and Brook
BT:	Bradley–Terry
C:	Center
CFP:	College Football Playoff
CLG:	Chiappori, Levitt, and Groseclose
CM:	Carter and Machol
CPAE:	Completion Percentage Above Expectation
CS:	Caught Stealing
CSW:	Cabot, Sagarin, and Winston
D:	Down
D:	Double
DBVD:	Decroos, Bransen, Van Haaren, and Davis
DFS:	Daily Fantasy Sports

DICE:	Defense-Independent Component ERA
DPY/A:	Defensive Passing Yards (allowed) per Attempt
DR:	Defensive Rebounds
DRP:	Defensive Rebounding Percentage
DRPM:	Defensive Real Plus Minus
DRY/A:	Defensive Rushing Yards (allowed) per Attempt
DTO:	Defensive Turnovers
DTPP:	Defense Turnover Percentage
E:	Error
EFF:	Efficiency
EFG:	Effective Field Goal Percentage
ENG:	Empty Net Goal
EPV:	Expected Points Value
ERA:	Earned Run Average
ESPN:	Entertainment and Sports Programming Network
F:	Forward
FG:	Field Goal
FGA:	Field Goal Attempted
FIBA:	Fédération Internationale de Basketball (International Basketball Federation)
FMBG:	Franks, Miller, Bornn, and Goldsberry
FOW%:	Face-off Win Percentage
FP:	Fantasy Points
FPI:	Football Power Index
FT:	Free Throw
FTA:	Free Throw Attempts
FTR:	Free Throw Rate
G:	Guard
GIDP:	Grounding Into Double Play
GLM:	Generalized Linear Model
GP:	Games Played
H:	Hit
HBP:	Hit by Pitch
HR:	Home Run
HTML:	Hypertext Markup Language

Innouts:	Number of innings a fielder faces
IP:	Internet Protocol
JSON:	Javascript Object Notation
K:	Strikeout
LC3:	Left Corner Three
LoL:	League of Legends
LW:	Linear Weights
MAD:	Mean Absolute Deviation
MC:	Monte Carlo
MCL:	Medial Collateral Ligament
MLB:	Major League Baseball
MLR:	Multiple Linear Regression
MLS:	Major League Soccer
MOBA:	Multiplayer Online Battle Arena
MPG:	Minutes per Game
NBA:	National Basketball Association
NCAA:	National Collegiate Athletic Association
NET:	NCAA Evaluation Tool
NFL:	National Football League
NHL:	National Hockey League
NMF:	Nonnegative Matrix Factorization
OAA:	Outs Above Average
OBP:	On-Base Percentage
OBSO:	Off-Ball Scoring Opportunities
OEFG:	Opponent Effective Field Goal
OFTR:	Opponent Free Throw Rate
OL:	Offensive Line
OLR:	Ordinal Logistic Regression
OppFGA:	Opponent Field Goal Attempts
OppFTA:	Opponent Free Throw Attempts
OppOREB:	Opponent Offensive Rebounds
OppTO:	Opponent Turnovers
OPS:	On-base Plus Slugging
OR:	Offensive Rebounds
OREB:	Offensive Rebound

ORP:	Offensive Rebounding Percentage
ORPM:	Offensive Real Plus Minus
OT:	Overtime
PAP:	Pitcher Abuse Points
PAT:	Points After Touchdown
PCA:	Principal Component Analysis
PENDIF:	Penalty Yards Difference
PER:	Player Efficiency Rating
PF:	Personal Fouls
PF:	Power Forward
PFF:	Pro Football Focus
PFR:	Pro Football Reference
PG:	Point Guard
PNR:	Pick-and-Roll
PO:	Putout
POT:	Probability a team wins in OT
PPP:	Points per Possession
PPW:	Pope, Price, and Wolfers
PTHREE:	Probability a three-point shot is made
PTS:	Points
PTS%:	Points Percentage
PTWO:	Probability a two-point shot is made
PY/A:	Passing Yards per Attempt
QB:	Quarterback
QBR:	Quarterback Rating
qSQ:	Quantified Shot Quality
R:	Correlation
RAPTOR:	FiveThirtyEight.com's NBA player rating system
RC:	Runs Created
RC3:	Right Corner Three
RD:	Rating Deviation
RET:	Return
RF:	Range Factor
RFID:	Radio Frequency Identification
ROY:	Rookie of the year

RPI:	Rating Percentage Index
RPM:	Real Plus-Minus
RSQ:	R-Squared Value
RY/A:	Rushing Yards per Attempt
SAT:	Shot Attempts
SB:	Stolen Base
SF:	Sacrifice Fly
SG:	Shooting Guard
SH:	Sacrifice Hit
SLG:	Slugging Percentage
SPG:	Sicilia, Pelechrinis, and Goldsberry
T:	Triple
TD:	Touchdown
TM:	Thaler–Massey
TO:	Turnovers
TPP:	Turnover Percentage
TPZSG:	Two-Person Zero-Sum Game
TS%:	True Shooting Percentage
U:	Utility
UBR:	Ultimate Base Running
URL:	Uniform Resource Locator
WAR:	Wins Above Replacement
wGDP:	Weighted Grounded into Double Play Runs
WINVAL:	Adjusted +/− system for rating NBA players
wOBA:	Weighted On-Base Average
WP:	Wolfers and Price
WPA:	Win Probability Added
wRAA:	Weighted Runs Above Average
wSB:	Weighted Stolen Base Runs
XERA:	Pitcher's estimated ERA if he had average luck
xG:	Expected Goals
XML:	Extensible Markup Language
YL:	Yard Line
YTG:	Yards to Gain
YVH:	Yurko, Ventura, and Horowitz

PART I

BASEBALL

BASEBALL'S PYTHAGOREAN THEOREM

The more runs that a baseball team scores, the more games the team should win. Conversely, the fewer runs a team gives up, the more games the team should win. Bill James, probably the most celebrated advocate of applying mathematics to analysis of Major League Baseball (often called sabermetrics), studied many years of Major League Baseball standings and found that the percentage of games won by a baseball team can be well approximated by the formula

$$\frac{\text{runs scored}^2}{\text{runs scored}^2 + \text{runs allowed}^2} = \begin{array}{c}\text{Estimate of percentage} \\ \text{of games won.}\end{array} \quad (1)$$

This formula has several desirable properties:

- Predicted win percentage is always between 0 and 1.
- An increase in runs scored increases predicted win percentage.
- A decrease in runs allowed increases predicted win percentage.

Consider a right triangle with a hypotenuse (the longest side) of length c and two other sides of length a and b. Recall from high school geometry that the Pythagorean Theorem states that a triangle is a right triangle if and only if $a^2 + b^2 = c^2$ must hold. For example, a triangle with sides of lengths 3, 4, and 5 is a right triangle because $3^2 + 4^2 = 5^2$. The fact that equation (1) adds up the squares of two numbers led Bill James to call the relationship described in (1) Baseball's Pythagorean Theorem.

Let's define $R = \dfrac{\text{runs scored}}{\text{runs allowed}}$ as a team's scoring ratio. If we divide the numerator and denominator of (1) by (runs allowed)^2, then the value of the fraction remains unchanged and we may rewrite (1) as equation (1′).

$$\frac{R^2}{R^2 + 1} = \text{Estimate of percentage of games won} \qquad (1')$$

Figure 1-1 (see file Mathleticschapter1files.xlsx for all of this chapter's analysis) shows how well (1′) predicts teams' winning percentages for Major League Baseball teams during the 2005–2016 seasons. For example, the 2016 Los Angeles Dodgers scored 725 runs and gave up 638 runs. Their scoring ratio was $R = \dfrac{725}{638} = 1.136$. Their predicted win percentage from Baseball's Pythagorean Theorem was $\dfrac{1.136^2}{1.136^2 + 1} = .5636$. The 2016 Dodgers actually won a fraction $\dfrac{91}{162} = .5618$ of their games. Thus (1′) was off by 0.18% in predicting the percentage of games won by the Dodgers in 2016.

For each team define Error in Win Percentage Prediction to equal Actual Winning Percentage minus Predicted Winning Percentage. For example, for the 2016 Atlanta Braves, Error $= .42 - .41 = .01$ (or 1.0%), and for the 2016 Colorado Rockies, Error $= .46 - .49 = -.03$ (or 3%). A positive error means that the team won more games than predicted while a negative error means the team won fewer games than predicted. Column J computes for each team the absolute value of the prediction error. Recall that absolute value of a number is simply the distance of the number from 0. That is, $|5| = |-5| = 5$. In cell J1 we average the absolute prediction errors for each team to obtain a measure of how well our predicted win percentages fit the actual team winning percentages. The average of absolute forecasting

	A	B	C	D	E	F	G	H	I	J
1						exp		2.000			MAD:	0.021
2	Year	Team	Wins	Losses	Runs	Opp Runs	Ratio	Pred W–L%	Act W–L%	Error
3	2016	ARI	69	93	752	890	0.845	0.42	0.43	0.009
4	2016	ATL	68	93	649	779	0.833	0.41	0.42	0.010
5	2016	BAL	89	73	744	715	1.041	0.52	0.55	0.030
6	2016	BOS	93	69	878	694	1.265	0.62	0.57	0.041
7	2016	CHC	103	58	808	556	1.453	0.68	0.64	0.043
8	2016	CHW	78	84	686	715	0.959	0.48	0.48	0.002
9	2016	CIN	68	94	716	854	0.838	0.41	0.42	0.007
10	2016	CLE	94	67	777	676	1.149	0.57	0.58	0.011
11	2016	COL	75	87	845	860	0.983	0.49	0.46	0.028
12	2016	DET	86	75	750	721	1.040	0.52	0.53	0.011
13	2016	HOU	84	78	724	701	1.033	0.52	0.52	0.002
14	2016	KCR	81	81	675	712	0.948	0.47	0.50	0.027
15	2016	LAA	74	88	717	727	0.986	0.49	0.46	0.036
16	2016	LAD	91	71	725	638	1.136	0.56	0.56	0.002
17	2016	MIA	79	82	655	682	0.960	0.48	0.49	0.008
18	2016	MIL	73	89	671	733	0.915	0.46	0.45	0.005
19	2016	MIN	59	103	722	889	0.812	0.40	0.36	0.033
20	2016	NYM	87	75	671	617	1.088	0.54	0.54	0.005
21	2016	NYY	84	78	680	702	0.969	0.48	0.52	0.034

FIGURE 1.1   Baseball's Pythagorean Theorem 2005–2016.

errors is called the MAD (mean absolute deviation).[1] We find that for our dataset the predicted winning percentages of the Pythagorean Theorem were off by an average of 2.17% per team.

Instead of blindly assuming win percentage can be approximated by using the square of the scoring ratio, perhaps we should try a formula to predict winning percentage, such as

$$\frac{R^{exp}}{R^{exp}+1}.$$ (2)

If we vary exp in (2) we can make (2) better fit the actual dependence of winning percentage on the scoring ratio for different sports.

1. Why didn't we just average the actual errors? Because averaging positive and negative errors would result in positive and negative errors canceling out. For example, if one team wins 5% more games than (1′) predicts and another team wins 5% less games than (1′) predicts, the average of the errors is 0 but the average of the absolute errors is 5%. Of course, in this simple situation estimating the average error as 5% is correct while estimating the average error as 0% is nonsensical.

	N	O
5		MAD
6		0.021
7	1.1	0.02812245
8	1.2	0.02617963
9	1.3	0.02441563
10	1.4	0.02289267
11	1.5	0.02160248
12	1.6	0.02069009
13	1.7	0.02014272
14	1.8	0.0199295
15	1.9	0.0201094
16	2	0.020513
17	2.1	0.02114432
18	2.2	0.02208793
19	2.3	0.02328749
20	2.4	0.02473436
21	2.5	0.02640258
22	2.6	0.02823811
23	2.7	0.03019355
24	2.8	0.03228514
25	2.9	0.03447043
26	3	0.03670606

FIGURE 1.2   Dependence of Pythagorean
Theorem Accuracy on Exponent.

For baseball, we will allow exp in (2) (exp is short for exponent) to vary between 1 and 3. Of course exp = 2 reduces to the Pythagorean Theorem.

Figure 1-2 shows how the MAD changes as we vary exp between 1 and 3. This was done using the Data Table feature in Excel.[2] We see that indeed exp = 1.8 yields the smallest MAD (1.99%). An exp value of 2 is almost as good (MAD of 2.05%), so for simplicity we will stick with Bill James's view that exp = 2. Therefore exp = 2 (or 1.8) yields the best forecasts if we use an equation of form (2). Of course, there might be another equation that predicts winning percentage better than the Pythagorean Theorem from runs scored and allowed. The Pythago-

2. See Chapter 1 Appendix for an explanation of how we used Data Tables to determine how MAD changes as we vary exp between 1 and 3. Additional information available at https://support.office.com/en-us/article/calculate-multiple-results-by -using-a-data-table-e95e2487-6ca6-4413-ad12-77542a5ea50b.

rean Theorem is simple and intuitive, however, and does very well. After all, we are off in predicting team wins by an average of $162 * .0205$, which is approximately three wins per team. Therefore, I see no reason to look for a more complicated (albeit slightly more accurate) model.

HOW WELL DOES THE PYTHAGOREAN THEOREM FORECAST?

To test the utility of the Pythagorean Theorem (or any prediction model) we should check how well it forecasts the future. We chose to compare the Pythagorean Theorem's forecast for each Major League Baseball playoff series (2005–2016) against a prediction based just on games won. For each playoff series the Pythagorean method would predict the winner to be the team with the higher scoring ratio while the "games won" approach simply predicts the winner of a playoff series to be the team that won more games. We found that the Pythagorean approach correctly predicted 46 of 84 playoff series (54.8%) while the "games won" approach correctly predicted the winner of only 55% (44 out of 80) playoff series.[3] The reader is probably disappointed that even the Pythagorean method only correctly forecasts the outcome of under 54% of baseball playoff series. We believe that the regular season is a relatively poor predictor of the playoffs in baseball because a team's regular season record depends a lot on the performance of five starting pitchers. During the playoffs, teams only use three or four starting pitchers, so a lot of the regular season data (games involving the fourth and fifth starting pitchers) are not relevant for predicting the outcome of the playoffs.

For anecdotal evidence of how the Pythagorean Theorem forecasts the future performance of a team better than a team's win-loss record, consider the case of the 2005 Washington Nationals. On July 4, 2005, the Nationals were in first place with a record of 50–32. If we had extrapolated this win percentage, we would have predicted

---

3. In four playoff series the opposing teams had identical win-loss records, so the "games won" approach could not make a prediction.

a final record of 99–63. On July 4, 2005, the Nationals' scoring ratio was .991. On July 4, 2005, equation (1) would predict the Nationals to win around half (40) of the remaining 80 games and finish with a 90–72 record. In reality, the Nationals only won 31 of their remaining games and finished at 81–81!

IMPORTANCE OF PYTHAGOREAN THEOREM

The Baseball Pythagorean Theorem is also important because it allows us to determine how many extra wins (or losses) will result from a trade. As an example, suppose a team has scored 850 runs during a season and also given up 800 runs. Suppose we trade an SS (Joe) who "created"[4] 150 runs for a shortstop (Greg) who created 170 runs in the same number of plate appearances. This trade will cause the team (all other things being equal) to score $170 - 150 = 20$ more runs. Before the trade, $R = \dfrac{850}{800} = 1.0625$, and we would predict the team to have won $\dfrac{162 * 1.0625^2}{1 + 1.0625^2} = 85.9$ games. After the trade, $R = \dfrac{870}{800} = 1.0875$, and we would predict the team to have won $\dfrac{162 * 1.0875^2}{1 + 1.0875^2} = 87.8$ games. Therefore, we estimate the trade makes our team $87.8 - 85.9 = 1.9$ games better. In Chapter 9, we will see how the Pythagorean Theorem can be used to help determine fair salaries for Major League Baseball players.

FOOTBALL AND BASKETBALL "PYTHAGOREAN THEOREMS"

Does the Pythagorean Theorem hold for football and basketball? Daryl Morey, currently the General Manager for the Houston Rockets NBA team, has shown that for the NFL, equation (2) with

---

4. In Chapters 2–4 we will explain in detail how to determine how many runs a hitter creates.

	A	B	C	D	E	F	G	H	I	J	K	L	M	N
1								Exp	2.370	MAD	0.051			
2	Year	Team	Wins	Losses	Ties	PF	PA	Ratio	Pred W–L%	Act W–L%	Error			
3	2015	Arizona Cardinals	13	3	0	489	313	1.56	0.742	0.813	0.071			
4	2015	Atlanta Falcons	8	8	0	339	345	0.98	0.490	0.5	0.010			MAD
5	2015	Baltimore Ravens	5	11	0	328	401	0.82	0.383	0.313	0.070		Exp	0.051130558
6	2015	Buffalo Bills	8	8	0	379	359	1.06	0.532	0.5	0.032		1.5	0.087458019
7	2015	Carolina Panthers	15	1	0	500	308	1.62	0.759	0.938	0.179		1.6	0.083786393
8	2015	Chicago Bears	6	10	0	335	397	0.84	0.401	0.375	0.026		1.7	0.080410576
9	2015	Cincinnati Bengals	12	4	0	419	279	1.50	0.724	0.75	0.026		1.8	0.077291728
10	2015	Cleveland Browns	3	13	0	278	432	0.64	0.260	0.188	0.072		1.9	0.074380834
11	2015	Dallas Cowboys	4	12	0	275	374	0.74	0.325	0.25	0.075		2	0.071698879
12	2015	Denver Broncos	12	4	0	355	296	1.20	0.606	0.75	0.144		2.1	0.069282984
13	2015	Detroit Lions	7	9	0	358	400	0.90	0.435	0.438	0.003		2.2	0.067048672
14	2015	Green Bay Packers	10	6	0	368	323	1.14	0.577	0.625	0.048		2.3	0.065010818
15	2015	Houston Texans	9	7	0	339	313	1.08	0.547	0.563	0.016		2.4	0.063455288
16	2015	Indianapolis Colts	8	8	0	333	408	0.82	0.382	0.5	0.118		2.5	0.062158811
17	2015	Jacksonville Jaguars	5	11	0	376	448	0.84	0.398	0.313	0.085		2.6	0.061279631
18	2015	Kansas City Chiefs	11	5	0	405	287	1.41	0.693	0.688	0.005		2.7	0.060819271
19	2015	Miami Dolphins	6	10	0	310	389	0.80	0.369	0.375	0.006		2.8	0.060758708
20	2015	Minnesota Vikings	11	5	0	365	302	1.21	0.610	0.688	0.078		2.9	0.060941558
21	2015	New England Patriots	12	4	0	465	315	1.48	0.716	0.75	0.034		3	0.061357921
22	2015	New Orleans Saints	7	9	0	408	476	0.86	0.410	0.438	0.028		3.1	0.061891886
23	2015	New York Giants	6	10	0	420	442	0.95	0.470	0.375	0.095		3.2	0.062648637
24	2015	New York Jets	10	6	0	387	314	1.23	0.621	0.625	0.004		3.3	0.063594958
25	2015	Oakland Raiders	7	9	0	359	399	0.90	0.438	0.438	0.000		3.4	0.06474528
26	2015	Philadelphia Eagles	7	9	0	377	430	0.88	0.423	0.438	0.015		3.5	0.065955742
27	2015	Pittsburgh Steelers	10	6	0	423	319	1.33	0.661	0.625	0.036			

FIGURE 1.3   Predicted NFL Winning Percentages: Exp = 2.37.

exp = 2.37 gives the most accurate predictions for winning percentage, while for the NBA, equation (2) with exp = 13.91 gives the most accurate predictions for winning percentage. Figure 1-3 gives the predicted and actual winning percentages for the 2015 NFL, while Figure 1-4 gives the predicted and actual winning percentages for the 2015–2016 NBA. See the file Sportshw1.xls

For the 2008–2015 NFL seasons we found MAD was minimized by exp = 2.8. Exp = 2.8 yielded a MAD of 6.08%, while Morey's exp = 2.37 yielded a MAD of 6.39%. For the NBA seasons 2008–2016 we found exp = 14.4 best fit actual winning percentages. The MAD for these seasons was 2.84% for exp = 14.4 and 2.87% for exp = 13.91. Since Morey's values of exp are very close in accuracy to the values we found from recent seasons we will stick with Morey's values of exp. See file Sportshw1.xls.

Assuming the errors in our forecasts follow a normal random variable (which turns out to be a reasonable assumption) we would

	A	B	C	D	E	F	G	H	I	J	K	L	M
1							Exp	13.910	MAD	0.0287			
2	Year	Team	Wins	Losses	Points	Opp Points	Ratio	Pred W-L%	Act W-L%	Error			
3	2015-16	Atlanta Hawks	48	34	8433	8137	1.04	0.622	0.585	0.037			
4	2015-16	Boston Celtics	48	34	8669	8406	1.03	0.606	0.585	0.021			
5	2015-16	Brooklyn Nets	21	61	8089	8692	0.93	0.269	0.256	0.013		Exp	0.0287
6	2015-16	Charlotte Hornets	48	34	8479	8256	1.03	0.592	0.585	0.007		12	0.0340286
7	2015-16	Chicago Bulls	42	40	8335	8456	0.99	0.450	0.512	0.062		12.2	0.0332135
8	2015-16	Cleveland Cavaliers	57	25	8555	8063	1.06	0.695	0.695	6E-05		12.4	0.0324282
9	2015-16	Dallas Mavericks	42	40	8388	8413	1	0.490	0.512	0.022		12.6	0.0317199
10	2015-16	Denver Nuggets	33	49	8355	8609	0.97	0.397	0.402	0.005		12.8	0.0310445
11	2015-16	Detroit Pistons	44	38	8361	8311	1.01	0.521	0.537	0.016		13	0.0304509
12	2015-16	Golden State Warriors	73	9	9421	8539	1.1	0.797	0.89	0.093		13.2	0.0298964
13	2015-16	Houston Rockets	41	41	8737	8721	1	0.506	0.5	0.006		13.4	0.0294269
14	2015-16	Indiana Pacers	45	37	8377	8237	1.02	0.558	0.549	0.009		13.6	0.0290408
15	2015-16	Los Angeles Clippers	53	29	8569	8218	1.04	0.641	0.646	0.005		13.8	0.0287533
16	2015-16	Los Angeles Lakers	17	65	7982	8766	0.91	0.214	0.207	0.007		14	0.0285995
17	2015-16	Memphis Grizzlies	42	40	8126	8310	0.98	0.423	0.512	0.089		14.2	0.0284997
18	2015-16	Miami Heat	48	34	8204	8069	1.02	0.557	0.585	0.028		14.4	0.0284481
19	2015-16	Milwaukee Bucks	33	49	8122	8465	0.96	0.360	0.402	0.042		14.6	0.0284727
20	2015-16	Minnesota Timberwolves	29	53	8398	8688	0.97	0.384	0.354	0.03		14.8	0.028568
21	2015-16	New Orleans Pelicans	30	52	8423	8734	0.96	0.377	0.366	0.011		15	0.0287573
22	2015-16	New York Knicks	32	50	8065	8289	0.97	0.406	0.39	0.016		15.2	0.0289692
23	2015-16	Oklahoma City Thunder	55	27	9038	8441	1.07	0.721	0.671	0.05		15.4	0.0292675
24	2015-16	Orlando Magic	35	47	8369	8502	0.98	0.445	0.427	0.018		15.6	0.0296178
25	2015-16	Philadelphia 76ers	10	72	7988	8827	0.9	0.200	0.122	0.078		15.8	0.0300081
26	2015-16	Phoenix Suns	23	59	8271	8817	0.94	0.291	0.28	0.011		16	0.0304529
27	2015-16	Portland Trail Blazers	44	38	8622	8554	1.01	0.528	0.537	0.009			

FIGURE 1.4   Predicted NBA Winning Percentages: Exp = 13.91.

expect around 95% of our NBA win forecasts to be accurate within
2.5 * MAD = 7.3%. Over 82 games this is about 6 games. So whenever
the Pythagorean forecast for wins is off by more than six games, the
Pythagorean prediction is an "outlier." When we spot outliers we try
and explain why they occurred. The 2006–2007 Boston Celtics had a
scoring ratio of .966, and Pythagoras predicts the Celtics should have
won 31 games. They won seven fewer games (24). During that season
many people suggested the Celtics "tanked" games to improve their
chance of having the #1 pick (Greg Oden and Kevin Durant went 1–2)
in the draft lottery. The shortfall in the Celtics' wins does not prove
this conjecture, but the evidence is consistent with the Celtics winning substantially fewer games than chance would indicate.

CHAPTER 1 APPENDIX: DATA TABLES

The Excel Data Table feature enables us to see how a formula changes as the values of one or two cells in a spreadsheet are modified. In this appendix we show how to use a one-way data table to determine how the accuracy of (2) for predicting team winning percentage depends on the value of exp. To illustrate let's show how to use a one-way data table to determine how varying exp from 1 to 3 changes our average error in predicting an MLB's team winning percentage (see Figure 1-2).

Step 1: We begin by entering the possible values of exp (1, 1.1, . . . , 3) in the cell range N7:N26. To enter these values we simply enter 1 in N7 and 1.1 in N8 and select the cell range N7:N8. Now we drag the cross in the lower right-hand corner of N8 down to N26.

Step 2: In cell O6 we enter the formula we want to loop through and calculate for different values of exp by entering the formula = J1. Then we select the "table range" N6:O26.

Step 3: Now we select Data Table from the What If section of the ribbon's Data tab.

Step 4: We leave the row input cell portion of the dialog box blank but select cell G1 (which contains the value of exp) as the column input cell. After selecting OK we see the results shown in Figure 1-2. In effect, Excel has placed the values 1, 1.1, . . . , 3 into cell G1 and computed our MAD for each listed value of exp.

CHAPTER 2

WHO HAD A BETTER YEAR: MIKE TROUT OR KRIS BRYANT?

The Runs Created Approach

At age 24, Los Angeles Angels outfielder Mike Trout won the 2016 American League Most Valuable Player award for the second time in his career. Also at age 24, Kris Bryant of the Chicago Cubs won the 2016 National League Most Valuable Player award. Table 2.1 shows their key statistics:

Recall that a batter's slugging percentage is given by

$$\text{Slugging Percentage} = \frac{\text{Total Bases}}{\text{At Bats}}, \quad \text{where}$$

Total Bases = Singles + 2 * (Doubles) + 3 * (Triples) + 4 * (Home Runs).

We see in Table 2-1 that Trout had a higher batting average than Bryant. However, Bryant had a slightly higher slugging percentage since he hit more doubles and home runs. Bryant also had 54 more at bats than Trout and three more hits. So, which player had a better hitting year?

TABLE 2.1

Mike Trout and Kris Bryant 2016 Statistics

Event	Trout (2016)	Bryant (2016)
At Bats	549	603
Batting Average	.315	.292
Slugging Percentage	.550	.554
Hits	173	176
Singles	107	99
Doubles	32	35
Triples	5	3
Home Runs	29	39
Walks + Hit by Pitcher	127	93

We know that when a batter is hitting, he can cause good things (like hits or walks) to happen or bad things (outs) to happen. To compare hitters, we must develop a metric which measures how the relative frequency of a batter's good events and bad events influences the number of runs the team scores.

In 1979 Bill James developed the first version of his famous Runs Created formula in an attempt to compute the number of runs "created" by a hitter during the course of a season. The most easily obtained data we have available to determine how batting events influence runs scored is season-long team batting statistics. A sample of the complete data from 2010–2016 found in the worksheet Fig 2-1 of the workbook Chapter2mathleticsfiles.xlsx is shown in Figure 2-1.

James realized there should be a way to predict the runs for each team from hits, singles, 2Bs, 3Bs, HRs, outs, and BBs + HBPs.[1]

---

1. Of course, this leaves out things like sacrifice hits, sacrifice flies, stolen bases, and caught stealings. See http://danagonistes.blogspot.com/2004/10/brief-history -of-run-estimation-runs.html for an excellent summary of the evolution of runs created.

	A	B	C	D	E	F	G	H	I	J	K	L	M	N
1	Team	Runs	At Bats	Hits	Singles	2B	3B	HR	BB+HBP					
2	ARI	752	5665	1479	948	285	56	190	513					
3	ATL	649	5514	1404	960	295	27	122	561	RC Formula	=(D5+I5)*(E5+2*F5+3*G5+4*H5)/(C5+I5)			
4	BAL	744	5524	1413	889	265	6	253	512					
5	BOS	878	5670	1598	1022	343	25	208	601	RC	916.9805	917		
6	CHC	808	5503	1409	887	293	30	199	752	Actual		878		
7	CHW	686	5550	1428	950	277	33	168	508	Error		−39		
8	CIN	716	5487	1403	929	277	33	164	504	% Error	−0.04442			
9	CLE	777	5484	1435	913	308	29	185	580					
10	COL	845	5614	1544	975	318	47	204	534					
11	DET	750	5526	1476	983	252	30	211	546					
12	HOU	724	5545	1367	849	291	29	198	601					
13	KCR	675	5552	1450	1006	264	33	147	427					

FIGURE 2.1   Selected Team Batting Data for 2016 Season.

Using intuition James came up with the following relatively simple formula:

$$\text{Runs Created} = \frac{(\text{Hits} + \text{Walks} + \text{HBPs}) * (\text{Total Bases})}{(\text{At Bats} + \text{Walks} + \text{HBPs})} \qquad (1)$$

As we will soon see, (1) does an amazingly good job of predicting how many runs a team scores in a season from these components. What is the rationale for (1)? To score runs you need to have runners on base and then you need to advance them toward the home plate. (Hits + Walks + HBPs) is basically the number of base runners the team will have in a season. $\dfrac{\text{Total Bases}}{\text{AB} + \text{Walks} + \text{HBP}}$ measures the rate at which runners are advanced per plate appearance. Therefore (1) is multiplying the number of base runners by the rate at which they are advanced. Using the information in Figure 2-1 we can compute Runs Created for the 2016 Boston Red Sox:

$$\text{Runs Created} = \frac{(1598+601) * (1022 + 2(343) + 3(25) + 4(208))}{(5670 + 601)} \cong 917$$

Actually, the 2016 Boston Red Sox scored 878 runs, so Runs Created overestimated the actual number of runs by around 4%. The TeamRC worksheet in the file Teams.xlsx calculates runs created for each team during the 2010–2016 seasons and compares runs created to actual runs scored. We find that Runs Created was off by an average of 21 runs per team. Since the average team scored 693 runs, this is an average error of about 3% when we try to use (1) to predict Team Runs Scored. It is amazing that this simple, intuitively appealing formula does such a good job of predicting runs scored by a team. Even though more complex versions of Runs Created more accurately predict actual runs scored, the simplicity of (1) has caused this formula to still be widely used by the baseball community.

BEWARE BLIND EXTRAPOLATION!

The problem with any version of Runs Created is that the formula is based on team statistics. A typical team has a batting average of .250, hits HRs on 3% of all plate appearances, and has a walk or HBP in around 10% of all plate appearances. Contrast these numbers to Miguel Cabrera's great 2013 season in which he had a batting average of .348, hit an HR on approximately 7% of all plate appearances, and received a walk or HBP during approximately 15% of his plate appearances. One of the first ideas we teach in business statistics courses is **to not use a relationship that is fit to a dataset to make predictions for data that is very different from the data used to fit the relationship**. Following this logic, we should not expect a Runs Created formula based on team data to accurately predict the runs created by a superstar such as Miguel Cabrera or a very poor player. In Chapter 4 we will remedy this problem with a different type of model.

TROUT VS. BRYANT

Despite this caveat, let's plunge ahead and use (1) to compare Mike Trout's 2016 season to Kris Bryant's 2016 season. For fun we also computed Runs Created for Miguel Cabrera's great 2013 season. See the worksheet Figure 2-2 of the workbook Chapter2mathleticsfiles.xlsx.

From our data, we calculated that Mike Trout created 134 runs and Kris Bryant created 129 runs. Cabrera created 148 runs in 2013.

	A	B	C	D	E	F	G	H	I	J	K	L	M
1	Player	At Bats	Hits	1B	2B	3B	HR	Estimated Outs	Other Outs	BB+HBP	Runs Created	Game Outs Used	Runs Created / Game
2	Bryant (2016)	603	176	99	35	3	39	416.146	11	93	129.09	427.146	8.10837519
3	Trout (2016)	549	173	107	32	5	29	366.118	17	127	134.02	383.118	9.385763732
4	Cabrera (2013)	555	193	102	26	1	44	352.01	21	95	147.54	373.01	10.61264318
10										J2 Formula	=(C2+J2)*(D2+2*E2+3*F2+4*G2)/(B2+J2)		
11										K2 Formula	=I2+H2		
12										L2 Formula	=K2/(L2/26.83)		

FIGURE 2.2   Runs Created for Trout, Bryant, and Cabrera.

This indicates that Trout had a slightly better hitting year in 2016 than Bryant. Miguel Cabrera's 2013 season was superior to both Trout and Bryant's 2016 year according to this Runs Created approach.

RUNS CREATED PER GAME

A major problem with any Runs Created metric is that a mediocre hitter with 700 plate appearances might create more runs than a superstar with 400 plate appearances.

	A	B	C	D	E	F	G	H	I	J	K	L	M
6	Player	At Bats	Hits	1B	2B	3B	HR	Outs		BB+HBP	Runs Created	Game Outs Used	Runs Created / Game
7	Christian	700	190	170	10	1	9	497.4		20	66.79	497.4	3.60
8	Gregory	400	120	90	15	0	15	272.8		20	60.00	272.8	5.90

FIGURE 2.3   Christian and Gregory's Fictitious Statistics.

As shown in Figure 2-3 (see worksheet Fig2_3 of the workbook Chapter2mathleticsfiles.xlsx), we have created two hypothetical players: Christian and Gregory. Christian had a batting average of .257 while Gregory had a batting average of .300. Gregory walked more often per plate appearance and had more extra base hits. Yet Runs Created says Christian was a better player. To solve this problem, we need to understand that hitters consume a scarce resource—outs. During most games a team bats for nine innings and gets $3 * 9 = 27$ outs.[2] We can now compute Runs Created Per Game.

2. Since the home team does not always bat in the ninth inning and some games go into extra innings, average outs per game is not exactly 27. For the years 2010–2016, average outs per game was 26.83.

To see how this works let's look at Trout's 2016 data. See Figure 2-2 and the PlayerRC worksheet in file Teams.xlsx.

How did we compute outs? Essentially all at bats except for hits and errors result in an out. Approximately 1.8% of all at bats result in errors. Therefore, we computed outs as at bats − hits − .018(at bats). Hitters also create "extra" outs through sacrifice hits, sacrifice bunts, caught stealings, and grounding into double plays. In 2016 Trout created 17 of these extra outs. As shown in cell L3 Trout "used" up 383.11 outs for the Angels. This is equivalent to $\frac{383.11}{26.83} = 14.28$ games. Therefore, Trout created $\frac{134.02}{14.28} = 9.39$ runs per game. More formally:

$$\text{Runs Created Per Game} = \frac{\text{Runs Created}}{\frac{.982(\text{At Bats}) - \text{Hits} + \text{GIDP} + \text{SF} + \text{SH} + \text{CS}}{26.83}} \qquad (2)$$

Equation (2) simply states that Runs Created per game is Runs Created by a batter divided by Number of Games worth of outs used by the batter. Figure 2-2 shows that Miguel Cabrera created 10.61 runs per game. Figure 2-2 also makes it clear that Trout was a more valuable hitter than Bryant in 2016. Specifically, Trout created 9.39 runs per game while Bryant created approximately 1.28 fewer runs per game (8.11 runs). We also see that runs created per game by the notional Gregory is 2.29 runs (5.88 − 3.59) better per game than fictitious Christian. This resolves the problem that ordinary runs created ranked Christian ahead of Gregory.

Our estimate of runs created per game of 9.39 for Mike Trout indicates that we believe a team consisting of nine Mike Trouts would score an average of 9.39 runs per game. Since no team consists of nine players like Trout, a more relevant question might be, how many runs would Mike Trout create when batting with eight "average hitters"? In his book *Win Shares* (2002) Bill James came up with a more complex version of runs created that answers this question. We will provide our own answer to this question in Chapters 3 and 4.

CHAPTER 3

EVALUATING HITTERS BY LINEAR WEIGHTS

In Chapter 2 we described how knowledge of a hitter's at bats, BBs + HBPs, singles, 2Bs, 3Bs, and HRs allows us to compare hitters via the runs created metric. As we will see in this chapter, the linear weights approach can also be used to compare hitters. In business and science, we often try and predict a given variable (called Y or the dependent variable) from a set of independent variables (call the independent variables $x_1, x_2, \ldots, a_n$). Usually we try to find weights $B_1, B_2, \ldots, B_n$ and a constant that make the quantity

$$\text{Constant} + B_1 x_1 + B_2 x_2 + \cdots + B_n x_n$$

a good predictor for the dependent variable.

Statisticians call the search for the weights and the constant that best predict Y a multiple linear regression. Sabermetricians (people who apply math to baseball) call the weights linear weights.

For our team batting data for year 2010–2016,

$$Y = \text{dependent variable} = \text{runs scored in a season.}$$

For independent variables we will use BBs + HBPs (walks + hits by pitcher), singles, 2Bs, 3Bs, HRs, SBs (stolen bases), and CSs (caught stealings). Thus, our prediction equation will look like

$$\text{Predicted runs for season} = \text{Constant} + B_1(\text{BB} + \text{HBP}) + B_2(\text{Singles}) + B_3(2B) + B_4(3B) + B_5(\text{HR}) + B_6(\text{SB}) + B_7(\text{CS}). \quad (1)$$

Let's see if we can use basic arithmetic to come up with a crude estimate of the value of an HR. For the years 2010–2016 in a game an average MLB team has 38 batters come to the plate and score 4.3 runs. So roughly one out of nine batters scores. During a game the average MLB team has around 12 batters reach base. Therefore 4.3/12 or around 36% of all runners score. If we assume an average of one base runner on base when an HR is hit then a Home Run creates "runs" in the following fashion,

- The batter scores all the time instead of 1/8 of the time, which creates 7/8 of a run.
- An average of one base runner will score 100% of the time instead of 37% of the time, which creates .63 runs.

This leads to a crude estimate that a home run is worth around .87 + .63 = 1.5 runs. We will soon see that our regression model provides a similar estimate for the value of a home run.

We can use the Regression tool in Excel to search for the set of weights and constant that enable (1) to give the best forecast for Runs Scored. See the chapter appendix for an explanation of how to use the Regression tool. Essentially Excel's Regression tool finds the constant and set of weights that minimize the sum over all teams of (actual runs scored − predicted runs scored)^2.[1] The results of our regression are in sheet MLR of workbook Chapter3.xlsx. See Figure 3-1.

---

1. If we did not square the prediction error for each team, we would find that the errors for teams that scored more runs than predicted would be cancelled out by the errors for teams that scored fewer runs than predicted.

	A	B	C	D	E	F	G
1	SUMMARY OUTPUT						
2							
3	Regression Statistics						
4	Multiple R	0.949366525					
5	R Square	0.9012968					
6	Adjusted R Square	0.897876392					
7	Standard Error	22.07547927					
8	Observations	210					
9							
10	ANOVA						
11		df	SS	MS	F	Significance F	
12	Regression	7	898893.5132	128413.359	263.505645	6.66264E–98	
13	Residual	202	98440.01059	487.3267851			
14	Total	209	997333.5238				
15							
16		Coefficients	Standard Error	t Stat	P–value	Lower 95%	Upper 95%
17	Intercept	–411.8133561	33.00675506	–12.47663866	7.3423E–27	–476.8953293	–346.731383
18	BB+HBP	0.326171191	0.026991877	12.08405016	1.1813E–25	0.272949219	0.37939316
19	1B	0.459107774	0.028209869	16.2747222	1.325E–38	0.403484193	0.51473135
20	2B	0.805141015	0.070539419	11.41405797	1.31E–23	0.666052984	0.94422905
21	3B	1.072129559	0.185083303	5.792686554	2.6244E–08	0.707186489	1.43707263
22	HR	1.428105264	0.052270693	27.32133795	9.1608E–70	1.325039094	1.53117143
23	SB	0.250044999	0.063490957	3.938277396	0.00011296	0.124854967	0.37523503
24	CS	–0.254380304	0.190576335	–1.334794818	0.18344599	–0.630154411	0.1213938

FIGURE 3.1   Regression Output with CS and SB Included.

Cells B17:B24 (listed under Coefficients) show that the best set of linear weights and constant (Intercept cell gives the constant) to predict runs scored in a season is given by

$$\text{Predicted Runs} = -411.81 + 46(\text{Singles}) + .81(2Bs) + 1.07(3Bs) + \\ 1.43(\text{HRs}) + .33(\text{BBs} + \text{HBPs}) + .25(\text{SBs}) - .25(\text{CSs}). \quad (2)$$

The highlighted R-squared value in cell B5 indicates that our independent variables (singles, 2Bs, 3Bs, HRs, BBs + HBPs, SBs, and CSs) explain roughly 90% of the variation in the number of runs a team actually scores during a season.

Equation (2) indicates that a single "creates" .46 runs, a double creates .81 runs, a triple creates 1.07 runs, a home run creates 1.43 runs, a BB or HBP creates .33 runs, and a stolen base creates .25 runs, while being caught stealing "eliminates" .25 runs. We see that our HR weight agrees with our simple calculation of 1.5, and the fact that a double is worth more than a single but less than two singles makes sense. We also observe that the fact that a single is worth more than a walk makes sense because singles often advance runners two bases. It is also reasonable to see that a triple is worth more than a double but less than an HR.

THE MEANING OF P-VALUES

When we run a regression, we should always check that each independent variable has a significant effect on the dependent variable. We do this by looking at each independent variable's p-value. These are shown in column E of Figure 3-1. Each independent variable has a p-value between 0 and 1. **Any independent variable with a** *small* **p-value (traditionally small means <.05, even though this choice has been arbitrary) is considered to be a useful predictor of the dependent variable (after adjusting for the other independent variables).** Essentially the p-value for an independent variable gives the probability that (in the presence of all other independent variables used to fit the regression) the independent variable **does not enhance our predictive ability (or equivalently the probability that the value of the weight is obtained purely by chance, and in reality the coefficient is 0).** For example, there is only around one chance in 10^{23} that doubles do not enhance our ability for predicting runs scored even after we know singles, triples, HRs, BBs + HBPs, CSs, and SBs. We find from Figure 3-1 that all independent variables except for CS have p-values that are very close to 0. For example, singles have a p-value of 1.33×10^{-38}. This means that singles almost surely help us predict team runs even after adjusting for all other independent variables. However, the high p-value for CS indicates that we should drop it from the regression and rerun the analysis. The resulting regression is shown in Figure 3-2 (see sheet MLRnoCS of workbook Ch3Data.xlsx).

All our independent variables have p-values <.05, so they all pass the test of statistical significance. We will now use the following

	A	B	C	D	E	F	G
1	SUMMARY OUTPUT						
2							
3	Regression Statistics						
4	Multiple R	0.948907909					
5	R Square	0.900426219					
6	Adjusted R Square	0.897483152					
7	Standard Error	22.11794065					
8	Observations	210					
9							
10	ANOVA						
11		df	SS	MS	F	Significance F	
12	Regression	6	898025.2542	149670.8757	305.948214	8.8346E–99	
13	Residual	203	99308.26962	489.2032986			
14	Total	209	997333.5238				
15							
16		Coefficients	Standard Error	t Stat	P–value	Lower 95%	Upper 95%
17	Intercept	-422.3214856	32.11582993	-13.14994775	5.654E–29	-485.6448728	-358.9980984
18	BB+HBP	0.328427033	0.026990732	12.16814092	6.1158E–26	0.275208898	0.381645169
19	1B	0.462425312	0.028154216	16.4247273	3.9961E–39	0.406913115	0.51793751
20	2B	0.809004928	0.070615562	11.45646795	9.2244E–24	0.669770893	0.948238964
21	3B	1.056646807	0.185074775	5.709296723	3.9868E–08	0.691731384	1.421562229
22	HR	1.432093994	0.052285581	27.38984579	4.1936E–70	1.329001529	1.535186459
23	SB	0.204454976	0.05362427	3.812732098	0.00018226	0.098722992	0.31018696

FIGURE 3.2   p-Values for Linear Weights Regression.

equation (derived from cells B17:B23 of Figure 3-2) to predict runs scored by a team in a season,

Predicted Runs =
$$-422.32 + .46(\text{Singles}) + .81(2\text{Bs}) + 1.06(3\text{Bs})$$
$$+ 1.43(\text{HRs}) + .33(\text{BBs} + \text{HBPs}) + .205(\text{SBs}). \qquad (3)$$

Note our R^2 is still 90%, even after dropping CS as an independent variable. This is unsurprising because the high p-values for CS indicated that it would not help us predict Runs Scored after we knew the other independent variables.

ACCURACY OF LINEAR WEIGHTS VS. RUNS CREATED

Do linear weights do a better job of forecasting runs scored than Bill James's original runs created formula? We see in cell E2 of Figure 3-3 (see sheet Accuracy LW of workbook Ch3Data.xlsx) that

	A	B	C	D	E	F	G	H	I	J	K
1					MAD	Linear Weights					
2					17.15642123	0.462425312	0.809	1.057	1.432	0.328427	0.204
3	Year	Team	Runs	Predicted Runs	Absolute Error	1B	2B	3B	HR	BB+HBP	SB
4	2016	ARI	752	774.3875946	−22.38759461	948	285	56	190	513	137
5	2016	ATL	649	663.0898878	−14.08988777	960	295	27	122	561	75
6	2016	BAL	744	743.8598697	0.140130263	889	265	6	253	512	19
7	2016	BOS	878	866.4120047	11.58799535	1022	343	25	208	601	83
8	2016	CHC	808	802.0454766	5.954523405	887	293	30	199	752	66
9	2016	CHW	686	699.1220277	−13.12202769	950	277	33	168	508	77
10	2016	CIN	716	695.0452205	20.95477948	929	277	33	164	504	139
11	2016	CLE	777	762.5111346	14.48886543	913	308	29	185	580	134
12	2016	COL	845	816.4903997	28.50960026	975	318	47	204	534	66
13	2016	DET	750	761.1626238	−11.16262382	983	252	30	211	546	58
14	2016	HOU	724	738.1344612	−14.1344612	849	291	29	198	601	102
15	2016	KCR	675	666.8202365	8.179763509	1006	264	33	147	427	121
16	2016	LAA	717	675.9107863	41.08921374	955	279	20	156	522	73
17	2016	LAD	725	704.664866	20.33513399	894	272	21	189	583	45
18	2016	MIA	655	670.7167317	−15.7167317	1031	259	42	128	501	71
19	2016	MIL	671	709.9591957	−38.95919567	837	249	19	194	636	181
20	2016	MIN	722	745.3214576	−23.32145761	886	288	35	200	557	91
21	2016	NYM	671	702.8567334	−31.8567334	865	240	19	218	579	42
22	2016	NYY	680	673.6639336	6.336066414	930	245	20	183	517	72
23	2016	OAK	653	639.032281	13.96771904	892	270	21	169	475	50
24	2016	PHI	610	616.0473558	−6.047355848	878	231	35	161	482	96
25	2016	PIT	729	733.814162	−4.814161981	964	277	32	153	642	110

FIGURE 3.3   Measuring Accuracy of Linear Weights.

for the Team Hitting Data (years 2010–2016) linear weights were off by an average of 17.15 runs (an average of 2.5% per team!) while, as we previously mentioned, runs created was off by 26 runs per game.

Thus, linear weights do a better job of predicting team runs than basic runs created.

THE HISTORY OF LINEAR WEIGHTS

We would be remiss if we did not briefly trace the history of linear weights (see Dan Agonistes's excellent summary, http://danagonistes .blogspot.com/2004/10/brief-history-of-run-estimation.html or Alan Schwarz's excellent book *The Numbers Game*, 2002). In 1916, F. C. Lane, editor of *Baseball Magazine*, used the results of how 1,000 hits advanced runners around the bases to come up with an estimate of linear weights. During the late 1950s and 1960s, military officer George Lindsay looked at a large set of game data and came up with a set of linear weights. Then in 1978, statistician Pete Palmer (see *The Hidden Game of Baseball*) used a Monte Carlo simulation model (see Chapter 4) to estimate the value of each type of baseball event. During 1989, *The Washington Post* reporter Thomas Boswell also came up with a set of linear weights (see the book *Total Baseball*). The weights obtained by these pioneers are summarized in Table 3-1 (a - indicates the author did not use the event in his model):

TABLE 3.1
The Historical Evolution of Linear Weights Estimates

Event	Lane	Lindsay	Palmer	Boswell	Our Regression
BB + HBP	.164	—	.33	.33	.35
Single	.457	.41	.46	.47	.63
2B	.786	.82	.8	.78	.71
Triple	1.15	1.06	1.02	1.09	1.26
Home Runs	1.55	1.42	1.4	1.4	1.49
Outs	—	—	− .25	—	—
SB	—	—	.3	.3	—
CS	—	—	− .6	—	—

For reasons that we will discuss in Chapter 4, we believe Monte Carlo simulation (as implemented by Palmer) is the best way to determine linear weights. Despite this, let's use our regression to evaluate hitters. Recall that (2) predicted runs scored given a team's statistics for an entire season.

USING LINEAR WEIGHTS TO DETERMINE RUNS CREATED BY A HITTER

How can we use (2) to predict how many runs we would score if we had a team consisting entirely of, say, Mike Trout 2016, Kris Bryant 2016, or Miguel Cabrera 2013. See Figure 3-4.

Trout 2016 made 366.118 outs (see cell I4). As explained in Chapter 2, we computed outs made by a hitter as .982ABs + sacrifice hits +

	A	B	C	D	E	F	G	H	I	J	K	L	M	N
1														
2					0.4624	0.809	1.057	1.432		0.328427033	0.204			
3	Player/Year		At Bats	Hits	1B	2B	3B	HR	Outs Used	BB+HBP	SB			
4	Trout16		549	173	107	32	5	29	366.118	127	30			
5	Bryant16		603	176	99	35	3	39	416.146	93	8			
6	Cabrera13		555	193	122	26	1	44	352.01	95	3			
7														
8		Scale Factor	At Bats	Hits	Singles	2B	3B	HR		BB+HBP	SB	Linear Weights Runs	Runs per game	
9	Trout16	11.82405672	6491.4	2045.561813	1265.2	378.4	59.12	342.9		1501.655204	354.7	1588.071256	9.802909	
10	Bryant16	10.40259909	6272.8	1830.857439	1029.9	364.1	31.21	405.7		967.4417152	83.22	1297.190499	8.007349	
11	Cabrera13	12.29794608	6825.4	2373.503594	1500.3	319.7	12.3	541.1		1168.304878	36.89	1709.31659	10.55134	

FIGURE 3.4 Linear Weights Estimates of Runs per Game Created by Trout, Bryant, and Cabrera.

sacrifice bunts + caught stealings + grounding into double plays. Given an average of 26.72 outs per game a team's season has $26.72 * 162 = 4,329$ outs. Now Trout hit 29 HRs. So, for each out he hit $29/366.118 = .079$ HRs. Thus, for a whole season we would predict a team of nine Mike Trouts to hit $4,329 * (29/366.118) = 342.9$ HRs. Now we see how to use (2) to predict runs scored in a team by a team consisting entirely of that player.[2] Simply "scale up" each of Trout's statistics by

$$4,329/366.118 = 11.824 = \text{Outs for Season/Player Outs.}$$

In rows 9 to 11 we multiply each player's statistics (from rows 4 to 6) by 4,329/(player's outs). We call this a player's "Scale Factor." Then in Column L we apply our linear weights regression model (equation 2) to the data in rows 9 to 11 to predict total season runs for a team consisting of the single player (see cells L9:L11). In cells M9:M11 we divide the predicted runs for a season by 162 to create a predicted runs per game. We predict a team of Trout 2016 to score 9.803 runs per game, a team of Bryant 2016 to score 8.007 runs per game, and a team of Cabrera 2013 to score 10.551 runs per game. Note that using runs created we estimate 9.39, 8.11, and 11.25 runs, respectively, for the three players. Thus, for the three players we have found that runs created and linear weights give similar predictions for the number of runs a player is responsible for during a game.

2. It might be helpful to note that both sides of this equation have the same units.

OBP, SLUGGING PERCENTAGE, OBP + SLUGGING, AND RUNS CREATED

As Michael Lewis brilliantly explains in his best seller *Moneyball*, during the 1980s and 1990s major league front offices came to realize the importance of on-base percentage (OBP) as a measure of a hitter's effectiveness. OBP is simply the fraction of a player's plate appearances where he reaches base on a hit, a walk, or an HBP. During the 2010–2016 seasons the average OBP was .319. OBP is a better measure of hitting effectiveness than ordinary batting average because a player with a high OBP uses less of a team's scarce resource (outs!). Unfortunately, many players with a high OBP (such as Ty Cobb and Rogers Hornsby) do not hit many home runs, so their value is overstated by simply relying on OBP. Therefore, baseball experts created a new statistic: OPS or on-base plus slugging, which is slugging percentage (total bases divided by at bats) added to OBP. The rationale is that by including slugging percentage in OPS we give proper credit to power hitters. In 2004, OPS "arrived" when it was included on Topps baseball cards.

Of course, OPS gives equal weight to slugging percentage and OBP. Is this reasonable? To determine the proper relative weight to give slugging percentage and OBP we used our 2010–2016 team data and ran a regression to predict team runs scored using as independent variables OBP and slugging percentage (SLG). See sheet OBP_SLG in workbook Ch3Data.xlsx and Figure 3-5.

We find that both OBP and SLG are highly significant (each has a p-value near 0). The R-squared in cell B5 indicates that we explain 88.5% of the variation in runs scored. This is similar to our best linear weights model, which had an R-squared of .90. Since this model seems easier to understand, it is easy to why OBP and slugging percentage are highly valued by baseball front offices. Note, however, that

We predict team runs scored as

$$-738.74 + 2{,}338.1(\text{OBP}) + 1{,}707(\text{SLG}).$$

This indicates that OBP is somewhat more important than SLG.

	A	B	C	D	E	F	G
1	SUMMARY OUTPUT						
2							
3	Regression Statistics						
4	Multiple R	0.940845892					
5	R Square	0.885190993					
6	Adjusted R Square	0.884081728					
7	Standard Error	23.51922548					
8	Observations	210					
9							
10	ANOVA						
11		df	SS	MS	F	Significance F	
12	Regression	2	882830.6526	441415.3263	797.9972172	5.09931E−98	
13	Residual	207	114502.8712	553.153967			
14	Total	209	997333.5238				
15							
16		Coefficients	Standard Error	t Stat	P–value	Lower 95%	Upper 95%
17	Intercept	−738.7520251	43.82154709	−16.85819133	1.04367E−40	−825.1457842	−652.3582661
18	OBP	2338.121668	191.8515917	12.18713719	4.14782E−26	1959.888099	2716.355238
19	SLG	1707.332494	92.94672979	18.3689356	2.39874E−45	1524.088909	1890.576078

FIGURE 3.5   Regression Predicting Team Runs
from OBP and Slugging Percentage.

RUNS CREATED ABOVE AVERAGE

One way to evaluate a player such as Bryant16 is to ask how many more runs an average MLB team would score if Bryant16 were added to the team. We answer this question in the AboveAverage sheet. See Figure 3-6. After entering a player's batting statistics in row 7, cell D11 computes the number of runs the player would add to an average MLB team. We now explain the logic underlying this spreadsheet.

In row 7 we enter the number of singles, doubles, triples, home runs, BBs + HBPs, sacrifice bunts, and total outs made by Bryant16. We see that Bryant created 416 outs. In row 6 we entered the same statistics for an average MLB team (based on 2010–2016 seasons).

If we add Bryant to an average team, the rest of the "average players" will create $4,328.64 − 416.15 = 3,912.49$ outs. Let $3,912.49/4,328.64 = .904$ be defined as teammult. Then the non-Bryant plate appearances by the remaining members of our average player + Bryant16 team will create teammult * 939.83 singles, teammult * 276.2 doubles, etc. Thus, our Bryant16 + average player team will create

	A	B	C	D	E	F	G	H	I	J	K
1						Outs Used					
2						416.146					
3											
4	teammult		Intercept	1B	2B	3B	HR	BB+HBP	SB		
5	0.903862	Linear Weights	−422.321	0.462425	0.809	1.0566468	1.432094	0.328427	0.204455	Outs	Predicted Runs Scored
6		Average Team		939.8286	276.2	29.161905	159.3571	544.5905	95.07619	4328.64	673.6088188
7		Bryant16		99	35	3	39	93	8	416.146	
8		Bryant16 Added to Average Team		948.4755	284.64	29.358343	183.0369	585.2347	93.93577	3912.494	731.9090705
9											
10				Bryant16 Runs Over Average							
11				58.30025							

FIGURE 3.6   Computing How Many Runs Bryant16 Would Add
to an Average Team.

$99 + \text{teammult} * 939.82 = 948.47$ singles, $35 + \text{teammult} * 276.2 = 284.64$ doubles, etc. This implies that our Bryant16 + average player team is predicted by linear weights to score $-422 + (.462) * (948.47) + (.809) * (284.64) + (1.056) * (29.36) + (1.432) * (183.04) + (.328) * (585.23) + (.204) * (93.94) = 731.91$ runs. Since an average team was predicted by linear weights to score 673.61 runs, we see in cell D11 that the addition of Bryant to an average team would add $731.91 - 673.61 = 58.30$ runs. Thus, we estimate that adding Bryant16 to an average team would add around 58 runs. This estimate of Bryant's hitting ability puts his contribution into the context of a typical MLB team, and therefore seems more useful than an estimate of how many runs would be scored by a team made up entirely of Bryant16.

In Chapter 4 we will use Monte Carlo simulation to obtain another estimate of how many runs a player adds to a particular team.

CHAPTER 3 APPENDIX: RUNNING REGRESSIONS IN EXCEL

To run regressions in Excel it is helpful to install the Analysis Toolpak Add-in.

INSTALLATION OF THE ANALYSIS TOOLPAK

To install the Analysis Toolpak in Excel, click on the File menu and choose the Options item at the bottom of the list. From the menu that appears choose Add-ins near the bottom. At the bottom of the panel is an option to Manage Excel Add-ins. Click on the Go button. Ensure that the Analysis Toolpak option (not the Analysis Toolpak—VBA option) is checked and click OK.

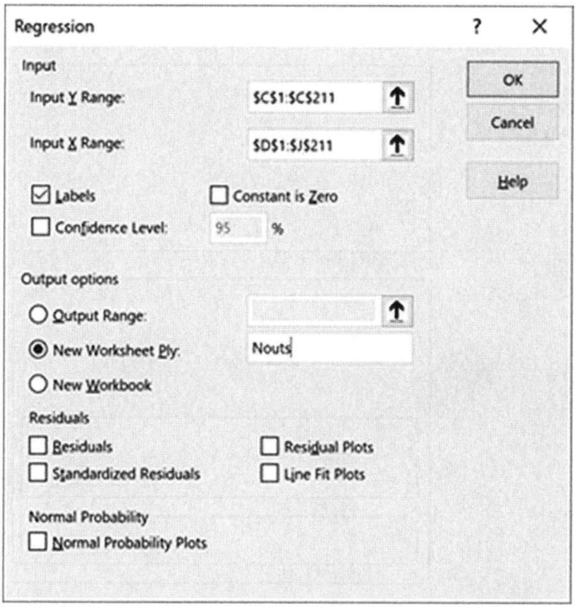

FIGURE 3.7   Regression of Runs on Various Statistics.

RUNNING A REGRESSION

The regression shown in Figure 3-1 predicts team runs scored from a team's singles, doubles, triples, HRs, BBs + HBPs, SBs, and CSs.

To run the regression, first go to the sheet Data of the workbook Ch3Data.xlsx. In Excel, bring up the Analysis Toolpak by selecting the Data Tab and then Choosing Data Analysis from the right-hand portion of the tab.

Now select the Regression option and fill in the dialog box as shown in Figure 3-7.

This tells Excel we want to predict the team runs scored (in cell range C2:C211) using the independent variables in cell range D2:J211 (singles, doubles, triples, HRs, BBs + HBPs, SBs, and CSs). We checked the Labels box so that our column labels shown in row 1 will be included in the regression output. The output (as shown in Figure 3-1) will be placed in the worksheet MLR.

EVALUATING HITTERS BY MONTE CARLO SIMULATION

In Chapters 2 and 3 we showed how to use Runs Created and Linear Weights to evaluate a hitter's effectiveness. These concepts were primarily developed to "fit" the relationship between Runs Scored by a team during a season and team statistics such as BB, singles, doubles, triples, and HRs. We pointed out that for players whose event frequencies differ greatly from typical team frequencies these metrics might do a poor job of evaluating a hitter's effectiveness.

A simple example (described by famed *USA Today* sports statistician Jeff Sagarin) will show how Runs Created and Linear Weights can be very inaccurate. Consider a player (let's call him Joe Hardy after the hero of the wonderful movie and play *Damn Yankees!*) who hits an HR during 50% of his plate appearances and makes an out during the other 50% of his plate appearances. Since Joe hits as many HRs as he makes outs, you would expect Joe "on average" to alternate HR, out, HR, out, HR, out and average three runs per inning. In the Appendix to Chapter 6 we will use the principle of conditional expectation to give a mathematical proof of this result.

	K	L	M	N	O	P
					Runs Created	Runs Created Per Game
3	Method	At Bats	Home Runs	Outs	Runs Created	Runs Created Per Game
4	Bill James	8748	4374	4374	8748	54
5	Linear Weights	8748	4374	4374	5957.26	36.77321

FIGURE 4.1   Runs Created and Linear Weights Predicted
Runs per Game for Joe Hardy.

In 162 nine-inning games our Joe Hardy will make on average
$162 * 27 = 4,374$ outs and hit 4,374 home runs. As shown in Figure 4-1
(see fileSimulationmotivation.xlsx), we find that runs created predicts that Joe Hardy would generate 54 runs per game (or six per
inning) and linear weights predicts Joe Hardy would generate 36.77
runs per game (or 4.01 runs per inning). Both these estimates are far
away from the true value of 27 runs per game!

INTRODUCTION TO MONTE CARLO SIMULATION

How can we show that our player generates three runs per inning
or 27 runs per game? By programming the computer to play out
many innings and averaging the number of runs scored per inning.
Developing a computer model to repeatedly play out an uncertain
situation is called Monte Carlo simulation.

Physicists and astronomers use Monte Carlo simulation to simulate the evolution of the universe. Biologists use Monte Carlo simulation to simulate the evolution of life on earth. Corporate financial
analysts use Monte Carlo simulation to evaluate the likelihood that
a new GM car model or a new Proctor and Gamble shampoo will be
profitable. Wall Street "rocket scientists" use Monte Carlo simulation to price exotic or complex financial derivatives. The term *Monte
Carlo simulation* was coined by the Polish born physicist Stanislaw
Ulam, who used Monte Carlo simulation in the 1930s to determine
the chance of success of the chain reaction needed for an atom bomb
to successfully detonate. Ulam's simulation was given the military

code name Monte Carlo, and the name Monte Carlo simulation has been used ever since.

How can we play out an inning? Simply flip a coin and assign a toss of heads to an out and a toss of tails to a home run. Or we could draw a card from a deck of cards and assign a red card to an out and a black card to an HR. Either the coin tossing or the card drawing method will assign a .5 chance to a home run and a .5 chance to an out. We keep flipping the coin or drawing a card (with replacement) until we obtain three outs. Then we record the number of HRs. We repeat this procedure 1,000 or so times and average the number of runs scored per inning. This average should closely approximate the average runs per inning scored by our hypothetical player. We will get very close to 3,000 total runs, which yields an estimate of three runs per inning. We choose to implement our simple Monte Carlo simulation using Microsoft Excel. See Figure 4-2 and file Simulationmotivator.xls. Excel contains a function RAND( ). If you type $= RAND( )$ in any cell and hit the F9 key the number in the cell will change. The RAND( ) function yields any number between 0 and 1 with equal probability. This means, for example, that half the time RAND( ) yields a number between 0 and .5 and half the time RAND( ) yields a number between .5 and 1. The results generated by the RAND( ) function are called random numbers. Therefore, we can simulate an inning for our player by assigning an outcome of an HR to a random number less than or equal to .5 and assigning an outcome of an out to a random number between .5 and 1. By hitting F9 in spreadsheet Simulationmotivator.xls we can see the results of a simulated inning. See Figure 4-2. For our simulated inning, each random number $\leq .5$ yielded an HR and any other random number yielded an out. For our simulated inning five runs were scored.

Cells J6:J1005 contain the results of 1,000 simulated innings while cell J3 contains the average runs per inning generated during our 1,000 hypothetical innings. (Note that rows 17 to 1,002 are hidden.) The chapter appendix explains how we used Excel's Data Table feature to perform our simulation 1,000 times. Whenever you hit F9

	B	C	D	E	F	G	H
2	Batter	Random Number	Result	Outs	Runs	Over?	Total Runs
3	1	0.9732	out	1	0	no	5
4	2	0.2423	HR	1	1	no	
5	3	0.7489	out	2	1	no	
6	4	0.3429	HR	2	2	no	
7	5	0.4219	HR	2	3	no	
8	6	0.0333	HR	2	4	no	
9	7	0.0767	HR	2	5	no	
10	8	0.9828	out	3	5	yes	
11	9					yes	
12	10					yes	
13	11					yes	
14	12					yes	
15	13					yes	
16	14					yes	

FIGURE 4.2   Simulating One Inning for Joe Hardy.

you will see cell J3 is very close to 3, sometimes a little lower and
sometimes a little higher, indicating that our player will generate
around three runs per inning or 27 runs per game (not 54 runs per
game as Runs Created predicts).

SIMULATING RUNS SCORED BY
A TEAM OF NINE TROUTS

Buoyed by the success of our simple simulation model, how could
we simulate the number of runs that would be scored by a team of,
say, nine Mike Trout 2016s? We need to follow through the progress
of an inning and track the runners on base, runs scored, and number
of outs. In our model the events that can occur on each plate appearance are displayed in Figure 4-3.

- We assume each error advances all base runners a single base.
- A long single advances each base runner two bases.
- A medium single scores a runner from second base but
  advances a runner on first one base.

	C	D
13		Event
14	1	Strikeout
15	2	Walk
16	3	HBP
17	4	Error
18	5	Long Single
19	6	Medium Single
20	7	Short single
21	8	Short double
22	9	Long double
23	10	Triple
24	11	Home run
25	12	Ground into Double Play
26	13	Normal Ground ball
27	14	Line drive or Infield fly
28	15	Long Fly
29	16	Medium Fly
30	17	Short Fly

FIGURE 4.3 Possible Batter Outcomes for Baseball Simulation.

- A short single advances all runners by one base.
- A short double advances each base runner two bases.
- A long double scores a runner from first.
- A GIDP (ground into double play) is a ground ball double play if there is a runner on first, first and second, first and third, or bases loaded. In other situations, the batter is out and the other runners stay where they are.
- A normal GO is a ground out that results in a force out with a runner on first, first and second, first and third, or bases loaded. We assume that with runners on second and third, the runners stay put; with a runner on third, the runner scores; and a runner on second advances to third.
- A long fly ball advances (if there are fewer than two outs) a runner on second or third base.
- A medium fly ball (if there are fewer than two outs) scores a runner from third.
- A short fly or a line drive or an infield fly does not advance any runners.

Our next step is to assign probabilities to each of these events. During recent seasons around 1.8% of all at bats result in an error. For a given player we input the information in the highlighted cells. Let's input Mike Trout's 2016 statistics. See Figure 4-4 and file Trout2016.xlsm. We note that our simulation omits relatively infrequent baseball events such as steals, caught stealings, passed balls, wild pitches, balks, etc.

	D	E	F
1	Outcome	Number	Probability
2	Plate Appearances	681	
3	At Bats+ Sacrifice Hits + Sacrifice Bunts	554	
4	Errors	10	0.0146843
5	Outs (in Play)	234	0.3436123
6	Strikeouts	137	0.2011747
7	Walks	116	0.1703377
8	Hit by Pitch	11	0.0161527
9	Singles	107	0.1571219
10	Doubles	32	0.0469897
11	Triples	5	0.0073421
12	Home Runs	29	0.0425844

FIGURE 4.4   Inputs to Trout Simulation.

For Trout, At bats + SH + SB = 554. He walked 116 times, hit 107 singles, etc.

Outs (in play) are plate appearances that result in non-strikeout outs.

Outs (in play) = (At Bats + SH + SB) − Hits − Errors − Strikeouts.

Historically, errors are 1.8% of At Bats + SH + SB, so we compute Errors = .018 * (At Bats + SH + SB).

Also,

Total plate appearances = BB + HBP + (At bats + SH + SB)
= 554 + 116 + 11 = 681.

We now compute the probability of various events as (Frequency of Event)/(Total Plate Appearances). For example, we estimate the probability of a Trout single as 107/681 = .157.

We need to also estimate probabilities for all possible types of singles, doubles, and outs in play. For example, what fraction of outs in play are GIDP? Using data from Earnshaw Cook's *Percentage Baseball* (1966) and discussions with *USA Today*'s Jeff Sagarin (who has built many accurate baseball simulation models) we estimated these fractions as follows.

- 30% of singles are long singles, 50% are medium singles, and 20% are short singles.
- 80% of doubles are short doubles and 20% are long doubles.
- 53.8% of outs in play are ground balls, 15.3% are infield flies or line drives, and 30.9% are fly balls.
- 50% of ground outs are GIDP and 50% are normal GOs.
- 20% of all fly balls are long fly balls, 50% are medium fly balls, and 30% are short fly balls.

To verify that these parameters provide an accurate representation of baseball, we simulated 50,000 innings using the composite major league statistics for the 2016 season. We found that our simulated runs per game were within 1% of the actual runs per game.

We now use the Excel simulation add-in @RISK to "play out" an inning thousands (or millions!) of times. You can download a free 15-day trial version of @RISK from Palisade.com. Basically, @RISK generates the event for each plate appearance based on the probabilities that we input (of course, these probabilities are based on the player we wish to evaluate). Essentially, for each plate appearance, @RISK generates a random number between 0 and 1. For example, for Trout a random number .157 would yield a single. This will cause 15.7% of Trout's plate appearances (just as in the actual 2016 season!) to result in a single. In a similar fashion, each other possible batter outcome will occur in the simulation with approximately the same probability as the outcome actually occurred!

One sample inning of our Trout 2016 simulation in which two runs were scored is shown in Figure 4-5.

	D	E	F	G	H	I	J
					Outs		
56	Outcome	State #	Outcome #	Runs	Made	Outs	Done?
57	Strikeout	1	1	0	1	1	no
58	Short double	1	8	0	0	1	no
59	Medium Fly	6	16	0	1	2	no
60	Long Single (advance 2 bases)	6	5	1	0	2	no
61	Walk	2	2	0	0	2	no
62	Long Single (advance 2 bases)	3	5	1	0	2	no
63	Line drive or Infield fly	4	14	0	1	3	yes

FIGURE 4.5 Example of Simulating a Single Inning.

The Entering State column tracks the runners on base; for example, 101 means runner on first and third while 100 means runner on first. The Outcome column tracks the outcome of each plate appearance using the codes shown in Figure 4-3. For example, event code 6 represents a medium single.

In the inning shown in Figure 4-5 our team of nine Trouts scored two runs. Playing out thousands of innings with @RISK enables us to estimate the average number of runs scored per inning by a team of nine Trouts. Then we multiply the average number of innings a team bats during a game (26.72/3) to estimate the number of runs created per game by Trout. Since we are playing out each inning using the actual probabilities corresponding to a given player, our Monte Carlo estimate of the runs per inning produced by nine Trouts (or any other player) should be a far better estimate than runs created or linear weights. Again, the Monte Carlo estimate of runs per game should be accurate for **any player, no matter how good or bad he is.** As we have shown with our Joe Hardy example, the accuracy of runs created and linear weights as a measure of hitting effectiveness breaks down for extreme cases.

SIMULATION RESULTS FOR TROUT, BRYANT, AND CABRERA (AND BONDS04)

For Trout16, Bryant16, Cabrera13, and Bonds04 our simulation yields the following estimates for Runs Created per game:

- Trout16: 9.38 runs per game.
- Bryant16: 7.95 runs per game.
- Cabrera13: 10.24 runs per game.
- Bonds04 : 21.02 runs per game!!!

There is a problem with our Bonds04 result. Barry Bonds received 232 walks during 2004. However, 120 of these walks were intentional. Of course, Bonds received intentional walks because pitchers would rather pitch to the other players (who were not as good at hitting as Bonds). For a team consisting of nine Bonds04 players there would be no point in issuing an intentional walk. Therefore, we reran our simulation after eliminating the intentional walks from Bonds's statistics. We found that the no Intentional Walks Bonds created 15.98 runs per game, which is still quite high in comparison to the other players.

HOW MANY RUNS DID TROUT ADD TO THE 2016 ANGELS?

To be honest, there will never be a team of nine Mike Trouts or nine Barry Bonds. What we really want to know is how many runs a player adds to his team. Let's try and determine how many runs Mike Trout 2016 added to the 2016 LA Angels. The hitting statistics for the 2016 LA Angels (excluding Mike Trout) are shown in Figure 4-6.

	D	E	F
	Outcome	Number	Probability
1	Plate Appearances	5360	
2	At Bats+Sacrifice Hits + Sacrifice Bunts	4962	
3			
4	Errors	89	0.0166045
5	Outs (in play)	2782	0.5190299
6	Strikeouts	854	0.1593284
7	Walks	355	0.0662313
8	HBP	40	0.0074627
9	Singles	848	0.158209
10	Doubles	247	0.0460821
11	Triples	15	0.0027985
12	Home Runs	127	0.023694

FIGURE 4.6 2016 LA Angels Statistics (sans Trout).

Note, for example, that 4.3% of Trout's plate appearances resulted in HRs but for the rest of the 2016 Angels only 2.4% of all plate appearances resulted in home runs. We can now estimate how many runs Trout added to the LA Angels. For the Angels without Trout we assume that each hitter's probabilities are governed by the data in Figure 4-6. Playing out 25,000 innings we found that, based on the runs per inning from our simulation, the Angels were projected to score an average of 626 runs without Trout. With Trout, the Angels actually scored 717 runs. So, compared to an average LA Angels hitter, how many wins can we estimate that Trout added? Enter the Pythagorean Theorem from Chapter 1. During 2016 the Angels gave up 727 runs. This yields a scoring ratio of $717/727 = 0.986$. The Pythagorean Theorem predicts they should have won $\dfrac{162 * .986^2}{.986^2 + 1} = 79.88$ games.

Without Mike Trout our simulation yielded a scoring ratio of $626/727 = .861$. Therefore without Mike Trout the Pythagorean Theorem predicts that the Angels would have won $\dfrac{161 * .861^2}{.861^2 + 1} = 68.97$ games.

Thus our model estimates that Trout added $79.88 - 68.97 = 10.91$ wins for the Angels (assuming that Trout plate appearances were replaced by an average non-Trout Angels hitter).

TROUT VS. THE AVERAGE MAJOR LEAGUER

In his *Historical Baseball Abstract* (2001), Bill James advocated comparing a player to an "average major leaguer." Let's try and determine how many extra runs an "average 2016" team would score if we replaced 681 of the average team's plate appearances by Trout's statistics (shown in Figure 4-4). The file Troutoveraverage.xlsx allows us to input two sets of player statistics. We input Trout's 2016 statistics in cells B2:B12. Then we input the average 2016 major league team's statistics in H2:I12. See Figure 4-7.

	H	I	J
1	Outcome	Number	Probability
2	Plate Appearances	6153	
3	At Bats+Sacrifice Hits + Sacrifice Bunts	5593	
4	Errors	101	0.01641476
5	Outs (in play)	2784	0.45246221
6	Strikeouts	1299	0.21111653
7	Walks	503	0.08174874
8	HBP	55	0.00893873
9	Singles	918	0.14919551
10	Doubles	275	0.04469365
11	Triples	29	0.00471315
12	Home Runs	187	0.03039168

FIGURE 4.7   Average 2016 Team Statistics.

We can see that Trout hit more HRs and had many more walks per plate appearance than the average 2016 batter. When simulating an inning, each batter's probabilities will be generated using either Trout's data from column D or the team data in column J. Since Trout had 681 plate appearances and the average team had 6,153 plate appearances, we choose each batter to be Trout (column D data) with probability $681/6{,}153 = .102$, and choose each batter to be an "average batter" (column J data) with probability $1 - .111 = .889$. After running 50,000 innings for the average team, and the team replacing 11.1% of the average team's at bats by Trout, we find the marginal impact is that Trout would increase the number of runs scored over an average team from 727 to 804. How many wins is that worth? With Trout, our scoring ratio is $804/727 = 1.106$. Using the Pythagorean Theorem of Chapter 1 we predict that the team with Trout would win $\dfrac{162 * (1.106)^2}{(1 + 1.106^2)} = 89.12$. Therefore, we estimate that adding Trout to an average team would lead to $89.12 - 81 = 8.12$ wins. We will see in Chapter 9 that an alternative analysis of Trout's 2016 batting record indicates that he added around 6.64 wins more than an average player.

CHAPTER 4 APPENDIX:
USING A DATA TABLE TO PERFORM
A SIMULATION IN EXCEL

In the cell range B2:H22 of the file Simulationmotivator.xlsx we have programmed Excel to "play out" an inning for a team in which each hitter who has a 50% chance of striking out or hitting an HR. Hit F9 and the number of runs scored by the team is recorded in cell H3. Note that whenever the Excel RAND( ) function returns a value <.5 the batter hits an HR, and otherwise the batter strikes out. To record the number of runs scored during many (say, 1,000) innings we enter the numbers 1 through 1,000 in the cell range I6:I1005. An easy way to accomplish this is to enter a 1 in cell I6 and, after choosing Fill from the Editing group on the Home tab, choose Series . . . and fill in the dialog box as shown in Figure 4-8.

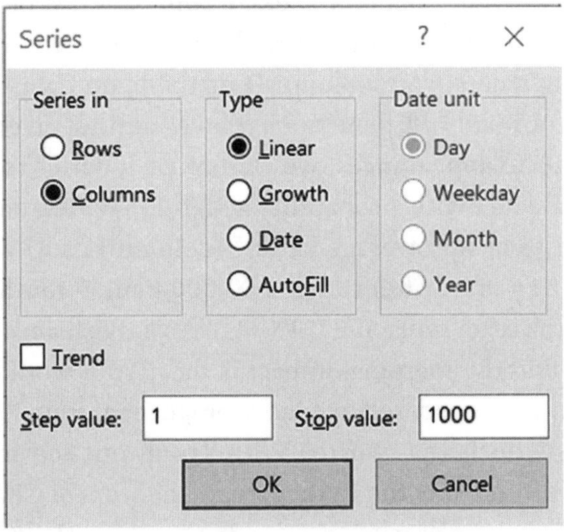

FIGURE 4.8   Entering the Numbers 1–1000 in a Column.

Next, we enter in cell J3 the formula (= H3) that we want to play out or simulate 1,000 times. Now we select the cell range I3:J1005 (this is called the Table Range). We select the Data tab and then

	I	J
2		**Average Runs per Inning**
3		3.01
4		
5	**Iteration Number**	1
6	1	6
7	2	6
8	3	5
9	4	4
10	5	0
11	6	0
12	7	7
13	8	1
14	9	2
15	10	6
16	11	1
17	998	1
18	999	8
19	1000	4

FIGURE 4.9 Simulating 1,000 Innings of Joe Hardy Hitting.

choose the What-If icon (the one with a question mark) and choose Data Table.

Next, we leave the row input cell blank and **choose any blank cell as your column input cell**. Then Excel puts the numbers 1, 2, . . . , 1000 successively in our selected blank cell. Each time cell H3 (runs in the inning) is recalculated as the RAND( ) functions in column C recalculate. Entering the formula = AVERAGE(I6:I1005) in cell J3 calculates the average number of runs scored per inning during our 1,000 simulated innings. For the 1,000 innings simulated in Figure 4-9, the mean number of runs scored per inning was 3.01.

CHAPTER 5

EVALUATING BASEBALL PITCHERS, FORECASTING FUTURE PITCHER PERFORMANCE, AND AN INTRODUCTION TO STATCAST

In Chapters 2–4 we have discussed three methods, runs created, linear weights, and Monte Carlo simulation, that can be used to evaluate the performance of a baseball hitter. We now turn our attention to evaluating the performance of baseball pitchers. As we will see, evaluating the performance of baseball pitchers is no easy matter.

Until recently, the most frequently used technique for evaluating the performance of pitchers was Earned Run Average. Let's consider a pitcher named Joe Hardy. Consider all the runners that Joe allows to reach base. Any of these base runners who scored or would have scored if Joe's team made no fielding miscues (such as an error, passed ball, etc.) results in Joe's being charged with an earned run. For example, if Joe gives up a triple with two outs in an inning and the next batter hits a single, Joe is charged with an earned run. Now suppose the next batter after the triple hits a ball to the SS that the shortstop misplays and is charged with an error. If the runner scores, this is an unearned run because without the error the runner would not have scored. A pitcher's earned run average (or ERA) is the number of earned runs he gives up per nine innings. For example, if Joe gives up 20 earned runs in 45 innings, he has given up $\dfrac{20*9}{45} = 4$ runs per nine innings. Then Joe's ERA is 4. In general, a pitcher's ERA is computed as $\dfrac{\text{(Earned runs allowed)}*9}{\text{Innings Pitched}}$.

PROBLEMS WITH ERA

There are several problems with evaluating pitchers by ERA.

1. Errors are subjective. Some official scorers are more reluctant than others to call a batted ball an error. Kalist and Spurr (2006) found slight evidence that official scorers were biased in favor of the home team.

2. When a starting pitcher is pulled from the game and there is at least one base runner, the number of earned runs he gives up depends greatly on the performance of the relief pitcher. For example, suppose that Joe leaves the game with two outs and the bases loaded. If the relief pitcher gets the next out, Joe is charged with no earned runs while if the relief pitcher gives up a grand slam, then Joe is charged with three earned runs.

3. A pitcher with good fielders behind him will clearly give up fewer earned runs than a pitcher with a leaky defense. We will discuss the evaluation of fielders in Chapter 7.

Starting pitchers are often evaluated on their win-loss record. This clearly depends on the batting support the pitcher receives.

For example, in 2018, New York Met pitcher Jacob DeGrom led the National League with a 1.70 ERA, and due to abysmal hitting support had a mediocre 10–9 win-loss record.

Relief pitchers are often evaluated on saves. Most saves credited to relief pitchers are given to a relief pitcher who faces a batter representing the tying run. The official definition of a Save is

> The official scorer shall credit a pitcher with a save when such pitcher meets all four of the following conditions:
> (1) He is the finishing pitcher in a game won by his team;
> (2) He is not the winning pitcher;
> (3) He is credited with at least a third of an inning pitched; and
> (4) He satisfies one of the following conditions:
>   (a) He enters the game with a lead of no more than three runs and pitches at least one inning;
>   (b) He enters the game, regardless of the count, with the potential tying run either on base, or at bat or on deck; or
>   (c) He pitches for at least three innings.

To paraphrase George Orwell: "All saves are created equal, but some saves are more equal than others." In looking at a relief pitcher's published statistics, all we see is the number of saves, so all saves seem equal. However, consider a relief pitcher who comes in with his team ahead 3–2 during the top of the ninth inning, and the other team has bases loaded and none out. If this pitcher holds the lead, he has done a fabulous job. Consider a second relief pitcher who enters the game with two outs in the ninth and a runner on first and a 4–2 lead. He strikes out the next batter and receives a save. Clearly the first relief pitcher deserves much more credit, but each pitcher receives a save.

In Chapter 8 we will show how the concept of win probability added (WPA) resolves many of the problems involved in evaluating pitcher performance. WPA also allows us to compare the value of relief pitchers and starting pitchers.

USING PAST ERA TO PREDICT FUTURE ERA DOES NOT WORK WELL!

Despite the problems with ERA, it seems important to be able to predict a pitcher's future ERA from his past performance. This would aid baseball management in their quest to improve their team's future pitching performance.

It seems logical to try and predict a pitcher's ERA for the next season from his last season's ERA. For a long time the baseball community thought that this approach would yield good predictions of next year's ERA. Let's check out this hypothesis. For all pitchers who pitched in both 2015 and 2016, Figure 5-1 (see the file XERA.xlsx) plots on the x-axis the pitcher's ERA during the 2015 season and on the y-axis we plot the pitcher's ERA during the 2016 season. We used Excel's Trend Curve feature to plot the line (see the chapter appendix for details on using the Trend Curve feature) that best fits this data.

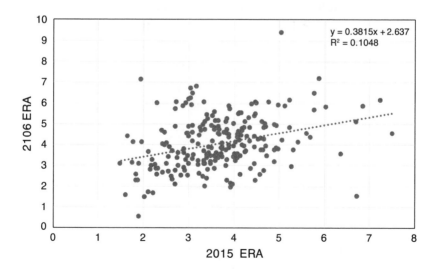

FIGURE 5.1   Predicting Next Year's ERA from Last Year's ERA.

This graph indicates that the best fitting[1] straight line that can be used to predict next year's ERA from last year's ERA is the equation

$$(2016 \text{ ERA}) = 2.637 + .3815 * (2015 \text{ ERA}). \tag{1}$$

For example, a pitcher who had an ERA of 4.0 last year would be predicted to have an ERA next year of $2.637 + .3815 * (4) = 4.16$ runs per game.

From the graph we see that the best fitting line does not fit the data very well. Many pitchers with predicted 2016 ERAs of around four runs actually have ERAs of over five or below three runs per game! Statisticians quantify how well a line fits a set of data using the R-squared and correlation as well as the mean absolute deviation of the regression forecasts.

R² AND CORRELATION

- R-squared value (RSQ for short): From Figure 5-1 we find the RSQ value for predicting 2016 ERA from 2015 ERA is .1048. This indicates that last year's ERA explains only 10.5% of the variation in next year's ERA. In other words, over 89% of the variation in next year's ERA is unexplained by last year's ERA. Statisticians also measure **linear association** by looking at the square root of RSQ (using the same sign as that of the slope of the least squares line), which is often called r or the correlation coefficient. We find that the correlation between next year's ERA and last year's ERA is $\sqrt{0.104} = 0.32$. Correlation is a unit-free measure of linear association between two datasets that is always between −1 and +1. Call our two datasets X and Y.
- The correlation (usually denoted by r) between two variables (call them X and Y) is a unit-free measure of the strength

---

1. Excel chooses as best fitting the line that minimizes the sum of the squared vertical distances of the points to the fitted line. This is called the least squares line.

of the linear relationship between X and Y. The correlation between any two variables is always between –1 and +1. The exact formula used to compute the correlation between two variables is not very important.[2] It is important, however, to be able to interpret the correlation between two variables X and Y.

- A correlation near +1 means that there is a strong positive linear relationship between X and Y. That is, when X is larger than average Y tends to be larger than average, and when X is smaller than average, Y tends to be smaller than average. Alternatively, when a straight line is fit to the data, there will be a straight line with positive slope that does a good job of fitting the points. As an example, for the data shown in Figure 5-2 (here X = units produced and Y = monthly production cost) X and Y have a correlation of +0.95.

- A correlation near –1 means that there is a strong negative linear relationship between X and Y. That is, when X is larger than average Y tends to be smaller than average, and when X is smaller than average, Y tends to be larger than average. Alternatively, when a straight line is fit to the data, there will be a straight line with negative slope that does a good job of fitting the points. As an example, for the data shown in Figure 5-3 (here X = price charged for a product and Y = product demand), X and Y have a correlation of –0.90.

- A correlation near 0 means that there is a weak linear relationship between X and Y. That is, knowing whether X is larger or smaller than its mean tells you little about whether Y will be larger or smaller than its mean. Figure 5-4 shows the dependence of Y = unit sales on X = years of sales experience. Years of experience and unit sales have a correlation of 0.003. In our dataset, average experience is 10 years. We see that when a person has more than 10 years of sales experience, his sales

2. If you wish, however, you can use Excel to compute correlations between two columns of numbers with the = CORREL function.

can either be low or high. We also see that when a person has
less than 10 years of sales experience, sales can be low or high.
Although there is little or no linear relationship between experience and sales, we see there is a strong nonlinear relationship
(see fitted curve) between years of experience and sales. Correlation does not measure the strength of nonlinear relationships.

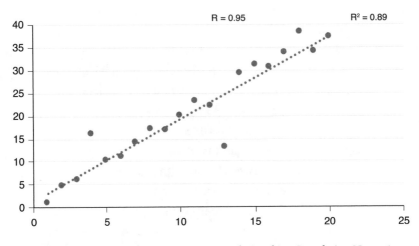

FIGURE 5.2   Strong Positive Linear Relationship: Correlation Near +1.

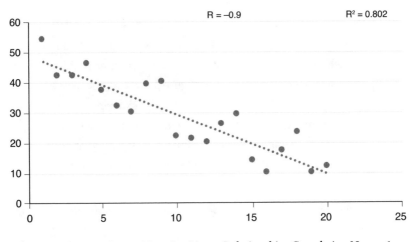

FIGURE 5.3   Strong Negative Linear Relationship: Correlation Near −1.

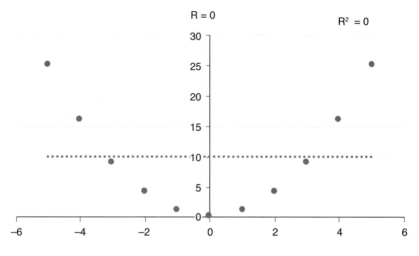

FIGURE 5.4 Weak Linear Relationship.

MEAN ABSOLUTE DEVIATION OF FORECASTS

- The mean absolute deviation (MAD for short) of the forecast errors) is a commonly used measure of forecast accuracy. For each pitcher we compute the predicted ERA from (1) and take the absolute value of (Predicted ERA) − (Actual ERA). We find the MAD for predicting next year's ERA from last year's ERA to be .93 runs. In other words, our error on average in predicting next year's ERA from last year's ERA is .93 runs.

VOROS McCRACKEN STUNS THE BASEBALL WORLD!

Voros McCracken (see http://www.baseballprospectus.com/article .php?articleid=878, January 2001) appears to be the first person to have successfully explained why future ERAs are hard to predict from past ERAs. McCracken observed that a pitcher's effectiveness is primarily based on the following:

1) The fraction of BFPs (batters faced by pitchers) that result in balls in play (a ball in play is a plate appearance that results in a ground out, error, single, double, triple, fly out, or line out).

2) The fraction of balls in play that result in hits (referred to as BABIP, or batting average on balls in play).

3) The outcome of BFPs that do not result in balls in play. What fraction of BFPs that do not yield a ball in play result in strikeouts, walks, HBPs, or home runs?

McCracken's brilliant insight was to realize that a pitcher's future performance on (1) and (3) can be predicted fairly well from past performance, **but it is very difficult to predict (2) from past performance.** Suppose we try and predict a pitcher's next season's fraction of BFPs resulting in strikeouts from the percentage of BFPs resulting in strikeouts the last season. We find $r = .78$. For BBs we find $r = .66$ and for HRs $r = .34$. McCracken called SOs, BBs, HBPs, and HRs defense-independent pitching statistics (DIPS for short) because these results are independent of the team's fielding ability. DIPS seem to be much more predictable from season to season than ERA. However, the fraction of balls in play resulting in an out or a hit seems to be very hard to predict.[3] For example, a pitcher's BABIP for the next season has only a .24 correlation with a pitcher's BABIP for the last season. **The unpredictability of BABIP is what makes it so difficult to predict next year's ERA from last year's ERA.** McCracken sums things up when he states:

> The pitchers who are the best at preventing hits on balls in play one year are often the worst at it the next. In 1998, Greg Maddux had one of the best rates in baseball, then in 1999 he had one of the worst. In 2000, he had one of the better ones again. In 1999, Pedro Martinez had one of the worst; in 2000, he had the best. This happens a lot.

I believe that, most of the time, luck and season-to-season differences in team fielding quality are major factors in causing the lack of predictability of BABIP. In some cases, however, players continu-

---

3. Later researchers found that for certain types of pitchers (particularly knuckleball pitchers) the outcome of balls in play is much easier to predict.

ally "ace the BABIP exam." Average BABIP is consistently around 0.30. Mike Trout has an average BABIP through 2016 of .360. This is the highest value in baseball since World War II. This incredible performance can't be due to chance! On the other hand, Clayton Kershaw's career BABIP is around .270. Again, this cannot be due to chance and indicates that when Kershaw does not strike out hitters, their balls in play are less likely to be hits than the balls hit by an average major leaguer.

DICE: A BETTER MODEL FOR PREDICTING A PITCHER'S FUTURE PERFORMANCE

So how can we use McCracken's insights to better predict a pitcher's future ERA? McCracken came up with a very complex method to predict future ERAs. Since DIPS are fairly well predictable from year to year it seems reasonable that there should be some simple combination of DIPS (BBs, SOs, HBPs, and HRs) that can be used to predict next year's ERA more accurately than our previous approach. Clay Dreslogh in 2001 (see https://web.archive.org/web/20070528164743 /http://www.sportsmogul.com/content/dice.htm) came up with a simpler formula, known as Defense-Independent Component ERA, or DICE, to predict next year's ERA.

$$DICE = Constant + (13 * HR + 3 * (BB + HBP) - 2 * K)/IP\ (2) \tag{2}$$

The website fangraphs.com lists a pitcher's DICE as FIP (Fielder Independent Pitcher) statistic. The constant for a given season is chosen so League Average FIP = League Average ERA. The constant is usually around 3.1. FanGraphs also lists a pitcher's FIP−, which is a park adjusted version of FIP. See FanGraphs's excellent glossary for more details.

DICE provides a "fielding independent" estimate of a pitcher's season performance. For example, in 2016, Clayton Kershaw had the following statistics:

- 11 BBs
- 2 HBPs
- 172 Ks
- 8 HRs
- 149 IPs

Substituting these numbers into (2) using a constant value of 3.10 we obtain a fielding-independent estimate of Kershaw's 2016 pitching effectiveness on an ERA scale as

$$DICE = 3.10 + (3 * (11 + 2) + 13 * 8 - 2 * 172)/149 = 1.75.$$

Kershaw's actual ERA in 2016 was 1.80. Kershaw's actual ERA probably outperformed his DICE number due to Kershaw's fantastic ability to reduce BABIP.

As another example, consider Kyle Hendricks's pitching statistics for the 2016 World Champion Chicago Cubs (I still can't write this without thinking it's a mirage!). Hendricks's DICE was 3.16, yet his actual ERA was much better (2.13). This stark difference is probably due to Hendricks's having an amazing BABIP of .250. During the three previous seasons, Hendricks's BABIP averaged .296, near the league average. This suggests that Hendricks's ERA would be likely to rise in future seasons, due to his BABIP reverting to a number closer to the major league average of .300. Sure enough, during the 2017–2019 seasons, Hendricks's combined ERA was 3.33.

DOES DICE PREDICT FUTURE ERA BETTER THAN PAST ERA?

For each pitcher who pitched at least 100 innings in 2019, we tried to predict his 2019 ERA from his 2018 DICE. The results (see file Dice18_19.xlsx) are shown in Figure 5-5.

Last year's DICE explains 13.5% of variation in next year's ERA (compared to last year's ERA, which explained only 10.5% of next year's ERA). The correlation between 2018 DICE and 2019 ERA is

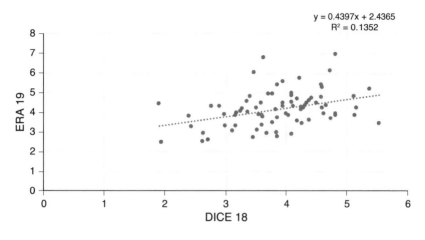

FIGURE 5.5 Predicting Future ERA using DICE.

.37 (last year's ERA had only a .34 correlation with next year's ERA). On average our predictions of next year's ERA from last year's DICE were on average off by only .64 runs (predictions based on last year's ERA were off by .93 runs).

Surprisingly, our study shows that we can more accurately predict next year's ERA from HRs BBs, HBPs, and Ks than from last year's ERA. In Chapter 8 we will use Win Probability to evaluate a pitcher's performance.

CAN STATCAST HELP US PREDICT FUTURE PITCHER PERFORMANCE?

In 2015 MLB introduced the fantastic Statcast system. Statcast is a combined radar- and camera-based system that collects lots of valuable data on each pitch and play. You can download a text file containing the dataset of interest from https://baseballsavant.mlb.com/statcast_search. For pitchers, Statcast tracks the following information:

Active Spin: For each type of pitch, Statcast tracks the percentage of spin that contributes to the movement of the ball.

Extension: Extension measures how much closer than 60 feet 6 inches the pitcher is when he releases the ball. A pitcher with an extension of 4 feet makes a 90 mph fastball with no extension look like a $90 * (60.5/56) = 97$ mph fastball.

Perceived Velocity: Incorporating extension, perceived velocity adjusts actual velocity of a pitch relative to the average MLB extension to show the speed perceived by a hitter. Of course, if two pitchers throw the ball at the same speed, the pitcher with more extension will have the higher perceived velocity.

Pitch Movement: Relative to an average pitcher, pitch movement is measured in inches horizontally and vertically. The statistic is adjusted for the fact that gravity pulls pitches down. During 2019, Trevor Bauer's curveball averaged 9.5 inches more movement than the average curveball.

Putaway Percentage: The fraction of two strike pitches that result in a strikeout.

Spin Rate: The number of revolutions per minute the baseball makes on a pitch. In 2020 Garrett Richards of the Padres had the highest average curveball spin rate (3,343 rpm).

XERA: We have discussed how luck often determines if a batted ball results in a hit or an out. XERA attempts to determine what a pitcher's ERA would be if he had average luck (based on where his balls pitched are hit and how fast they are hit). The file XERA.xlsx contains the ERAs and XERAs for all MLB pitchers during the 2015 and 2016 seasons. We used Excel's RSQ function to determine how well a pitcher's 2015 ERA or XERA predicts the same 2016 statistics. We found the following.

- 2015 ERA explains only 10.5% of variation in 2016 ERA, while 2015 XERA explains a much higher percentage (18%) of 2016 ERA.

- 2015 XERA explains 23% of variation in 2016 XERA. Since XERA attempts to take luck out of the equation, it is not surprising that XERA is more predictable than ERA.

TABLE 5.1

Type of Hit by Launch Angle

Launch Angle	Type of Hit
0–10 degrees	Ground Ball
10–25 degrees	Line Drive
25–50 degrees	Fly Ball
> 50 degrees	Popup

STATCAST BATTING METRICS

In Chapters 2–4 we discussed various methods for evaluating hitters. Statcast tracks many interesting batting metrics that can provide further insight into a hitter's ability.

> **Launch Angle:** The average angle at which the ball leaves the bat. Table 5-1 summarizes how launch angle determines the result of the at bat.
> **Exit Velocity:** The average speed at which the baseball comes off the bat.
> **Barrel:** A barrel is any batted ball that is expected to lead to a batting average of at least .500 and a slugging percentage of at least 1.5. A barrel requires an exit velocity of at least 98 mph. At a 98 mph exit velocity, a barrel requires a launch angle between 26 and 30 degrees. At higher speeds a wider-angle range yields a barrel. For example, with an exit velocity of 100 mph, a batted ball is considered a barrel if the launch angle is between 24 and 33 degrees.

A HOLY GRAIL OF SPORTS ANALYTICS

In this chapter we have briefly discussed the problems involved in developing a model to predict future performance by a baseball pitcher. There are many other important sports problems which involve predicting future performance of a player or team from past performance. Some examples follow:

- Predict performance of an NBA, an NFL, or an MLB draft pick from its high school, college, or international performance.
- Predict a batter's future runs created.
- Predict a running back's, quarterback's, or wide receiver's future performance from his past professional performance.
- Predict a team's record for the next season based on the past season performance and player trades.
- Predict likelihood of a pitcher needing a Tommy John surgery.

Many people have developed such forecasting models. For example, each season, Ron Shandler and Bill James put out projections for each MLB's next season performance. Each year, Aaron Schatz's *Football Prospectus* (and dozens of fantasy football magazines) "*predict*" NFL team and player performance for the next season. Each year, ESPN.com's Kevin Pelton and FiveThirtyEight.com's Nate Silver predict NBA team and player performance for the next season.

While these forecasts are fascinating, what is really needed is a comparison of the forecasting accuracy of various methods. Then we can judge which forecasts to use in our fantasy leagues, draft decisions, or player personnel decisions. For example, who predicts future baseball performance better: Ron Shandler or Bill James? Toward this end, for example, FiveThirtyEight has opened its various forecasts of game outcomes to the public (see https:// fivethirtyeight.com/features/when-we-say-70-percent-it-really -means-70-percent/), and hence, anyone can evaluate their accuracy. Hopefully in the future databases of accuracy of sports forecasts (and stock pickers!!) will be commonplace on the Internet.

CHAPTER 5 APPENDIX:
USING THE EXCEL TREND CURVE

The Excel Trend Curve feature enables us to plot the line that best fits a set of data. In the file XERA.xlsx we are given the ERAs of pitchers who pitched during the 2015 and 2016 seasons. We would like to plot on the x-axis each pitcher's 2015 ERA, and plot on the y-axis the same pitcher's 2016 ERA. Then we want to graph the line (and obtain the equation of the line) that best fits this relationship. The best fitting line is shown in Figure 5-1.

To obtain Figure 5-1, proceed as follows:

- Select the data to be graphed (cell range F5:G235 of the worksheet 2016).
- Then from the Insert tab choose the first option from the Scatter icon (the icon showing the x- and y-axes and some dots). Then right-click on any of the charted points and select Add Trendline . . . and choose the Linear option. Then check the Display Equation and Display R-squared options. You will then see the graph shown in Figure 5-1.

We find that the straight-line relationship based on last year's ERA that best predicts a pitcher's ERA during 2016 is (2016 ERA) = 2.637 + .3815 * (2015 ERA).

This equation explains 10.5% of the variation in 2016 ERA.

BASEBALL DECISION MAKING

During the course of a season managers make many crucial decisions including the following:

- With a man on first and nobody out, should we attempt a sacrifice bunt to advance the runner to second base?
- With a man on first and one out, should we attempt to steal second base?
- If we are the home team and the score is tied in the top of the ninth inning, and the opposing team has a man on third base and none out, should we play the infield in? This increases the chance of a hit (most people think bringing the infield in makes a .250 hitter a .300 hitter), but bringing the infield in ensures that a ground out will not score the runner from third.

Let's focus first on bunting with nobody out and a runner on first. If the bunt succeeds, we have a runner on second who is now one base closer to scoring, but we have given up a precious out.

Is the benefit of the extra base worth giving up the precious out? If we steal successfully with a runner on first, we have moved the runner one base closer to scoring and we still have one out, so we are clearly better off than we were before the stolen base. If the runner is caught stealing, however, we have basically killed off our inning because we have two outs and nobody on base.

POSSIBLE STATES DURING
A BASEBALL GAME

The key to developing a framework for baseball decision making is to realize that during an inning a team is in one of the 24 situations (often called states) listed in Figure 6-1

Each state is denoted by four numbers. The first number is the number of outs (0, 1, or 2). The second number lets us know if first base is occupied (1 = base occupied, 0 = base not occupied). Similarly, the third and fourth numbers tell us if second and third base, respectively, are

State	Outs	Runner on first?	Runner on second?	Runner on third?
0000	0	No	No	No
1000	1	No	No	No
2000	2	No	No	No
0001	0	No	No	Yes
1001	1	No	No	Yes
2001	2	No	No	Yes
0010	0	No	Yes	No
1010	1	No	Yes	No
2010	2	No	Yes	No
0011	0	No	Yes	Yes
1011	1	No	Yes	Yes
2011	2	No	Yes	Yes
0100	0	Yes	No	No
1100	1	Yes	No	No
2100	2	Yes	No	No
0101	0	Yes	No	Yes
1101	1	Yes	No	Yes
2101	2	Yes	No	Yes
0110	0	Yes	Yes	No
1110	1	Yes	Yes	No
2110	2	Yes	Yes	No
0111	0	Yes	Yes	Yes
1111	1	Yes	Yes	Yes
2111	2	Yes	Yes	Yes

FIGURE 6.1 Possible States during an Inning.

occupied. For example, 1010 means there is one out and a runner on second base and 2001 means there are two outs and a runner on third base. Intuitively we know the best state is 0111 (bases loaded, nobody out) and the worst state is 2000 (two outs, nobody on). How can we explicitly measure how much better one state is than another? Simply look at many past games and look at the **average number of runs scored in each situation**. Thanks to the website Bats and Stats (https://batsandstats.com/2016/08/03/one-of-my-favorite-baseball-stats/), we always have access to the latest run expectancy matrix. Figure 6-2 shows the expected runs scored during 2016 in each of the 24 possible states.

Situation	0 Outs	1 Out	2 Out
000	0.5062	0.2737	0.1028
001	1.3163	0.9225	0.3638
010	1.0932	0.668	0.3174
011	1.9033	1.3168	0.5784
100	0.8744	0.5263	0.2199
101	1.6845	1.1751	0.4809
110	1.4614	0.9206	0.4345
111	2.2715	1.5694	0.695

FIGURE 6.2   Expected Runs.

For example, with a runner on first and second and one out (state 1110), teams scored an average of .92 runs. The information in our table of expected runs is vital to proper baseball decision making. To see why, let's look at the decision of whether to bunt with a runner on first and none out. This is state 0100. Since our table aggregates data over all teams and batters, we will assume the numbers in Figure 6-2 refer to the expected numbers of runs scored given that "an average" hitter is at bat. In state 1000 we see that an "average" team is expected to score .87 runs. Now if we attempt a sacrifice the bunt might succeed in advancing the runner on first to second with the batter being out (leading us to state 1010) or fail by having the lead runner out and the batter reaching first (state 1100). These are by far the most common outcomes when we bunt. If, on average, we score more runs with a bunt than without a bunt, then bunting is a good idea. So how do we figure out the average number of runs that our team will score after bunting? Before proceeding further, a brief introduction to some important concepts in mathematical probability theory is in order.

EXPERIMENTS AND RANDOM VARIABLES

First let's define the important concepts of an **experiment** and a **random variable**. An experiment is any situation whose outcome is uncertain. Examples of experiments include

- Tossing a pair of dice (outcomes of 1, 2, 3, 4, 5, or 6 are possible for each die).
- A batter's plate appearance (the many possible outcomes include HR, single, strikeout, etc.).
- Shooting a free throw (outcomes include made free throw, missed free throw rebounded by us, and missed free throw rebounded by the other team).
- A QB throws a pass (outcomes include incomplete, interception, completion for 10 yards, completion for 15 yards, etc.).
- An FG kicker attempts a field goal (outcome is either a made or a missed field goal).

Random variables can be associated with experiments. Here are some examples:

- The sum of the total on two dice (possible values include 2, 3, . . . , 10, 11, 12).
- Number of runners batted in during a batter's plate appearance (0, 1, 2, 3, and 4 are possible values).
- Number of points scored on free throw (possible values are 0 and 1).
- Number of points scored on passing play (0 and 6 are possible).
- Number of points scored on a field goal (0 and 3 are possible).

EXPECTED VALUE

We will denote random variables by capitalized boldface letters. In our analysis of baseball, basketball, and football we will often need to determine the **expected value** of a random variable. The expected value of a random variable is the average value of the random variable we can expect if an experiment is performed many times. In general, we find the expected value of a random variable as follows:

$$\sum_{\text{all outcomes}} (\text{Probability of outcome}) *$$

(Value of random variable for outcome).

For example, if we toss a die, each possible outcome has probability $1/6$. Therefore if we define the random variable $\mathbf{X} =$ number of dots showing up when die is tossed, then $E(\mathbf{X}) =$ expected value of $\mathbf{X} = \frac{1}{6}(1) + \frac{1}{6}(2) + \frac{1}{6}(3) + \frac{1}{6}(4) + \frac{1}{6}(5) + \frac{1}{6}(6).$

Therefore if we were to toss a die many times and average the total number of dots showing, we would expect to get a number near 3.5.

In baseball, we will compare various decisions based on expected runs scored. For example, if expected runs scored is higher if we do not bunt than if we bunt, then we should not bunt. In football or basketball, we will usually compare decisions based on expected number of points by which we beat our opponent. For example, suppose that early in a football game we face fourth and three on our opponent's 35-yard line. If an FG means we beat the opposition on average by .5 points during the rest of the game, and going for the first down means we beat the opponents by an average of 1.5 points during the rest of the game, then we should eschew the field goal and go for the first down.[1]

---

1. If we are near the end of the game, however, we should maximize the probability of winning the game. If we are not near the end of game, then maximizing choosing decisions based on maximizing expected number of points by which we beat an opponent is virtually equivalent to maximizing our probability of victory. We will use this approach in Chapter 22 to study the basis for making important football decisions, such as, should I go for a field goal on fourth down?

In computing expected values of random variables we will often use the **Law of Conditional Expectation:**

$$\text{Expected Value of Random Variable} =$$
$$\sum_{\text{all outcomes}} (\text{Probability of outcome}) *$$
$$(\text{Expected value of random variable given outcome})$$

For example, suppose in football a running play gains an average of five yards if an opponent plays a passing defense and an average of three yards if an opponent plays a rushing defense. Also assume the opponent plays a rushing defense 40% of the time and passing defense 60% of the time. Then we can use the law of conditional expectation to compute the expected number of yards gained on a running play as (probability of pass defense) * (expected yards gained given pass defense is played) + (probability of run defense) * (expected yards gained given run defense is played) = ((.6)(5) + (.4) * (3)) = 4.2 yards per play.

TO BUNT OR NOT TO BUNT: THAT IS THE QUESTION

We are now ready to determine whether bunting with a man on first and nobody out is a good play. In his book *Baseball Hacks* (O'Reilly, 2006), Joseph Adler tabulated (for the 2004 season) the results of bunts with a runner on first and found the results shown in Figure 6-3.

From Figure 6-2 we find that in the current state (0100) we will score on average .87 runs. Since this is based on data from all teams and players, this number essentially assumes an average batter is at the plate. If a great hitter is up our expected runs would be more than .87 runs while if a poor hitter is up our expected run would be fewer than .93 runs. Applying the law of conditional expectations to the data in Figure 6-3 we find the expected number of runs scored after the bunt is given by .10(1.46) + .70(.67) + .02(.10) + .08(.53) + .10(.53) = .71 runs. This calculation shows us that bunting makes us on aver-

Result	Resulting State	Probability	Expected Runs (from Figure 6-2)
Batter is safe and runner advances to second base	0111	0.1	1.46
Runner advances to second base and batter is out	1010	0.7	0.67
Both runners are out	2000	0.02	0.1
Runner is out at second base and batter reaches first base	1100	0.08	0.53
Batter is out and runner remains on first base	1100	0.1	0.53

FIGURE 6.3   Results of Bunt with Runner on First.

age .16 run (.71 − .87 = −.16) worse off than we are if we do not bunt. Therefore, bunting is not a good idea if an average hitter is up and the goal is to maximize the expected number of runs in an inning.

HOW ABOUT IF BATTER IS A POOR HITTER?

What if a bad hitter is up? Let's assume a weak hitting pitcher, Joe Hardy, is up. Joe strikes out 85% of the time, hits a single 10% of the time, and walks 5% of the time. We will assume the single always advances a runner on first to third base. If we do not bunt with Joe at the plate, the law of conditional expectation tells us the expected number of runs we will score in the inning is given by $.85 * E(1100) + .10 * E(0101) + .05 * E(0110)$, where E(state) is expected runs scored in that state. Substituting in the expected runs for each state from Figure 6-2, we find that with our weak hitter up we can expect to score $.85(.53) + .10(1.68) + .05(1.46) = .69$ runs.

Therefore, for Joe Hardy, bunting would increase the number of expected runs.

IS BUNTING WORTHWHILE WITH THE SCORE TIED?

Is bunting worthwhile when we only need to score one run? For example, suppose the score is tied in the bottom of the ninth inning and we have a runner on first with none out. If we score a run, we win the game. Should we bunt? We know that unless a very weak hitter is up, bunting will decrease the expected number of runs we score. In this situation, however, we want to look at **the probability of scoring at least one run.** In his excellent book *Baseball Between the Numbers* (2006), Jonah Keri tabulated the probability of scoring at least one run for all 24 states. With a runner on first and none out the probability of scoring at least one run is .417. The other probabilities that are germane to our example are summarized in Figure 6-4.

Now the law of conditional expectation tells us that bunting will yield a probability $.10(.625) + .70(.41) + .02(.071) + .08(.272) + .10(.272) = .40$ of scoring at least one run. Therefore, bunting and not bunting give us just about the same probability of scoring at least one run. Therefore, if an average hitter is at bat, we are virtually indifferent between bunting and not bunting.

TO STEAL OR NOT TO STEAL (A BASE!)?

Let's now examine the stolen base decision. Let's suppose we have a runner on first base and none out. Let p = probability of a successful steal of second base. For which values of p should we steal? In our current state (0100) we expect to score .87 runs. If

Result	Resulting State	Probability	Probability of scoring at least one run
Batter is safe and runner advances to second base	110	0.1	0.625
Runner advances to second base and batter is out	1010	0.7	0.41
Both runners are out	2000	0.02	0.071
Runner is out at second base and batter reaches first base	1100	0.08	0.272
Batter is out and runner remains on first base	1100	0.1	0.272

FIGURE 6.4 Probabilities of Scoring at Least One Run.

the steal is successful, the new state is 0010, in which we expect to score 1.09 runs. If the steal is unsuccessful, the new state is 1000, in which we expect to score .27 runs. Therefore, if we steal, the law of conditional expectation tells us our expected runs scored is $1.09p + .27(1-p)$. As long as $1.09p + .27(1-p) > .87$, we should steal second base. Solving this inequality, we find we should steal if $.82p > .60$ or $p > .60/.82 = .73$. Therefore, if our chance of stealing second base exceeds 73%, then trying to steal second base is a good idea. During 2016 71% of all stolen base attempts were successful. This success rate (if it is for stealing with a runner on first and none out) is not consistent with good decision making. Of course, the breakeven success rate depends on the score, opposing pitcher, and hitters coming up.

To determine the breakeven success probability for other states, define

	I	J	K	L	M
7	Status quo	Success	Failure		Probability
8	0.87	1.09	0.27	1st	0.7317073
9	1.09	1.32	0.27	2nd	0.7809524
10	1.32	1.51	0.27	3rd	0.8467742

FIGURE 6.5 Breakeven Percentages for Stolen Base Attempts.

- S = Expected runs if stolen base attempt is successful
- F = Expected runs if stolen base attempt fails
- SQ = Expected runs given the current situation.

If we let P = chance stolen base attempt succeeds, then the stolen base increases expected runs if and only if $P * S + (1 - P) * F > SQ$ or $P > (SQ - F)/(S - F)$. The workbook Stolenbase.xlsx and Figure 6-5 show the breakeven success rates in each situation.

ARE BASE RUNNERS TOO CONSERVATIVE WHEN TRYING TO ADVANCE ON A SINGLE OR DOUBLE?

If there is a runner on first and a single is hit, the coaches and runner must decide whether to try and advance to third base or stop at second base. If a runner is on second and a single is hit, the runner and coaches must decide whether .to stop at third base or try and score. If a runner is on first and a double is hit, the runner and coaches must decide whether to try and score or stop at third base. We will now show that most major league teams are much too conservative when deciding whether the base runner should try and "go for the extra base."

Let's suppose we have a runner on first and nobody out. If we hit a single and the runner makes it to third base, we find from Figure 6-2 we are in a situation (first and third nobody out) worth on average 1.68 runs. If the runner is out we now have a runner on first with one out (assuming the batter does not take second on the throw). In this situation, we average scoring .53 runs. If the runner stops at second, we are in a situation (runners on first and second with none out) that yields an average of 1.46 runs. Let P = probability runner makes it from first to third. Then we will maximize our expected runs by trying to advance if and only if $P * (1.68) + (1 - P) * .53 \geq 1.46$. We find this inequality is satisfied for $P \geq .93/1.15 = 81\%$. This implies that a runner on first with nobody out should try and go for third if his chance of success is at least 81%. According to data from the 2005 MLB season (see http://baseballanalysts.com/archives/2005 /10/can_baserunning.php), base runners trying to go from first to third are thrown out only 3% of the time. This means there are probably many situations[2] in which base runners had an 80–90% chance of making it from first to third in which they did not try to advance (and they should have!). The "breakeven" probability of a successful taking of the extra base that is needed to justify attempting to go for the extra base is summarized in Figure 6-6. See file Baserunners.xls for the calculations.

		Breakeven Probability
		J
5	Situation	
6	first 0 outs	0.81
7	first 1 out	0.73
8	first 2 outs	0.90
9	second 0 outs	0.86
10	second 1 out	0.73
11	second 2 outs	0.39

FIGURE 6.6   Breakeven Probability
Needed to Justify Trying for the Extra Base.

The probability of being thrown out when trying to score on a single from second base is around 5%, so, with none out, runners are behaving in a near-optimal fashion because they should advance if their chance of being thrown out is < 14%. In all other situations, however, runners are not trying to advance in many situations where

2. Of course, there might be a problem of self-selection with these percentages, i.e., runners that attempt to go from first to third are not "average" runners.

they should be trying for the extra base. For example, a runner on second with two outs should try and score as long he has at least a 39% chance of scoring.

TAGGING UP FROM THIRD?

When the legendary sabermetric guru Pete Palmer reviewed the first edition of *Mathletics*, he mentioned that many people have no idea how to answer the following question. A runner is on third with one out. A fly ball is hit to the outfield. The third base coach must decide whether to let the runner tag up and try and score. What probability of success justifies the decision to have the runner tag up?

To answer this question, we note that if the tag up succeeds, we score one run and have nobody on with two outs (worth .10 runs). If we fail, we score 0 run. If we do not tag up, we have a runner on third with two outs and expect to score .36 runs. If we let P = probability tagging up is successful, then we should tag up if

$$P * 1.1 + (1 - P) * 0 > .36 \quad \text{or} \quad P > .36/1.1 = .33.$$

This answer shocked me! I would have guessed 65%. The logic behind the decision is that the next batter is much more likely to make an out than get a hit, so tagging up is worth the gamble. The problem is that when the runner is thrown out, the manager and coaches get second-guessed. Since runners tagging up from third are safe almost 95% of the time, we see that third base coaches are much too conservative in making this decision.

In summary, to choose between strategic options like bunting or not bunting, or stealing or not stealing a base, simply choose the decision which, on average, makes us better off with respect to expected runs scored or (if the game is tied late) with respect to the probability of scoring at least one run.

CHAPTER 6 APPENDIX:
AVERAGE RUNS PER INNING AND
CONDITIONAL PROBABILITY

Recall in Chapter 4 we stated (without proof) and verified by simulation that a team in which each batter had a 50% chance of hitting an HR and a 50% chance of striking out would average three runs per inning. We now use conditional expectation to prove this result.

Let R_i = expected runs scored by this team in an inning in which i outs are allowed. Then $R_1 = .5(0) + .5(1 + R_1)$. This follows because with probability .5 the first batter makes an out and the inning ends with us scoring 0 runs. Also, with probability .5 the first batter hits an HR and we can expect to score $1 + R_1$ runs, because we still have one out left. Solving this equation we find $R_1 = 1$.

Now we can solve for R_2 from the following equation.

$R_2 = .5(R_1) + .5(1 + R_2)$. This equation follows because with probability .5 the hitter makes an out and we now expect to score R_1 runs. Also, with probability .5 we hit an HR and can now expect to score a total of $1 + R_2$ runs because we have two outs remaining. After substituting in $R_1 = 1$, we find that $R_2 = 2$.

Now we can solve for R_3 using the equation $R_3 = .5(R_2) + .5(1 + R_3)$. After substituting $R_2 = 2$ in that, we find that $R_3 = 3$, as we claimed. Generalizing this logic, you can easily show that $R_n = n$.

CHAPTER 7

EVALUATING FIELDERS

Surprisingly, until the late 1990s little progress had been made in determining how to evaluate the effectiveness of fielders and the relative importance of fielding (as compared to batting and hitting). In this chapter we discuss the never-ending effort to accurately evaluate a player's fielding ability. We also discuss the huge impact that shifting fielders has had on the game. We close with a discussion of the benefits catchers create by "framing" balls outside the strike zone so umpires call them strikes.

FIELDING PERCENTAGE: THE TRADITIONAL, FATALLY FLAWED METRIC

During the first 100 years of Major League Baseball, the only measure of fielding effectiveness available was **Fielding Percentage.** For any fielder,

$$\text{Fielding Percentage} = \frac{PO + A}{PO + A + E}. \tag{1}$$

Here PO = putouts made by fielder. For example, an SS gets credit for a putout when he catches a fly ball or line drive, tags a runner out, or receives the ball and steps on second base to complete a force-out.

A = Assists made by fielder. For example, an SS gets credit for an assist when he throws to first base and the batter is put out.

E = Errors made by fielder. Again, whether a batted ball is scored an error is a subjective decision made by the official scorer.

Essentially, fielding percentage computes the percentage of balls in play that a fielder handles without screwing up. Figure 7-1 gives fielding data for Hall of Famer Derek Jeter for the 2000–2006 seasons. Derek Jeter won five Gold Gloves (emblematic of the best fielder at each position), and most casual baseball fans thought of Jeter as a great fielder. As we will soon see, this is not the case.

	B	C	D	E	F	G	H	I
13		Year	InnOuts	PO	A	E	RF	FA
14	Jeter	2000	3836	236	349	25	0.9081	0.9605
15	Jeter	2001	3937	211	344	15	0.8394	0.9736
16	Jeter	2002	4150	219	367	14	0.8408	0.9766
17	Jeter	2003	3101	160	271	14	0.8276	0.9685
18	Jeter	2004	4025	273	392	13	0.9838	0.9808
19	Jeter	2005	4058	262	454	15	1.0506	0.9794
20	Jeter	2006	3877	214	381	15	0.9138	0.9754

FIGURE 7.1 Derek Jeter's Fielding Statistics 2000–2006.

We will define Innouts and RF later in the chapter. To illustrate the computation of fielding percentage (listed in the FA column), let's compute Jeter's 2004 fielding percentage:

$$2004 \text{ Jeter FA} = \frac{273 + 392}{273 + 392 + 13} = .981.$$

Therefore, Jeter properly handled 98.1% of his fielding chances. By the way, during the years 2000–2006 the average fielding percentage for an SS was .974, so Jeter's 2004 performance looks pretty good. During the years 2002–2006 Jeter's fielding percentage was below average only during 2003 (Jeter was injured during much of the 2003 season). This superficial analysis indicates that Jeter is a better than average shortstop. Not so fast! The problem with fielding percentage is that you cannot make an error on a ball you do not get to. If an SS does not move, he will field easy balls and make few errors. Unfortunately, an immobile shortstop will allow many more base hits than a shortstop with a vast range.

THE RANGE FACTOR: AN IMPROVED MEASURE OF FIELDING EFFECTIVENESS

So how can we measure whether a shortstop has great or poor range? Enter Bill James, who developed an ingenious yet simple measure of fielding effectiveness, the **Range Factor (RF)**. James defines a fielder's range factor as the sum of Putouts and Assists a fielder gets per game played. Then James normalizes this statistic relative to all players playing a given position. It turns out that shortstops during 2000–2006 averaged 4.483 PO + A per game. Thus, an SS who had 5 PO + A per game would have a range factor of $5/4.48 = 1.11$. This shortstop fields 11% more balls than a typical SS. SSs with range factors larger than 1 have above-average range and shortstops with range factors less than 1 have below-average range. Let's compute Derek Jeter's 2006 range factor. We assume 8.9 innings per game. The column Innouts gives the number of defensive outs for which Jeter was on the field. Thus in 2006 Jeter was on the field for $3877/(8.9 * 3) = 145.2$ games. Jeter had PO + A = $214 + 381 = 595$. This implies that Jeter successfully handled $\frac{595}{145.2} = 4.098$ chances per game. This is lower than the average SS, who successfully handled 4.48 chances per game. Thus, Jeter's normalized range factor is $\frac{4.098}{4.483} = .91$. This implies that in 2006 Jeter successfully handled 9% fewer chances than an average shortstop.

By the way, Ozzie Smith is generally considered the greatest fielding shortstop of all time. Ozzie had a lifetime fielding average of .978 (slightly above average), but **he had several years where his range factor exceeded 1.3!** For Ozzie Smith, the Range Factor metric shows his true greatness.

How much did the balls that Jeter did not field cost the Yankees? We will discuss this later in the chapter.

PROBLEMS WITH RANGE FACTOR

There are several problems with Range Factor. Suppose that SS1 plays for a team whose pitchers strike out an average of eight hitters per game and SS2 plays for a team whose pitchers strike out only five batters a game. On average, SS2's team will face three more balls in play than SS1's team, so even if both shortstops have equal ability, SS2 will have a higher range factor. Also suppose SS1's team has primarily left-handed pitchers and SS2's team has primarily right-handed pitchers. Then most managers will stack their lineups against SS1's team with right-handed batters to take advantage of the platoon effect (see Chapter 12). Right-handed batters are believed to hit more ground balls to shortstop than left-handed batters. SS2's opponents will use primarily left-handed batters (who are believed to hit to shortstop less often than right-handed batters). In such a situation SS1 would have more balls hit near him and would tend to have a larger RF than SS2.

When evaluating the range factor of outfielders, we must realize that the park dimensions have a significant effect on the number of chances that an outfielder can successfully field. For example, in the spacious Dodger Stadium, the left fielder will be able to make many more putouts than in Fenway Park (whose Green Monster prevents many fly balls from being caught). Suffice it to say that sabermetricians understand these problems and have created adjusted range factors to account for these and other problems.

THE FIELDING BIBLE: A GREAT LEAP FORWARD

I believe that John Dewan (author of *The Fielding Bible,* 2006) deserves the most credit for improving the evaluation of fielders. Dewan and his colleagues at Baseball Info Solutions watched videotapes of every Major League Baseball play and determined how hard each ball was hit and to which "zone" of the field the ball was hit. Then they determined the chance (based on all plays during a season) that a ball hit at a particular speed to a zone would be successfully fielded. For example, they might have found that 20% of all balls hit softly over second base are successfully fielded by shortstops. A shortstop who successfully fields such a ball prevented one hit. An average fielder would have successfully fielded this ball 20% of the time so our shortstop prevented $1 - .2 = .8$ hits more than an average player. In this case the shortstop receives a score of $+.8$ on the batted ball. If the shortstop failed to make the play, he prevented $0 - .2 = -.2$ hits and received a score of $-.2$ on the batted ball. Note if the shortstop successfully fielded one in five chances in this zone, his net score would be $.8 - 4(.2) = 0$, as we would hope. If over the course of a season a shortstop had a net score of -20, then he had effectively given up 20 more hits than an average fielder. A shortstop with a score of $+30$ had effectively prevented 30 more hits than an average fielder.

CONVERTING FIELDER'S SCORES TO RUNS

Can we convert a fielder's score to runs (and possibly games won or lost due to fielding!)? To convert an extra hit allowed or saved to runs we must look at Figure 6-2, which gives expected runs in all possible states. Suppose an SS fails to field a ball with 0 outs and bases empty. Before this ball was hit the state was 0000 and an average team was expected to score .51 runs. Now if the SS gives up a hit the new state is 0100 and the average batting team is expected to score .87 runs. If the SS turns a potential hit into an out, then the new state is now 1000 and the average batting team is expected to score only .27 runs. In this situation the SS failing to prevent a hit cost his team $.87 - .27 = .60$ runs. If we average the cost of allowing an out to become a hit over all possible states (weighting each state by the fraction of the time it occurs), we find that a hit allowed costs a team around .8 runs.

CONVERTING RUNS SAVED BY A FIELDER INTO WINS

How do we convert runs saved by a good fielder or extra runs allowed by a bad fielder into wins? Let's look back at the Pythagorean Theorem of Chapter 1. In 2016 an average team scored 725 runs and gave up 725 runs. (Of course, an average team wins 81 games.) If a fielder saves 10 runs, then our average team now outscores its opponent 725 − 715, which yields a scoring ratio of 725/715 = 1.014. This translates by Pythagoras to our team now winning $\frac{162(1.014^2)}{1.014^2 + 1} = 82.12$ games. Therefore 10 runs translate into around one game won. This implies a fielder whose *Fielding Bible* rating was −12.5 would cost his team around one game a year. Most front-line players have a *Fielding Bible* rating between +20 and −20. Therefore, few fielders cost their team or save their team two more wins than an average fielder. In Chapter 8 we will learn how to combine a batter's fielding and hitting abilities to obtain a measure of overall player effectiveness.

STATCAST IMPROVES ON THE *FIELDING BIBLE*!

Beginning with the 2017 season, Statcast determined (based on the location of the batted ball and fielder and the time an average outfielder would need to catch the ball) an estimate of the chance an average outfielder would catch each fly ball. Based on this probability Statcast computes outs above average (OAA). For example, suppose an outfielder had two balls hit to him and an average fielder had a 40% chance to catch the first ball and an 80% chance to catch the second ball. Suppose the fielder caught the first ball but did not catch the second ball. Then the outfielder gained $(1 - 0.4) + (0 - 0.8) = -0.2$ OAA. The calculation of OAA for infielders is more complex (see http://m.mlb.com/glossary /statcast/outs-above-average). The file OAA19.xlsx (downloaded

from https://baseballsavant.mlb.com/leaderboard/outs_above
_average#:~:text=Outs%20Above%20Average%20(OAA)%20
is,%25%20Out%20Probability%20play%20gets%20%2B) shows
OAA for all fielders. Figure 7-2 shows the 10 leaders and 10 stragglers in OAA. For example, we find that the Nationals' center
fielder Victor Robles led the majors with 23 OAA, which translates
to 19 runs saved.

	A	B	C	D	E	F	G
1	last_name	first_name	Team	year	Position	Runs Prevented	Outs Above Average
2	Robles	Victor	Nationals	2019	CF	20	23
3	Baez	Javier	Cubs	2019	SS	14	18
4	Kiermaier	Kevin	Rays	2019	CF	15	17
5	Arenado	Nolan	Rockies	2019	3B	12	16
6	Simmons	Andrelton	Angels	2019	SS	12	16
7	Ahmed	Nick	Diamondbacks	2019	SS	11	15
8	Cain	Lorenzo	Brewers	2019	CF	12	14
9	Story	Trevor	Rockies	2019	SS	11	14
10	Bader	Harrison	Cardinals	2019	CF	11	13
11	Buxton	Byron	Twins	2019	CF	11	12
258	Jimenez	Eloy	White Sox	2019	LF	-10	-11
259	Pham	Tommy	Rays	2019	LF	-10	-11
260	Choo	Shin-Soo	Rangers	2019	LF	-11	-12
261	Santana	Domingo	Mariners	2019	LF	-12	-13
262	Tatis Jr.	Fernando	Padres	2019	SS	-10	-13
263	Villar	Jonathan	Orioles	2019	2B	-10	-13
264	Gregorius	Didi	Yankees	2019	SS	-11	-14
265	Guerrero Jr.	Vladimir	Blue Jays	2019	3B	-12	-16
266	Polanco	Jorge	Twins	2019	SS	-12	-16
267	Rosario	Eddie	Twins	2019	LF	-16	-18

FIGURE 7.2 OAA 2019 Leaders and Stragglers.

DEREK JETER'S UNDESERVED GOLD GLOVES

I love Derek Jeter. A true Hall of Famer on and off the field. Derek
Jeter won five Gold Gloves (during 2004, 2005, 2006, 2009, and
2010). In the pre-Statcast days, FanGraphs extended the work of the
Fielding Bible computed with its ultimate zone rating (UZR). UZR
approximates the number of runs above average saved by a fielder.

TABLE 7.1

Jeter's UZR: 2004–2010

Year	UZR
2004	−1
2005	−15
2006	−7
2009	6
2010	−4

Based on UZR, Jeter should not have won those Gold Gloves. Table 7-1 shows Jeter's UZRs during his Gold Glove years.

In four of his five Gold Glove years, Jeter was a below-average fielder! The fact that it is now virtually impossible for a fielder with a negative UZR to win a Gold Glove is testimony to the increased influence of sabermetrics in the sports journalism community.

HOW THE TAMPA BAY RAYS WON THE 2008 AL

Investors who exploit market efficiencies can outperform the stock market. For example, in the time period 1990–2018 the Renaissance Technologies Medallion fund used proprietary algorithms to achieve an average annual return of 41%. In a similar fashion, teams that are ahead of the curve on analytics can outperform their peers. For example, during the 2007–2008 time frame most teams did not understand the importance of fielding. In 2007, the Rays' (run by former Bear Stearns employees) UZRs summed to −47 runs, around five wins below average. In 2008 the Rays' UZR summed to +72 runs, 7.2 wins above average. The improvement in fielding between the 2007 and 2008 seasons was worth 12 wins. In 2007, Tampa won 66 games, and in 2008, 97 games! Fielding improvement was not the whole story but fielding certainly played a significant role in the Rays' improvement.

THE SHIFT!

The great left-handed hitter Ted Williams usually hit the ball to the right side of the field. On July 14, 1946, the Cleveland Indian player-manager Lou Boudreau decided to play all four infielders, the right fielder, and the center fielder on the right side of the field. During the rest of Williams's great career (interrupted by military service in the Korean War), teams used the shift during most of his plate appearances. Williams estimated that the shift cost him 15 points in batting average.

After a 50+year hiatus, the shift has returned. In 2010, teams shifted on 3% of plate appearances play, and in 2019, teams shifted on 15% of all plate appearances (41% of the time against left-handed hitters!). Now that teams know where each hitter tends to hit the ball against different types of pitchers, it is a relatively easy mathematical exercise to determine where to place fielders to minimize the batter's effectiveness. As an example, consider the worksheet Shift in the file Shift.xlsx in Figure 7-3, which contains made-up data about where a hitter hits the ball on his ground ball at bats. A 1 in column C means the ball is hit between first and second, and a 2 in column C means the ball is hit between second and third. For example, on the first at bat the ball was hit 67.65 feet of the way between first and second, and on the second at bat the ball was hit 64.76 feet of the way between second and third. In L6:L9 we determined where to put the four infielders to maximize the chance of successfully fielding the ball. We assumed the play results in an out if the fielder is positioned within five feet of where the ball is hit. We added the constraint that the first baseman must be positioned within 10 feet of first base. By using Excel Solver (to be discussed later in the book) we found that placing the first baseman 6.57 feet from first base, the second baseman 50 feet from first base, the shortstop 60 feet from first base, and the third baseman 69 feet from first base, we can successfully field 76% of all ground balls. Note that 87% of all balls were hit between first and second.

	C	D	E	F	G	H	I	J	K	L	M	N
4								0.7621	First baseman within 10 feet of bag			
5	Base	distance	changed distance	Distance to 1	Distance to 2	Distance to 3	Distance to 4	Fieldit		Optimal Position	Default	
6	1	67.64780626	67.65	61.07	1.27	7.81	17.76	1	1	6.57903	8	
7	2	64.76419856	154.76	148.19	85.84	94.92	104.88	0	2	68.9195	78	
8	1	61.70228162	61.70	55.12	7.22	1.86	11.82	1	3	59.8392	130	
9	1	57.92893392	57.93	51.35	10.99	1.91	8.04	1	4	49.8871	170	
10	1	47.12941883	47.13	40.55	21.79	12.71	2.76	1				
11	1	52.00030574	52.00	45.42	16.92	7.84	2.11	1				
12	2	54.77674419	144.78	138.20	75.86	84.94	94.89	0				
13	2	59.68765089	149.69	143.11	80.77	89.85	99.80	0				
14	2	41.28925462	131.29	124.71	62.37	71.45	81.40	0				

FIGURE 7.3   The Power of the Shift.

	C	D	E	F	G	H	I	J	K	L	M	N
4								0.1931	First baseman within 10 feet of bag			
5	Base	distance	changed distance	Distance to 1	Distance to 2	Distance to 3	Distance to 4	Fieldit		Optimal Position	Default	
6	1	67.65	67.65	61.07	2.35	62.35	102.35	1	1	6.57903	8	
7	2	64.76	154.76	148.19	84.76	24.76	15.24	0	2	70	78	
8	1	61.70	61.70	55.12	8.30	68.30	108.30	0	3	130	130	
9	1	57.93	57.93	51.35	12.07	72.07	112.07	0	4	170	170	
10	1	47.13	47.13	40.55	22.87	82.87	122.87	0				
11	1	52.00	52.00	45.42	18.00	78.00	118.00	0				
12	2	54.78	144.78	138.20	74.78	14.78	25.22	0				
13	2	59.69	149.69	143.11	79.69	19.69	20.31	0				
14	2	41.29	131.29	124.71	61.29	1.29	38.71	1				

FIGURE 7.4   Normal Fielder Alignment Fails!

As shown in the worksheet No Shift in Figure 7-4, if the infielders were positioned in their normal position, only 19% of the ground balls would have been fielded.

It is difficult to determine how well shifting works, but the shift surely destroyed Ryan Howard's career. The great writer Rob Arthur of FiveThirtyEight.com (https://fivethirtyeight.com/features/ryan-howards-career-is-dead-the-shift-killed-it/) gives us the sad details. During 2010–2016 the great left-handed Philly slugger recorded an OPS of .975 when teams did not shift and a mediocre .643 OPS when teams shifted.

Also, through June 2017, the Cubs' World Series hero and left-handed hitter Kyle Schwarber only hit .180 on ground balls, largely

	A	B	C	D
2	Last Name	First Name	Pitches	Runs from Extra Strikes
3	Hedges	Austin	2684	15
4	Flowers	Tyler	2252	15
5	Perez	Roberto	3036	12
6	Grandal	Yasmani	3701	12
7	Vazquez	Christian	3186	11
8	Realmuto	J.T.	3730	8
9	McCann	Brian	2269	8
10	Stassi	Max	1206	7
11	Posey	Buster	2675	7

FIGURE 7.5   2019 Catcher Framing Leaders.

due to the shift (see http://wrigleyville.locals.baseballprospectus
.com/2017/06/04/schwarber-and-the-shift/).

CATCHERS AND PITCH FRAMING

On any plate appearance, the ball-strike count has a significant impact on the result. For example, with a 1–0 count, hitters average
a 1.064 OPS, while with a 0–1 count, hitters average a .813 OPS.
Catchers try hard to convince umpires that pitches that are actually
out of the strike zone be called strikes. This ability is called "pitch
framing." Beginning with the 2015 season, Statcast evaluated the
number of runs each catcher saved his teams with extra strikes. The
file Catcherframing.xlsx contains the number of pitches and runs
saved by each catcher. The top 10 catchers in runs saved are shown
in Figure 7-5.

CHAPTER 8

WIN PROBABILITY ADDED (WPA)

A player's objective should be to help his or her team win games. Therefore, it seems reasonable to measure how much a professional athlete's efforts help his team win or cause his team to lose games. As we will see for basketball and football, this is a very difficult task. For baseball, however, the brothers Harlan and Eldon Mills (1970) came up with a simple yet elegant way to measure how a baseball player changes the chance of his team winning a game. To illustrate the method, consider perhaps the most famous hit in baseball history: Bobby Thomson's HR in the 1951 playoffs. If you have time, watch Russ Hodges's immortal call on https://www.youtube.com/watch?v=3T0drh8i4Tw. Thomson came to bat for the New York Giants in the bottom of the ninth inning of the deciding game of the 1951 playoff against the Brooklyn Dodgers. The Giants were down 4–2 and had runners on second and third with one out. Assuming the two teams were of equal ability, you can use Greg Stoll's outstanding Win Expectancy Finder (see Figure 8-1 and https://gregstoll.com/~gregstoll/baseball/stats.html#V.0.1.0.1.0.0) to estimate the chance of the Giants winning the game. In all past games during the 1957–2019 seasons where this situation occurred the visiting team won 73% of the games, so the Giants (the home

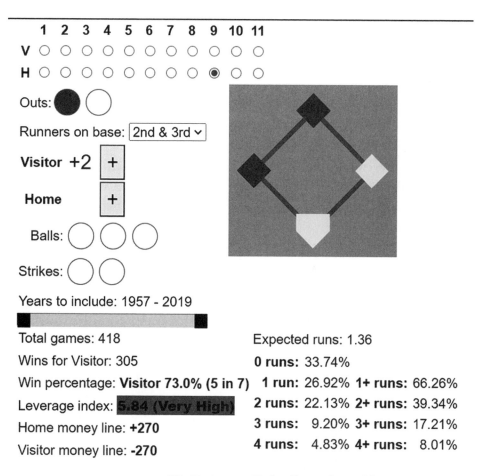

	1	2	3	4	5	6	7	8	9	10	11
V	○	○	○	○	○	○	○	○	○	○	○
H	○	○	○	○	○	○	○	○	◉	○	○

Outs: ● ○

Runners on base: 2nd & 3rd ⌄

Visitor +2 [+]

Home [+]

Balls: ○ ○ ○

Strikes: ○ ○

Years to include: 1957 - 2019

Total games: 418

Wins for Visitor: 305

Win percentage: **Visitor 73.0% (5 in 7)**

Leverage index: **5.84 (Very High)**

Home money line: **+270**

Visitor money line: **-270**

Expected runs: 1.36

0 runs: 33.74%

1 run: 26.92% **1+ runs:** 66.26%

2 runs: 22.13% **2+ runs:** 39.34%

3 runs: 9.20% **3+ runs:** 17.21%

4 runs: 4.83% **4+ runs:** 8.01%

FIGURE 8.1 Win Expectancy Finder (figure obtained from https://gregstoll.com/~gregstoll/baseball/stats.html#V.0.1.0.1.0.0).

team) would be estimated to have a $1 - 0.73 = 27\%$ chance of winning the game.

Thomson hit his historic home run and the Giants won. Of course, the Giants now had a 100% chance of winning. So how can we measure the credit Thomson should obtain for this batter-pitcher interaction? At the start of the game we assume each team has a 50% chance of winning. The metric we track for each plate appearance is the change in the batting team's chance of winning. Let's call this measure win probability added (WPA). After each game event (batter outcome, stolen base, pickoff, etc.) the batter and pitcher receive equal credit for changing the probability of their team winning. By the way, the great Willie Mays was on deck when Thomson's HR broke Brooklyn's heart.

Before Thomson's HR, the Giants had a 27.89% chance of winning. Therefore, Thomson's HR earns $1 - .2789 = .7211$ WPA and costs the pitcher (poor Ralph Branca!) .7211 WPA. FanGraphs tabulates each player's WPA. Figure 8-2 shows the leading hitters for the 2019 season.

	A	B	C
1	Name	Team	WPA
2	Christian Yelich	Brewers	7.34
3	Cody Bellinger	Dodgers	5.41
4	Mike Trout	Angels	5.17
5	Anthony Rendon	Nationals	4.95
6	Matt Olson	Athletics	4.82
7	Max Muncy	Dodgers	4.82
8	Bryce Harper	Phillies	4.71
9	Freddie Freeman	Braves	4.65
10	Anthony Rizzo	Cubs	4.54
11	Ronald Acuna Jr.	Braves	4.5
12	Alex Bregman	Astros	4.32
13	Xander Bogaerts	Red Sox	4.09
14	Mookie Betts	Red Sox	3.84
15	Pete Alonso	Mets	3.78
16	Matt Chapman	Athletics	3.67
17	Nolan Arenado	Rockies	3.66
18	Juan Soto	Nationals	3.57
19	Charlie Blackmon	Rockies	3.43
20	Kris Bryant	Cubs	3.35
21	Michael Brantley	Astros	3.32

FIGURE 8.2   2019 WPA Batting Leaders.

For example, Mike Trout's plate appearances added 5.17 wins for the Angels. FanGraphs gives the guide shown in Table 8-1 for interpreting WPA for a player who plays in most of a team's games.

Here are the 2019 WPA pitching leaders in Figure 8-3.

TABLE 8.1

Guide for Interpreting WPA

Rating	WPA
Excellent	+6.0
Great	+3.0
Above Average	+2.0
Average	+1.0
Below Average	0.0
Poor	−1.0
Awful	−3.0

	A	B	C
1	Name	Team	WPA
2	Justin Verlander	Astros	5.19
3	Hyun-Jin Ryu	Dodgers	4.33
4	Gerrit Cole	Astros	4.31
5	Jacob deGrom	Mets	4.21
6	Jack Flaherty	Cardinals	3.97
7	Shane Bieber	Indians	3.82
8	Max Scherzer	Nationals	3.47
9	Mike Soroka	Braves	3.45
10	Charlie Morton	Rayes	3.4
11	Zach Greinke	---	3.36
12	Mike Minor	Rangers	3.32
13	Stephen Strasburg	Nationals	3.28
14	Sonny Gray	Reds	2.88
15	Lucas Giolito	White Sox	2.74
16	Lance Lynn	Rangers	2.64
17	Patrick Corbin	Nationals	2.6
18	Kyle Hendricks	Cubs	2.58
19	Luis Castillo	Reds	2.52
20	Clayton Kershaw	Dodgers	2.48
21	Walker Buehler	Dodgers	2.25

FIGURE 8.3   2019 WPA Pitching Leaders.

	K	L	M
3	Season	Team	WPA
4	1995	Yankees	-0.86
5	1996	Yankees	5.26
6	1997	Yankees	1.99
7	1998	Yankees	4.65
8	1999	Yankees	3.38
9	2000	Yankees	2.53
10	2001	Yankees	3.23
11	2002	Yankees	1.38
12	2003	Yankees	3.69
13	2004	Yankees	4.93
14	2005	Yankees	3.15
15	2006	Yankees	3.33
16	2007	Yankees	2.25
17	2008	Yankees	4.26
18	2009	Yankees	3.94
19	2010	Yankees	2.01

FIGURE 8.4   Mariano Rivera's WPA.

Although relief pitchers rarely pitch over 100 innings, they often pitch during "high leverage" situations in which one plate appearance can result in a large change in a team's chance of winning. This makes it common for a top reliever to build up a great WPA in a few innings. Figure 8-4 shows the great Yankee reliever Mariano Rivera's WPA during his best seasons. Only during 1996 did Rivera pitch more than 100 innings, but 14 of his best years would have ranked in the Top 20 WPAs for 2016 Pitchers!

INCORPORATING FIELDING RATINGS INTO WPA

Using OAA, we can adjust a player's WPA to incorporate a player's fielding ability. Assume again that 10 runs = 1 win. For example, Cody Bellinger's 2019 WPA is 5.41 and his fielding prevented five runs more than average, so after including Bellinger's fielding, his 2019 WPA would be 5.41 + 0.5 = 5.91.

INCORPORATING BASE RUNNING ABILITY INTO WPA

A good base runner can certainly help his team win some games. Most good base runners help their team in the following ways:

- Stealing bases often and rarely getting caught when they try and steal.
- Rarely grounding into double plays.
- Taking the extra base when the batter gets a hit. For example, a fast base runner will score from second on a higher percentage of singles than an average runner.

FanGraphs's UBR (Ultimate Base Running) rating measures how many runs a hitter adds or subtracts (relative to an average base runner) by his base running (excluding stolen bases and grounded into double play tendencies). For example, a runner who scores from second on 100% of all singles hit would garner a positive contribution to UBR, while a runner who never scored from second base on a single would incur a negative contribution to his UBR.

The 2019 team UBR ratings are shown in Figure 8-5. The wSB (weighted stolen bases) number for a player or team measures the number of runs above or below average a player's stolen bases contribute to a team. Finally, the wGDP (weighted ground into double play) statistic measures the number of runs above or below average contributed by a player's tendency to ground into double plays.

As an example of these statistics, consider the great baserunner Billy Hamilton of the Cincinnati Reds. In 2016 Hamilton has UBR = 4.1 runs, wGDP = 0.6 runs, and wSB = 8.1 runs. Therefore, Hamilton's baserunning was worth 12.8 runs and would contribute 1.28 wins to his WPA.

	A	B	C	D
1	Team	UBR	wGDP	wSB
2	Yankees	7.3	0.3	0.6
3	Rockies	6.6	0.5	2.8
4	Cubs	6.2	−1.8	−1.2
5	Diamondbacks	5.3	0.5	−0.1
6	Athletics	5.2	1.9	2.2
7	Mariners	4	0.3	1.4
8	Phillies	2.8	2.7	1.8
9	Rangers	1.7	−2.2	2.3
10	Braves	1.1	0.8	1.1
11	Twins	1.1	0.3	−1.8
12	Pirates	0.7	0.8	−3
13	Cardinals	0.3	0.8	−2.4
14	Rays	0.2	1.2	4
15	Tigers	0.2	−0.1	−0.3
16	Marlins	0.1	−0.1	2.4
17	Royals	0	2.4	−0.5
18	White Sox	−0.1	−0.9	−1.2
19	Indians	−0.5	1.4	−1.1
20	Astros	−0.7	1.5	−2
21	Giants	−1	−0.4	−1.4
22	Red Sox	−1.2	0	0.6
23	Angels	−1.3	−0.7	−1.1
24	Reds	−2.3	−2.9	0.1
25	Orioles	−2.9	2.5	−3.9
26	Dodgers	−3.6	−1.9	0.6
27	Padres	−3.9	0.9	3.7
28	Blue Jays	−5	0.4	2.3
29	Nationals	−5.2	−0.8	−0.3
30	Mets	−7.4	−1.7	−2.2
31	Brewers	−7.6	−6.2	−3.4

FIGURE 8.5   2019 Team UBR Ratings.

COMPUTING WPA WITHOUT THE WIN EXPECTANCY FINDER

The Win Expectancy Finder estimates the chance of a team winning in any given situation based on past game results. In *The Book*, 2014, by Tom Tango, Mitchel Lichtman, and Andrew Dolphin, the authors estimate win probabilities in various situations using Markov chain analysis (an extensive discussion of Markov chains is given in our Chapter 39 discussion of Soccer Analytics) to estimate the probability of the home team winning in all possible situations (there are thousands of possible situations by the way).[1] Essentially, using Markov chain analysis or our simulation model you can play out any situation thousands or even millions of times. For example, there are not many games in which the home team has been down, say, 12 runs after one inning, so it is hard to accurately estimate the probability of the home team winning in this situation. Using simulation or Markov chain analysis we can play out this situation millions of times and obtain a more accurate estimate of the home team's winning probability. For Bobby Thomson's "shot heard round the world" (apologies to Ralph Waldo Emerson and his *Concord Hymn*!), *The Book* estimated the Giants' chance of winning as 30.1%.

---

1. If we assume the run margin is always between −10 and +10 runs, then there are $24 * 18 * 21 = 9{,}072$ possible situations. We obtained this result by multiplying the number of states during an inning (24) by the number of possible half innings ($9 * 2$), and then multiplied by the 21 possible run differentials.

CHAPTER 9

WINS ABOVE REPLACEMENT (WAR) AND PLAYER SALARIES

People often try to summarize important quantities with a single statistic. For instance, the stacked ranking system popularized (and thankfully on the decline) by the late former GE CEO Jack Welch gave each employee an A, B, or C rating. Baseball general managers would love to have a single number that summarizes the quality of a baseball player's performance. Currently, Wins Above Replacement (WAR) is the best metric available. In this chapter we will primarily focus on the computation of WAR for non-pitchers. In the current lower-scoring post-steroid (?) environment it is probably true that nine runs, rather than 10 runs equals one win, but for simplicity we will assume 10 runs = 1 win.

Suppose that, when you consider batting, baserunning, and fielding, a center fielder has added 50 runs above average to his team's performance. Do we know his true value to the team? As Woody Allen said in the Oscar-winning *Annie Hall*, "Showing up is 80% of life" (see http://quoteinvestigator.com/2013/06/10/showing-up/). Baseball is no different than the rest of life. **To determine a player's true value in any sport you need to understand that each (in baseball) plate appearance has value because (unless the player** is really a way below-average player) he has replaced a "replacement player." FanGraphs assumes that a replacement non-pitcher performs 20 runs below average for a season. This turns out (when pitchers are included) to imply that a team of replacement players would win around 48 games in a season. Consider a non-pitcher who is completely average in all aspects of the game and has 1/9 of his team's plate appearances (we assume we are in the AL, which has a designated hitter, so pitchers do not hit). This player's plate appearances add 20 runs (or two wins) above a replacement player who took those same plate appearances. If the average player had 1/18 (half a season) of his team's plate appearances, then he would have "saved" 10 runs and would have earned one win above replacement. So, Woody was right: there is value in showing up (actually, I think *Manhattan* is a better movie than *Annie Hall*). After discussing Tom Tango's wOBA (weighted on-base average) we will be ready to describe the computation of WAR and illustrate the computations by computing Mike Trout's 2016 WAR.

WEIGHTED ON-BASE AVERAGE (WOBA)

Tom Tango developed wOBA to be a version of linear weights that summarize a hitter's batting ability. The linear weights given to any event vary slightly with the year (see Figure 9-1), but the following weights are approximately correct. As we will see, a player's wOBA can easily be converted to his batting runs above average.

- BB 0.69
- HBP 0.72
- 1B 0.89
- 2B 1.27
- 3B 1.62
- HR 2.1

For example, wOBA sets the value of three BBs equal to one HR.

We see that in 2016 an average batter had a wOBA of 0.318. Figure 9-2 shows the 2016 wOBA leaders.

	A	B	C	D	E	F	G	H	I
1	Season	wOBA	wOBAScale	wBB	wHBP	w1B	w2B	w3B	wHR
2	2017	0.321	1.185	0.693	0.723	0.877	1.232	1.553	1.98
3	2016	0.318	1.212	0.691	0.721	0.878	1.242	1.569	2.015
4	2015	0.313	1.251	0.687	0.718	0.881	1.256	1.594	2.065
5	2014	0.31	1.304	0.689	0.722	0.892	1.283	1.635	2.135
6	2013	0.314	1.277	0.69	0.722	0.888	1.271	1.616	2.101
7	2012	0.315	1.245	0.691	0.722	0.884	1.257	1.593	2.058
8	2011	0.316	1.264	0.694	0.726	0.89	1.27	1.611	2.086
9	2010	0.321	1.251	0.701	0.732	0.895	1.27	1.608	2.072

FIGURE 9.1   wOBA Linear Weights by Season.

	A	B	S
1	Name	Team	wOBA
2	David Ortiz	Red Sox	0.419
3	Mike Trout	Angels	0.418
4	Joey Votto	Reds	0.413
5	Daniel Murphy	Nationals	0.408
6	Josh Donaldson	Blue Jays	0.403
7	Freddie Freeman	Braves	0.402
8	Miguel Cabrera	Tigers	0.399
9	Kris Bryant	Cubs	0.396
10	Charlie Blackmon	Rockies	0.394
11	DJ LeMahieu	Rockies	0.391
12	Anthony Rizzo	Cubs	0.391
13	Jose Altuve	Astros	0.391
14	Nolan Arenado	Rockies	0.386
15	J.D. Martinez	Tigers	0.384
16	Nelson Cruz	Mariners	0.383
17	Paul Goldschmidt	Diamondbacks	0.382
18	Mookie Betts	Red Sox	0.379
19	Ryan Braun	Brewers	0.378
20	Matt Carpenter	Cardinals	0.375
21	Brandon Belt	Giants	0.374

FIGURE 9.2   2016 wOBA Leaders.

COMPUTING A BATTER'S WAR

We can now outline the steps needed to compute a batter's WAR.

Step 1: Compute the batter's wRAA (weighted runs above average) as wRAA = ((wOBA − lgwOBA)/wOBA Scale) ∗ PA. Here PA = plate appearances, lgwOBA is league average wOBA, and wOBA Scale serves to normalize wOBA for the relevant season. FanGraphs also makes a park adjustment to this number before computing WAR. For example, if there were two hitters with the same wRAA and one played for the Padres (a pitcher's park) and one played for the Rockies (a hitter's park) then clearly the Padres batter is a better hitter. Park factors are discussed in Chapter 10.

Step 2: Compute the batter's base running runs above average as UBR + wSB + wGDP.

Step 3: For non-catchers use the batter's UZR as the player's fielding runs above average. Things are trickier for catchers (see http://www.fangraphs.com/library/war/war-position-players/).

Step 4: For certain positions, MLB quality players are scarcer, so a positional adjustment is needed. The adjustments shown in Figure 9-3 are based on the assumption that the batter has 1/9 of the team's plate appearances.

Position	Full Season Adjustment
C	12.5
1B	−12.5
2B	2.5
SS	7.5
3B	2.5
LF	−7.5
CF	2.5
RF	−7.5
DH	−17.5

FIGURE 9.3 WAR Positional Adjustments.

This table shows that catchers are relatively scarce, and designated hitters are not.

Step 5: As previously discussed, the batter gets credit for being better than a replacement player. This adjustment is replacement level runs = 20 ∗ (fraction of team's plate appearances)/(1/9). Thus, a player who played roughly half a team's games would have 1/18 of his team's plate appearances and earn 10 replacement level runs.

Step 6: Add together wRAA + baserunning runs + defensive runs + replacement level runs to determine the batter's total runs above a replacement player.

Step 7: To obtain the batter's WAR, divide total runs above a replacement player by the number of runs per win. Before converting total runs above a replacement player to WAR, FanGraphs uses a very small league adjustment to ensure that for each league, total runs above average balances out to 0. We will ignore this league adjustment.

To compute the player's WAR we will use 10 runs per win, but in lower scoring environments, the number of runs per win decreases. This is because by the Pythagorean Theorem of baseball, a scoring ratio of 1.0124 translates to 82 wins. Therefore .0124 ∗ (average runs per season) translates to one win, and this number clearly decreases as average runs per season decreases.

We are now ready to compute Mike Trout's 2016 WAR.

- Trout's 2016 wOBA = 0.418 and PA = 681. Therefore, before a park adjustment, Trout's wOBA is computed as 681 ∗ (.418 − .318)/1.21 = 56.3 runs (after park adjustment, FanGraphs gives 58.3).
- Trout's baserunning runs above average are given by UBR + wGDP + wSB = 3.2 + 3.5 + 2.6 = 9.3 runs.
- Trout's UZR = −0.3 runs.
- Trout had 11.3% of the Angels' plate appearances (let's call it 1/9!), so as a positional adjustment for center field we get 2.5 runs.
- Since Trout had 1/9 of the Angels' plate appearances he gets 20 replacement runs.
- In total, Trout's runs above replacement is given by 56.3 + 9.3 − 0.3 + 2.5 + 20 = 87.8 runs.
- Using 10 runs per win, we obtain an 8.8 WAR for Trout. Using nine runs per win, we obtain a 9.8 WAR for Trout. FanGraphs reports a WAR of 9.4, so we are pretty close.

We will not discuss computing WAR for pitchers, but we note that instead of using wOBA as the key statistic, FanGraphs uses DICE and innings pitched as the key inputs to the WAR calculation.

FanGraphs suggests interpreting WAR as shown in Table 9-1.

Figure 9-4 shows the 2016 WAR leaders for hitters.

Figure 9-5 shows the 2016 WAR leaders among pitchers.

TABLE 9.1

Interpretation of WAR

Scrub	0–1 WAR
Role Player	1–2 WAR
Solid Starter	2–3 WAR
Good Player	3–4 WAR
All-Star	4–5 WAR
Superstar	5–6 WAR
MVP	6+ WAR

	D	E	F	G
1	Rank	Name	Tream	WAR
2	1	Mike Trout	Angels	9.4
3	2	Kris Bryant	Cubs	8.4
4	3	Mookie Betts	Red Sox	7.8
5	4	Josh Donaldson	Blue Jays	7.6
6	5	Corey Seager	Dodgers	7.5
7	6	Jose Altuve	Astros	6.7
8	7	Manny Machado	Orioles	6.5
9	8	Francisco Lindor	Indians	6.3
10	9	Freddie Freeman	Braves	6.1
11	10	Adrian Beltre	Rangers	6.1
12	11	Adam Eaton	White Sox	6
13	12	Robinson Cano	Mariners	6
14	13	Brian Dozier	Twins	5.9
15	14	Ian Kinsler	Tigers	5.8
16	15	Brandon Crawford	Giants	5.8
17	16	Justin Turner	Dodgers	5.6
18	17	Daniel Murphy	Nationals	5.5
19	18	Kyle Seager	Mariners	5.5
20	19	Dustin Pedroia	Red Sox	5.2
21	20	Nolan Arenado	Rockies	5.2

FIGURE 9.4 2016 WAR Leaders: Non-pitchers.

	J	K	L	M
5	Rank	Name	Team	WAR
6	1	Noah Syndergaard	Mets	6.5
7	2	Jose Fernandez	Marlins	6.1
8	3	Max Scherzer	Nationals	5.6
9	4	Johny Cueto	Giants	5.5
10	5	Rick Porcello	Red Sox	5.2
11	6	Justin Verlander	Tigers	5.2
12	7	Chris Sale	White Sox	5.2
13	8	Corey Kluber	Indians	5.1
14	9	Madison Bumgarner	Giants	4.9
15	10	Jose Quintana	White Sox	4.8
16	11	Masahiro Tanaka	Yankees	4.6
17	12	David Price	Red Sox	4.5
18	13	Kyle Hendricks	Cubs	4.5
19	14	Jon Lester	Cubs	4.3
20	15	Aaron Sanchez	Blue Jays	3.9
21	16	Jake Arrieta	Cubs	3.8
22	17	Jon Gray	Rockies	3.7
23	18	Marcus Stroman	Blue Jays	3.6
24	19	Carlos Martinez	Cardinals	3.3
25	20	Kenta Maeda	Dodgers	3.3

FIGURE 9.5   2016 WAR Leaders: Pitchers.

WAR AND FAIR PLAYER SALARIES

We can use a player's 2016 WAR to determine a "fair" 2016 salary based on the player's performance. WAR assumes that a team made up of 25 players, each paid the major league minimum salary of $500,000, will win 48 games. That means $12.5 million can buy 48 wins. An average MLB team wins 81 games and had a 2016 payroll of $114 million. This implies that 33 wins above replacement = $101.5 million, or each win above replacement is worth $3.08 million. Using this number, we can compute a "value-based salary" for a 2016 major leaguer. Figure 9-6 shows the value-based salaries for many top 2016 players.

Note that many young players like Francisco Lindor, Corey Seager, and Kris Bryant are vastly underpaid. Barring injury, their big paydays will arrive soon. The shock is that many successful, older players such as Jon Lester and Robinson Cano are vastly overpaid.

	E	F	G	H	I
	Name	**Team**	**WAR**	**Actual Salary**	**Fair Salary**
5	**Name**	**Team**	**WAR**	**Actual Salary**	**Fair Salary**
6	Jon Lester	Cubs	4.3	$25.00	$13.24
7	Robinson Cano	Mariners	6	$24.00	$18.48
8	Adrian Beltre	Rangers	6.1	$18.00	$18.79
9	Mike Trout	Angels	9.4	$16.08	$28.95
10	Freddie Freeman	Braves	6.1	$12.36	$18.79
11	Josh Donaldson	Blue Jays	7.6	$11.65	$23.41
12	Manny Machado	Orioles	6.5	$5.00	$20.02
13	Jose Altuve	Astros	6.7	$3.69	$20.64
14	Adam Eaton	White Sox	6	$2.75	$18.48
15	Kris Bryant	Cubs	8.4	$0.70	$25.87
16	Mookie Betts	Red Sox	7.8	$0.60	$24.02
17	Corey Seager	Dodgers	7.5	$0.50	$23.10
18	Francisco Lindor	Indians	6.3	$0.50	$19.40

FIGURE 9.6   Fair 2016 Salaries Based on 2016 WAR.

This is probably largely since, when seeking free agents, teams tend to overpay due to the "winner's curse." The winner's curse states that people often pay more for an item than it is worth. The curse was first observed in bidding for offshore oil leases, when analysis showed that on average the winning bid greatly exceeded the value of the lease.

To understand why the winner's curse is a common occurrence, consider a piece of art with an actual worth of $5 million. Nobody really knows the actual worth of the art, but one assumes the average bid will equal the true value. If this is the case, five bidders might bid for the art $3, $4, $5, $6, and $7 million. The average bid equals the true value of the art, but the winner pays $2 million more than fair value!

HOW WAR HELPED THE RED SOX BREAK "THE CURSE"

We close by noting that the book *Mind Game* (Goldman, 2005) details how the Boston Red Sox, led by savvy GM Theo Epstein, used WAR to evaluate many of the player transactions, which resulted in the Red Sox 2004 World Series title. For example, *Mind Game* describes how relief pitcher Keith Foulke's high WAR for the 1999–2003 seasons (average WAR of +2.2) led the Red Sox to sign Foulke for 2004. Foulke recorded 32 saves during the Red Sox 2004 championship season and a WAR of 1.6. He earned only $4 million in 2004. The Hall of Famer Mariano Rivera was paid $11 million in 2004 and achieved a WAR of 2.5. This shows that Epstein's signing of Foulke was an efficient use of resources, which is one of the essential components of playing Moneyball!

CHAPTER 10

PARK FACTORS

During 2016 the Colorado Rockies scored 845 runs and the Florida Marlins scored 655 runs. This would seem to indicate that the Rockies had a much better offense than the Marlins. The Rockies, however, play in Denver's thin air and the ball will travel farther, so it should be easier to score runs in Colorado than in other places. To determine if the Rockies had a better offense, we need to understand the concept of park factors.

Once again, the brilliant Bill James was the first to develop the concept of park factors. In every NBA arena the court is the same size and the baskets are 10 feet high. In every NFL stadium the fields are the same dimensions (although, as previously stated, Denver's thin air and domed stadiums and inclement weather may affect performance). In baseball, however, each stadium has different dimensions. This will certainly influence the runs scored in the park. Park factors are an attempt to measure how the park influences runs scored, home runs, etc.

We will discuss the simplest version of park factors. How much easier is it to score runs or hit an HR in Coors Field than in an average park? How much harder is it to score runs in Marlins Park than in an average stadium? In Figure 10-1 we see ESPN's list of 2016 park factors. As we will soon show, the Marlins' runs park factor of .834 means 83.4% as many runs are scored in Marlins Park stadium as in an average park and the Rockies' runs park factor of 1.368 means 36.8% more runs are scored in Coors Field than in an average park.

To compute the runs park factor for Coors Field, calculate runs scored per game in Coors Field divided by runs scored per game during the Rockies' road games. Using the data in Figure 10-2, we compute the Rockies and Marlins park run factors shown in Figure 10-1 (see the file Parkfactors.xlsx). In both the Rockies' road and home games, runs are equally affected by the Rockies offense, Rockies defense, average NL (ignoring interleague play) offense, and

	D	E	F	G	H	I	J	K
6	PARK NAME	PARK NAME	RUNS	HR	H	2B	3B	BB
7	1	Marlins Park (Miami, Florida)	0.834	0.793	0.868	0.963	0.667	0.967
8	2	Citi Field (New York, New York)	0.988	1.09	0.887	0.838	0.455	1.168
9	3	Tropicana Field (St. Petersburg, Florida)	0.889	0.877	0.901	0.861	0.952	1.048
10	4	Dodger Stadium (Los Angeles, California)	0.813	0.914	0.908	0.919	0.382	0.895
11	5	Oakland Coliseum (Oakland, California)	0.829	0.727	0.921	0.954	1.043	0.872
12	6	Citizens Bank Park (Philadelphia, Pennsylvania)	0.84	1.149	0.924	0.821	0.889	0.969
13	7	Minute Maid Park (Houston, Texas)	0.808	0.822	0.926	0.886	1.152	0.918
14	8	Wrigley Field (Chicago, Illinois)	0.874	0.819	0.928	0.93	1.045	1.066
15	9	Angel Stadium of Anaheim (Anaheim, California)	0.91	1.056	0.936	0.828	0.6	0.958
16	10	Safeco Field (Seattle, Washington)	0.941	1.158	0.953	0.963	0.538	1.029
17	11	Guaranteed Rate Field (Chicago, Illinois)	0.927	1.101	0.96	0.89	1.071	0.933
18	12	Miller Park (Milwaukee, Wisconsin)	0.972	1.126	0.961	1	1.037	0.977
19	13	Oriole Park at Camden Yards (Baltimore, Maryland)	0.953	1.009	0.967	0.834	0.846	0.975
20	14	Great American Ball Park (Cincinnati, Ohio)	0.99	1.175	0.97	0.949	0.645	1.019
21	15	Busch Stadium (St. Louis, Missouri)	0.921	0.901	0.972	0.914	0.543	0.914
22	16	Nationals Park (Washington, D.C.)	0.956	1.023	0.974	0.891	0.806	0.972
23	17	Yankee Stadium (New York, New York)	1.035	1.377	0.982	0.871	0.5	1.033
24	18	Comerica Park (Detroit, Michigan)	1.019	1.138	0.984	0.884	1.904	0.945
25	19	Turner Field (Cumberland, Georgia)	1.059	0.77	1.009	0.9	0.936	1.135
26	20	PNC Park (Pittsburgh, Pennsylvania)	1.007	0.8	1.013	1.034	1.769	1.007
27	21	Petco Park (San Diego, California)	1.014	0.957	1.042	1.062	0.75	0.977
28	22	Target Field (Minneapolis, Minnesota)	1.044	1.014	1.064	1.081	1.031	0.875
29	23	Rogers Centre (Toronto, Ontario)	1.156	1.01	1.083	1.3	1.1	1.039
30	24	AT&T Park (San Francisco, California)	1.012	0.704	1.084	1.101	1.526	1.076
31	25	Kauffman Stadium (Kansas City, Missouri)	1.171	0.783	1.092	1.256	1.571	1.039
32	26	Globe Life Park in Arlington (Arlington, Texas)	1.156	1.049	1.1	1.052	1.813	1.029
33	27	Progressive Field (Cleveland, Ohio)	1.207	1.168	1.115	1.299	0.507	1.119
34	28	Chase Field (Phoenix, Arizona)	1.225	1.292	1.139	1.145	2.032	1.042
35	29	Fenway Park (Boston, Massachusetts)	1.199	1.065	1.161	1.424	1.667	0.977
36	30	Coors Field (Denver, Colorado)	1.368	1.265	1.231	1.405	1.39	1.078

FIGURE 10.1   ESPN 2016 Park Factors.

	Z	AA	AB	AC
13	**Rockies**			
14		Games	Runs	RA
15	Home	81	508	477
16	Road	81	337	383
17	**Marlins**			
18		Games	Runs	RA
19	Home	80	302	302
20	Road	81	353	380

FIGURE 10.2   Rockies and Marlins Home-Away Run Splits.

average NL defense. Therefore, the only difference between the runs scored per game in Coors Field and the Rockies' away games must be due to the influence of Coors Field.

FIGURE 10-2 2016 ROCKIES' AND MARLINS' HOME-AWAY RUNS SPLITS

We find that the Rockies' park factor may be computed as $\frac{(508+477)/81}{(337+383)/81} = 1.368$ and the Marlins' park factor may be computed as $\frac{(302+302)/80}{(353+380)/81} = .834$.

Now we "convert" Rockies' and Marlins' scoring to runs scored in an average park. We have shown that 136.8 runs at Coors Park = 100 runs at an average park. Therefore to convert runs scored at Coors Park to runs scored at an average park, we should multiply by 100/136.8. Thus we find that the Rockies' offense scored the equivalent of 508/1.368 + 337 = 708 runs or 708/162 = 4.37 runs per game. In a similar fashion we find that the Marlins' offense scored the equivalent of 302/.834 + 353 = 715 runs or 715/161 = 4.44 runs per game. Our analysis shows the 2016 Marlins had a better offense than the 2016 Rockies!

As pointed out in Chapter 9, before computing a hitter's WAR, a hitter's weighted average runs above average must be park adjusted. For example, Trevor Story of the 2016 Rockies had a .380 wOBA.

Before a park adjustment, this translates to 21 weighted average runs above average. After park adjustment, FanGraphs reduces Story's weighted average runs above average to 10.4 runs. This reduces Story's WAR by 1.

In closing, we note that the HR, 3B, 2B, H, and BB columns of Figure 10-1 give "park factors" for home runs, triples, doubles, hits, and walks. For example, for Coors Field in 2016,

- 26.5% more HRs were hit in Coors Park than in an average park.
- 23.1% more hits occurred in Coors Park than in an average park.
- 40.5% more doubles occurred in Coors Park than in an average park!
- 39% more triples occurred in Coors Park than in an average park.
- 7.8% more walks occurred in Coors Park than in an average park.

FLAWS IN THE ESPN APPROACH

The ESPN method of computing park factors assumes that in any park the home team faces the same opponents. Due to interleague play and the fact that teams in the same league do not play each opponent the same number of times, this assumption is invalid.

STREAKINESS IN SPORTS

We have all heard Marv Albert tell us that LeBron James is "on fire" or Jack Buck tell us that Mike Trout is "red hot" and nobody can get him out. We also hear announcers tell us the Warriors are on a hot streak and are unbeatable, etc. Is it true that athletes and teams encounter hot streaks or are the observed patterns of player and team performance just randomness at work?

WHAT DOES A RANDOM SEQUENCE LOOK LIKE?

Let's first examine how a random sequence of 162 wins and losses appears. Let's suppose a team wins 60% of its games. Then to generate a random sequence of 162 games we should make sure that in each game the team has a .60 chance of winning **and the chance of winning a game does not depend on the team's recent past history.** For example, whether the team lost its last five games or won its last five games, its chance of winning the next game should still be .6. Figure 11-1 shows a randomly generated sequence of wins and losses for a baseball team that went 91–71 for 162 games. We created this sequence by "shuffling" 91 +1 and 71 −1 so that there was no dependence between successive game results. Note that this "team" had 10 and nine game winning streaks. When shown a randomly generated

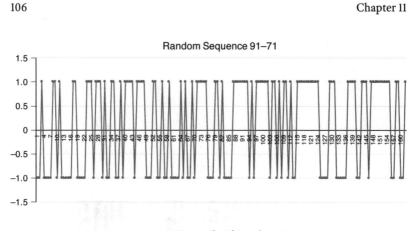

FIGURE 11.1   Example of Random Sequence.

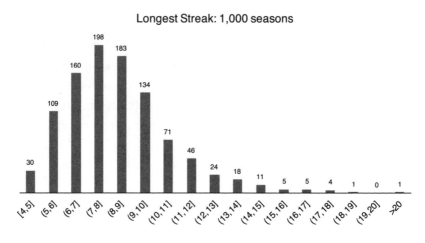

FIGURE 11.2   Distribution of Longest Winning Streaks in a Random Sequence.

sequence as in Figure 11-1, people invariably believe they are looking at a "streaky team," even though these sequences were generated by assumptions that involve no streakiness.

Most people think the occurrence of winning streaks indicates momentum or a "hot team" effect, but here we see long winning streaks are simply a consequence of randomness. Tversky and Kahneman (1971) found that subjects had great difficulty in distinguishing between random sequences and non–randomly generated sequences.

Suppose a team had a 60% chance of winning each game and past performance had no influence on the chance of winning a game. In the file Longeststreak.xlsx we simulated 1,000 seasons and found the frequency of longest streaks shown in Figure 11-2. We found that the chance of the longest winning streak being at least eight games is over 50%.

DOES THE HOT HAND EXIST?

Most people believe in the "hot hand" or streak hitter. In support of this view Gilovich, Vallone, and Tversky (1985) found that 91% of all surveyed basketball fans felt that basketball players were more likely to make a shot if the last shot was good than if the last shot was missed.

We can now look at whether the evidence supports the existence of the hot shooter, or streak hitter.

The most common technique used to test for streakiness is the Wald–Wolfowitz runs test (WWRT for short). Before discussing WWRT, we need to give the reader a quick tutorial in basic probability and statistics.

ENTER THE NORMAL RANDOM VARIABLE

A quantity such as a batting average or IQ of a randomly selected player or person is uncertain. We refer to an uncertain quantity as a random variable. Given a value of a random variable, such as a person's IQ or a player's batting average, can we determine how unusual the observation is? For example, which is more unusual: a person with an IQ of 140 or a 1980 major league batter who hit .360? The usual way to define the "unusualness" of an observation is to assume the data comes from a normal distribution. For example, IQs are known to follow a normal distribution which can be described by the probability density function (pdf) shown in Figure 11-3.

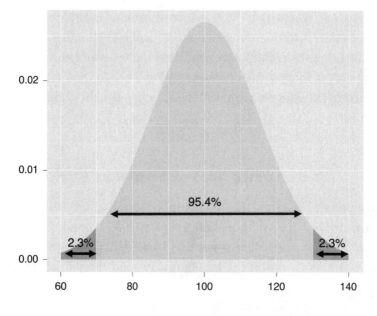

FIGURE 11.3 Normal Distribution Density for IQs.

The height of the IQ probability density function for an IQ value x is proportional to the likelihood that a randomly chosen person's IQ assumes a value near x. For example, the height of the pdf at 82 is approximately half of the height of the pdf at 100. This indicates that roughly half as many people have IQs near 82 as near 100. A normal density is characterized by two numbers:

- The mean μ or average value.
- The standard deviation σ, which measures the spread of a random variable about its mean.

IQs are known to have μ = 100 and σ = 15. If a random variable follows a normal pdf, then

- The mean is the most likely value of the random variable.
- 68% of the time the random variable will assume a value within σ of its mean. The area under a pdf represents probabil-

ity, so the total area under the pdf equals 1 and the area under the pdf between 85 and 115 for IQs is 68%.

- 95% of the time the random variable will assume a value within 2σ of its mean. Thus 95% of all people have IQs between 70 and 130, and the area under the IQ pdf between 70 and 130 is .95.
- The pdf is symmetric. That is, it is just as likely that a normal random variable will assume a value near $\mu + x$ as near $\mu - x$. Thus, roughly the same number of people have IQs near 120 as near 80, etc.

Z-SCORES

If a histogram or bar graph of our data tells us that the symmetry assumption fits our data, then statisticians measure how unusual a data point's value is by looking at how many standard deviations above or below average the data point is. This is called a z-score. Thus,

$$\text{z-score} = \frac{\text{value} - \text{mean}}{\text{standard deviation}}.$$

When averaged over all data points, z-scores have an average of 0 and a standard deviation of 1. Therefore computing a z-score is often called standardizing an observation. A z-score of 2 in absolute value indicates that an observation more extreme than our data point has roughly a 5% chance of occurring. A z-score of 3 in absolute value indicates that an observation more extreme than our data point has roughly three chances in 1,000 of occurring. As an example of the use of z-scores, we might ask what was more unusual: the stock market dropping 22% on October 19, 1987, or seeing a person shorter than Vern Troyer (30 inches tall) walk down the street. We are given the following information:

Mean daily stock return = 0%, sigma of daily stock returns = 1.5%
Mean height American male = 69 inches, sigma of height
    of American male = 4 inches

Then, October 19, 1987 z $-$ score $= \dfrac{-22-0}{1.5} = -14.67$ while Vern

Troyer's z-score is $\dfrac{30-69}{4} = -9.75$. Therefore a 22% market

drop in a day is much more unusual than the next person you
see walking down the street being shorter than Vern Troyer.

As another example of z-scores, look at Rogers Hornsby's .424
batting average in 1924 and George Brett's .390 batting average in
1980. If we assume that the pitching and fielding in both years were
of equal quality, which was the more outstanding performance?
From Figure 11-4 (see the file BA 1980.xlsx) we see that a histogram
of 1980 batting averages (including players with 300 or more at bats)
shows that the symmetry assumption is reasonable for batting averages. With Excel 2016 or newer, we can create beautiful histograms
by selecting the data and then choosing the Histogram icon from the
Insert Statistic Chart group on the Insert tab. After right-clicking
on the x-axis of a histogram and selecting Format Axis, we can set
overflow (we set .320) and underflow (we set .220) values and bin
widths (we chose 0.01.)

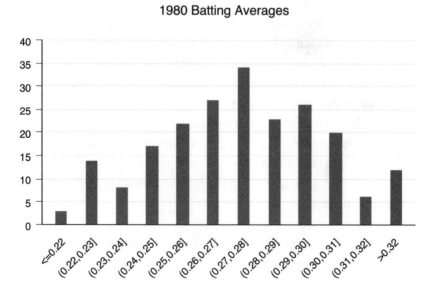

FIGURE 11.4 Histogram of 1980 Batting Averages.

Now, for 1980, mean batting average[1] = .274 standard deviation of batting averages = .0286. Then, George Brett's z-score = $\frac{.390 - .274}{.0286}$ = 4.04. In 1924, mean batting average = .299 standard deviation of batting averages = .0334. Then, Hornsby's z-score = $\frac{.424 - .299}{.0334}$ = 3.66. Thus even though Brett hit 34 points lower than Hornsby, relative to the overall performance during their respective seasons, Brett's performance was more outstanding.

BACK TO THE WALD–WOLFOWITZ RUNS TEST

We now return to our discussion of the WWRT. To motivate the WWRT, consider a team that is 5–5. The "streakiest" way to go 5–5 would be to have the sequence of wins and losses be as follows: WWWWWLLLLL. The "least streaky" way to go 5–5 would be to have the following sequence WLWLWLWLWL. Define a "run" to be an uninterrupted sequence of Ws or Ls. In this case WWWWWLLLLL has two runs (WWWWW and LLLLL), while the sequence WLWLWLWLWL has 10 runs. This example makes it clear that few runs in a sequence is an indication of streakiness while many runs is an indication of a lack of streakiness. The key question is, given a random sequence (that is, a sequence in which whether an observation is W or L does not depend on the prior events in the sequence), how many runs can we expect and how much spread is there about this average number of runs? Wald and Wolfowitz determined that for a random sequence consisting of S successes and F failures, the mean and standard deviation of the number of runs is as follows. In these formulas

$N = S + F$ is the total sequence length.

$$\mu = \text{Mean Number of Runs} = \frac{2FS}{N} + 1.$$

---

1. We restrict attention to players having at least 300 at bats during the season.

$$\sigma = \text{Standard deviation of number of runs} = \sqrt{\frac{(\mu-1)(\mu-2)}{N-1}}$$

For example, if a team goes 5–5 on average we would expect to see (if Ws and Ls are randomly sequenced) $\dfrac{2(5)(5)}{10}+1=5.5$

runs with a standard deviation of $\sqrt{\dfrac{(5.5-1)(5.5-2)}{10-1}}=1.32.$

Therefore the sequence WWWWWLLLLL has a z-score of $\dfrac{2-5.5}{1.32}=-2.65$, while the sequence WLWLWLWLWL has a z-score of $\dfrac{10-5.5}{1.32}=3.41.$

AN INTRODUCTION TO
HYPOTHESIS TESTING

Statisticians often set up a null and alternative hypothesis. The null hypothesis is to be accepted unless we observe a great deal of evidence in favor of the alternative hypothesis (this is as in the US justice system, in that the null hypothesis is "innocent" until proven guilty). What is a great deal of evidence against the null hypothesis? Most statisticians[2] believe that if the data indicates (under the assumption that the null hypothesis is true) that a **result at least as extreme as what we have observed has less than a 5% chance of occurring** then we should reject the null hypothesis.[3] Recall that if a random variable follows a normal distribution then there is only a 5% chance that the random variable's z-score will exceed 2 (actually 1.96, to be

2. In many court cases (such as Castaneda vs. Partida (1977)) the US Supreme Court has accepted the 5% level of significance or two standard deviations rule as the level of evidence needed to shift the burden of proof from plaintiff to defendant or vice versa.

3. Now this 5% "rule" is arbitrary and can be *easily* manipulated to obtain "statistically significant results," leading to what is known as, p-hacking. p-Hacking is not necessarily done on purpose, but many times due to not having a clear understanding of what p-value represents. I would urge you to clearly understand what p-value means. This will help you better understand the results of your analysis.

exact) in absolute value. This implies that if we define a test statistic based on our data (given that the null hypothesis is true), which follows a normal distribution, then there is only a 5% chance that the test statistic's z-score will exceed 2 in absolute value. There-fore, **if the test statistic's z-score exceeds 2 in absolute value we should reject the null hypothesis.** In this case we say the data is significant at the .05 level, because if the null hypothesis is true, then the chance of seeing a test statistic at least as extreme as what we observed is <.05.

In our situation, the null hypothesis = the Ws and Ls are ran-domly distributed within the sequence. That is, the past history of the sequence does not influence the chance that the next event is a W or an L.

Alternative hypothesis: The history of the sequence has some effect on whether the next event is a W or an L.

When trying to determine whether a sequence of Ws and Ls is random, the appropriate test statistic is the number of runs. Suppose we observed WWWWWLLLLL or WLWLWLWLWL. In either case each z-score exceeds 2 in absolute value, so either sequence would indicate that there is less than a 5% chance that the team's wins and losses came from a random sequence, and we would reject the null hypothesis and conclude that the sequence is not random.

BACK TO GVT AND THE HOT HAND

We can now discuss Gilovich, Vallone, and Tversky's (GVT's, for short) 1985 analysis of whether basketball players' shooting exhibits the "hot hand" or "streak shooting." GVT analyzed for each home game of the 1980–1981 Philadelphia 76ers season the results of suc-cessive FG attempts. For example, GGGMMG would mean the player made his first three shots, missed his next two, and made his sixth shot. They performed a WWRT on each player's sequence of field goal attempts. The results are shown in Table 11-1.

Since only Daryl Dawkins has a z-score for runs exceeding 2 in absolute value, he is the only player exhibiting significant streakiness.

TABLE 11.1

Data for Determining if Hot Hand Exists for 1980–1981

Philadelphia 76ers

	F	G	H	I	J	K
	Player	Good FGs	Missed FGs	Actual number of runs	Expected number of runs	Z–score
5	Chris Richardson	124	124	128	125.0	−0.38
6	Julius Erving	459	425	431	442.4	0.76
7	Lionel Hollins	194	225	203	209.4	0.62
8	Mo Cheeks	189	150	172	168.3	−0.41
9	Caldwell Jones	129	143	134	136.6	0.32
10	Andrew Toney	208	243	245	225.1	−1.88
11	Bobby Jones	233	200	227	216.2	−1.04
12	Steve Mix	181	170	176	176.3	0.04
13	Daryl Dawkins	250	153	220	190.8	−3.09

Perhaps this is because most of Dawkins's baskets were dunks (his nickname was "Chocolate Thunder"), and the fact that he made his last shot may indicate that he was being guarded by a player that he could overpower. This would indicate that the next shot would more likely be a success.

How can we look at streakiness aggregated over all the 76er players? Simply average their z-scores. Then we use a result from statistics that the standard deviation of the average of N independent z-scores[4] is $N^{-.5}$. The average z-score for the nine players is −0.56. The standard deviation of the average of nine z-scores is $9^{-.5} = 1/3 = .333$. In this case the z-score for "the average of z-scores" is $\dfrac{-0.56-0}{.333} = -1.68$, which is not statistically significant at the .05 level. Therefore, we conclude that as a whole the 76ers players do not exhibit significant streakiness or hot hand shooting behavior.

4. A set of random variables is independent if knowledge of the value of any subset of the random variables tells us nothing about the distribution of the values of the other random variables. In our case knowing the z-scores of any subset of 76er players clearly tells us nothing about whether or not the other players are more or less likely to exhibit hot hand shooting behavior, so it is appropriate to assume the z-scores of individual players are independent random variables.

RECENT "HOT HAND" DEVELOPMENTS

Recently the debate about the existence of the hot hand has become very "hot." Research by Joshua Miller and Adam Sanjurjo (2018) claims to contradict GVT and show that the hot hand (particularly in basketball shooting) exists. In this section, we will try and lead the reader through a nontechnical discussion of the hot hand debate.

Most of us (including the great NBA and WNBA players) who have played ball think we get in a zone and "can't miss." The easiest way to put a quantitative face on the hot hand is to claim that inequalities such as the following are true:

- Prob(Make next shot | Made last shot) > Prob(Make next shot | Missed last shot).
- Prob(Make next shot | Made last two shots) > Prob(Make next shot | Missed last two shots).
- Prob(Make next shot | Made last three shots) > Prob(Make Next shot | Missed last three shots).

For the 1980–1981 76ers, GVT found the following shooting percentages in these situations.

The lesson of Table 11-2 is that in each listed situation, the players tend to shoot better when their recent history is worse. Certainly Table 11-2 gives no evidence for the existence of the hot hand.

In a terrific paper, Andrew Bocskocsky, John Ezekowitz, and Carolyn Stein ("Heat check: New evidence on the hot hand," available https://papers.ssrn.com/sol3/papers.cfm?abstract_id =2481494) added to our knowledge of the hot hand. They examined every shot taken during the 2012–2013 NBA season. BES found two key results:

- If you try and predict a player's shooting percentage on the next shot using as independent variables the player's overall shooting

TABLE 11.2

1980–1981, 76ers Shooting Percentages

Results of Last Three Shots	Shooting Percentage on Next Shot
Make Make Make	46%
Miss Miss Miss	56%
Make Make	50%
Miss Miss	53%
Make	51%
Miss	54%

ability and the player's shooting percentage on his last four shots, then the player's shooting percentage on his last four shots does not have a significant effect on the chance that the next shot is made. This result is consistent with the GVT results.

- **After having success, shooters tend to take more difficult shots,** and **after adjusting for the difficulty of the next shot and the difficulty of the last four shots taken, a player's success on his last four shots does have a significant positive influence on the success of the next shot.** BES determined the difficulty of each shot considering factors such as length of the shot and the distance between the defender and the shooter. They concluded that there is sort of a "hot hand" effect: a player's raw shooting percentage does not increase with recent success, but if he has shot better recently, he has a hot hand in the sense that the player's shooting will improve on the next shot relative to that shot's difficulty. In a sense, the increased difficulty of the next shot and the improved shooting cancel each other out.

Miller and Sanjurjo's 2018 paper set the math, psychology, and sports world abuzz with a stimulating analysis of the hot hand issue. To motivate their work, consider Figure 11-5 and the file Broadie.xlsx (the author gratefully acknowledges the cogent insights on the hot hand graciously shared by Professor Mark Broadie of Columbia!).

We assume we are looking at a 50% shooter whose shooting on a given shot is totally independent of all other shots he takes. We assume the shooter plays 16 games and takes four shots in each game, and the outcomes replicate the 16 equally likely sequences of make and miss that can occur on four shots. We see that given the last shot was made, the player shot 12 for 24. This is exactly what we would expect in the absence of the hot hand; the success on the last shot has no impact on the next shot. However, suppose we consider each row as a sample of shots from a separate player who shoots 50% on average, and the outcome of any shot does not influence other shots. We consider a row of our spreadsheet to be evidence for the hot hand if in that sequence we find a better shooting percentage after a made shot than the shooter's overall success rate. Conversely, we consider a row of our spreadsheet to be evidence against the hot hand if in that sequence we find a worse shooting percentage after a made shot than the shooter's overall success rate. The rows in light font provide evidence for the hot hand and the shaded rows provide evidence against the hot hand. Surprisingly, we find that only three of the nine relevant rows provide evidence for the hot hand. **Thus, if half of all players provided similar evidence for the hot hand, this would be a higher percentage than we expected if there were no hot hand, and we should conclude that the hot hand is real.**

In Figure 11-6 we analyze the same 16 shot sequences, looking at success after two consecutive made shots. We see that after the last two shots were made the player shot four for eight (50%) as we might expect. However, in the five relevant rows where shooting percentage after two consecutive made shots differs from the overall shooting percentage, one row provides evidence for the hot hand and four rows provide evidence against the hot hand. Thus, if half of all players provided similar evidence for the hot hand after two consecutive made shots, this would be evidence that the hot hand is real.

	A	B	C	D	E	F	G	H	I	J	K	L
5									3 of 9 players have hot hand			
6							24	12	After Make, Make more likely than overall percentage.			
7	Sequence	Toss 1	Toss 2	Toss 3	Toss 4	P(HIH)	Attempt	Success	Make after make percentage	Actual shoot percentage	Payoff	
8	1	Miss	Miss	Miss	Miss		0	0	none	0	0	
9	2	Miss	Miss	Make	Miss	0	1	0	0	0.25	-1	
10	3	Miss	Make	Miss	Miss	0	1	0	0	0.25	-1	
11	4	Make	Miss	Miss	Miss	0	1	0	0	0.25	-1	
12	5	Miss	Make	Make	Miss	0.5	2	1	0.5	0.5	0	
13	6	Make	Miss	Make	Miss	0	2	0	0	0.5	-1	
14	7	Make	Make	Miss	Miss	0.5	2	1	0.5	0.5	0	
15	8	Make	Make	Make	Miss	0.666667	3	2	0.666666667	0.75	1	
16	9	Miss	Miss	Miss	Make		0	0	none	0.25	0	
17	10	Miss	Miss	Make	Make	1	1	1	1	0.5	1	
18	11	Miss	Make	Miss	Make	0	1	0	0	0.5	-1	
19	12	Make	Miss	Miss	Make	0	1	0	0	0.5	-1	
20	13	Miss	Make	Make	Make	1	2	2	1	0.75	1	
21	14	Make	Miss	Make	Make	0.5	2	1	0.5	0.75	0	
22	15	Make	Make	Miss	Make	0.5	2	1	0.5	0.75	0	
23	16	Make	Make	Make	Make	1	3	3	1	1	0	

FIGURE 11.5   Four Shots: Chance of Success after Made Shot.

	A	B	C	D	E	F	G	H	I	J	K
4						1 out of 5 players have hot hand					
5			as defined as shooting better after Make Make than total percentage								
6							8	4			
7	Sequence	Toss 1	Toss 2	Toss 3	Toss 4	P(HIHH)	Attempt	Success	Make after Make Make Percentage	Actual shot percentage	Payoff
8	1	Miss	Miss	Miss	Miss		0	0		0	0
9	2	Miss	Miss	Make	Miss		0	0		0.25	0
10	3	Miss	Make	Miss	Miss		0	0		0.25	0
11	4	Make	Miss	Miss	Miss		0	0		0.25	0
12	5	Miss	Make	Make	Miss	0	1	0	0	0.5	−1
13	6	Make	Miss	Make	Miss		0	0		0.5	0
14	7	Make	Make	Miss	Miss	0	1	0	0	0.5	−1
15	8	Make	Make	Make	Miss	0.5	2	1	0.5	0.75	−1
16	9	Miss	Miss	Miss	Make		0	0		0.25	0
17	10	Miss	Miss	Make	Make		0	0		0.5	0
18	11	Miss	Make	Miss	Make		0	0		0.5	0
19	12	Make	Miss	Miss	Make		0	0		0.5	0
20	13	Miss	Make	Make	Make	1	1	1	1	0.75	1
21	14	Make	Miss	Make	Make		0	0		0.75	0
22	15	Make	Make	Miss	Make	0	1	0	0	0.75	−1
23	16	Make	Make	Make	Make	1	2	2	1	1	0

FIGURE 11.6   Four Shots: Chance of Success after
Two Consecutive Made Shots.

MS (2018) also show that for a 50% shooter **who exhibits the property that no shot influences the outcome of any other shot,** the chance that the shooter taking 100 shots will make a shot after making three in a row is not 50%, **but 46%.** The file Hundredshots. xlsx uses a simulation to confirm this result. We assumed a player had a 50% chance of making each shot and took 100 shots and the result of each shot was independent of the results on the previous shots. Using @RISK, we simulated this situation 10,000 times, and found that after three successive made shots, there was indeed a 46% (not 50%) chance that the next shot was good. Given this result, MS argue that if a shooter makes half his shots after making three in a row, then he exhibits a hot hand, because in the absence of the hot hand, he should make only 46% of his shots after making three shots in a row.

EMPIRICAL EVIDENCE FOR THE HOT HAND

Two of this book's authors (Pelechrinis and Winston, 2021) examined actual NBA shot data from the 2013–2014 and 2014–2015 seasons to examine whether actual shot data supports the existence of the hot hand. We examined the 21 players who took at least 1,000 shots during the 2014–2015 season. To begin, we used a neural network (fitted to the 2013–2014 season) to estimate (based on distance of shot, distance of closest defender, and other factors) for each shot the chance that the shot would be good. Assuming no dependence between the shots, we simulated each player's sequence of 2014–2015 shots 500 times. For example, if we predicted Chris Paul to have a 40% chance to make his first shot and a 55% chance to make his second shot a RAND( ) $\leq$ 0.4 would result in the first shot being good, and a RAND( ) $\leq$ 0.55 would result in the second shot being good. As shown in Figure 11-7, for eight of the 21 players their **actual chance of making a shot after making the last shot was statistically different from the simulated chance of making a shot after making the last shot (p-value <0.05).** For example, Damian Lillard actually made 42% of his shots after making a shot, while the simulation indicated that with complete

Player π	$Pr[M_{\pi,i}|M_{\pi,i-1}]_{data}$	$Pr[M_{\pi,i}|M_{\pi,i-1}]_{sim}$	p-val
LaMarcus Aldridge	0.47	0.44	0.14
Rudy Gay	0.49	0.41	< 0.001
Pau Gasol	0.46	0.46	0.46
Chris Paul	0.48	0.41	0.01
Anthony Davis	0.51	0.52	0.70
Damian Lillard	0.42	0.37	0.04
Victor Oladipo	0.43	0.42	0.32
LeBron James	0.46	0.48	0.72
Dwyane Wade	0.47	0.46	0.26
Monta Ellis	0.44	0.42	0.18
Russell Westbrook	0.42	0.37	0.04
Andrew Wiggins	0.39	0.43	0.93
Blake Griffin	0.52	0.47	0.02
James Harden	0.44	0.40	0.10
Tyreke Evans	0.48	0.45	0.17
Stephen Curry	0.46	0.47	0.58
Dirk Nowitzki	0.42	0.39	0.15
Kyrie Irving	0.42	0.38	0.10
Klay Thompson	0.49	0.44	0.02
Markieff Morris	0.48	0.41	< 0.001
Nikola Vucevic	0.50	0.45	0.01

FIGURE 11.7   Empirical Evidence for the Hot Hand.

independence between shots, he was only expected to make 37% of his shots after making a shot. The chance of a discrepancy at least this large was only 0.04. With no hot hand, we would expect only one of our 21 players to have a p-value <0.05. Assuming there is a 0.05 chance for each player to have a p-Value <0.05, the Excel function = BINOM. DIST.RANGE(21,0.05,8,21) returns a small 0.000004 chance that eight or more of our players would have a p-value <0.05. A more in-depth analysis is provided in the working paper.

DOES STREAK HITTING IN BASEBALL EXIST?

S. C. Albright (1993) analyzed the streak hitting of Major League Baseball players. For each player, he looked at the player's sequence of hits and outs during the season. Albright began by computing the expected number of runs of hits or outs for each player and

	A	B	C	D
1	Cal Ripken 1987			
2				
3	Actual Runs	Hits	Outs	
4	233	150	446	
5				
6	Mean Runs	225.4966443	=2*B4*C4/(B4+C4)+1	
7	Sigma Runs	10.61843284	=SQRT((B6-1)*(B6-2)/(C4-1))	
8				
9	Z Score Runs	0.706634945	=(A4-B6)/B7	
10				
11	P-Value	0.479793367	=2*(1-NORM.S.DIST(B9,TRUE))	

FIGURE 11.8   Cal Ripken Is Not a Streak Hitter.

comparing it to the actual number of runs (for example, sequence HHOOHO has four runs). Then he would compute a z-score based on the actual number of runs. For example, Cal Ripken in 1987 (see the file Ripken.xlsx) displayed the results shown in Figure 11-8.

Ripken's positive z-score indicates more runs than were expected by chance. Therefore, Ripken in 1987 exhibited behavior slightly less streaky than expected. For all 501 players during the 1987 season the average z-score was -0.256. The standard deviation of the average of 501 z-scores is $\frac{1}{\sqrt{501}} = 0.045$. Then the z-score for the average of all 1987 z-scores is $\frac{-.256 - 0}{.045} = -5.68$, which is significant evidence that players are streakier than average. Albright went on and used a more sophisticated technique, logistic regression, to predict the probability that a player would get a hit on an at bat based on his recent history **as well as other variables, including**

- Is pitcher left- or right-handed?
- Pitcher's ERA.
- Is game home or road?
- Is game on grass or artificial turf?

After adjusting for these variables, the evidence of streakiness disappears. Albright also found that the players who exhibited significant streak hitting behavior during a season were no more likely than a randomly chosen player to be streak hitters during the next season. This indicates that streak hitting behavior does not persist from year to year.

WHAT ABOUT "HOT TEAMS"?

How about teams exhibiting momentum? We analyzed the 2002–2003 NBA season to see if team performance exhibited streaky behavior. We cannot look at simply the sequence of wins and losses. Often during the NBA season, a team will play six straight road games against tough teams. This will often result in a long losing streak. The losing streak would likely be due to the strength of the opposition, not a variation in the team's performance. Therefore, for each game we created a "point spread' by including the strength of the team, the strength of its opponent, and home court edge. To illustrate the idea, suppose the Kings are playing the Bulls in Chicago. Sacramento played 7.8 points better than average during the 2002–2003 season and the Bulls played 8.5 points worse than average. The NBA home edge is around three points. Since the Kings are 16.3 points $(7.8 - (-8.5) = 16.3)$ better than the Bulls, and the Bulls get three points for home edge, we would predict the Kings to win by 13.3. Therefore, if the Kings win by more than 13 points, they receive a W, and if the Kings win by less than 13 or lose, they receive an L. In short, a W indicates that a team played better than its usual level while an L indicates that a team played worse than its usual level. We used the WWRT to analyze the streakiness of the sequence of Ws and Ls for each team. The results are shown in Figure 11-9 (see file Teammomentum.xlsx). We see that only Portland exhibited significant streakiness. Note, however, that with 29 teams we would (even if there were no streakiness) expect, by chance, $.05(29) = 1.45$ teams on average to have z-scores exceeding 2 in absolute value. The average of the 29 team z-scores is $-.197$. The standard deviation of the average of 29 z-scores is $\dfrac{1}{\sqrt{29}}. = .186$, so the z-score for the average of all 29 team z-scores is $\dfrac{-.197 - 0}{.186} = -1.05$ which is not significant

	D	E	F	G	H	I	J
14	Team	z score	mean runs	sigma	actual runs	W	L
15	Atl	0.888957	42.0000	4.499657	46	41	41
16	Bos	0.45017	41.9756	4.496947	44	40	42
17	Charlotte	1.562035	41.9756	4.496947	49	42	40
18	Chicago	0.467286	41.9024	4.488816	44	43	39
19	Cleveland	0.672543	41.9756	4.496947	45	40	42
20	Dallas	−1.02608	40.4390	4.326202	36	49	33
21	Denver	0.005424	41.9756	4.496947	42	42	40
22	Detroit	−0.39785	41.7805	4.475265	40	44	38
23	Golden St.	0.222239	42.0000	4.499657	43	41	41
24	Houston	0.27368	40.8049	4.366855	42	48	34
25	Indiana	−1.55567	42.0000	4.499657	35	41	41
26	LAC	0.044682	40.8049	4.366855	41	34	48
27	LAL	−1.10644	41.9756	4.496947	37	40	42
28	Memphis	−0.66672	42.0000	4.499657	39	41	41
29	Miami	−0.64659	41.9024	4.488816	39	43	39
30	Milwaukee	0.467286	41.9024	4.488816	44	39	43
31	Minnesota	−1.03444	41.6098	4.456293	37	37	45
32	NJN	−0.39785	41.7805	4.475265	40	44	38
33	NYK	0.888957	42.0000	4.499657	46	41	41
34	Orlando	0.495951	41.7805	4.475265	44	38	44
35	Philadelphia	−0.20104	41.9024	4.488816	41	39	43
36	Phoenix	−0.31369	41.3902	4.431901	40	36	46
37	Portland	−2.21831	41.9756	4.496947	32	42	40
38	Sacramento	−0.1744	41.7805	4.475265	41	44	38
39	SAN	−0.13683	41.6098	4.456293	41	45	37
40	Seattle	−0.21695	41.9756	4.496947	41	42	40
41	Toronto	−0.39785	41.7805	4.475265	40	38	44
42	Utah	−0.13683	41.6098	4.456293	41	37	45
43	Washington	−1.5151	41.7805	4.475265	35	38	44

FIGURE 11.9   No Momentum for NBA Teams!

at the .05 level. Therefore, we conclude that the variation in team performance during the 2002–2003 NBA season is well explained by random variation. Our small study gives no support to the view that teams have momentum or encounter more hot streaks than a random sequence.

CHAPTER 12

THE PLATOON EFFECT

For most right-hand pitchers, their curve ball is an important part of their pitching repertoire. A right-hand pitcher's curve ball curves in toward a left-hand batter and away from a right-hand batter. In theory, when facing a right-handed pitcher, a left-handed batter had an edge over a right-handed batter. Similarly, when a left-handed pitcher is on the mound, a right-handed batter has the edge. Managers take advantage of this alleged result by platooning batters. That is, managers tend to start right-handed batters more often against left-handed pitchers and start left-handed batters more often against right-handed pitchers. Ignoring switch hitters, Joseph Adler (2006) found that 35% of batters against left-handed pitchers are left-handed and 41% of batters against right-handed pitchers are left-handed. This shows that platooning does indeed exist. As the great American statistician and quality guru W. Edwards Deming said, "In God we trust; all others must bring data." Does actual game data confirm that batters have an edge over pitchers who throw with a different hand than the batter hits with? As shown in Figure 12-1 (see the file Platoon.xlsx), for the 2019 season, FanGraphs gives the OPS for each pitcher-hitter hand combination.

	Y	Z	AA	AB
8	Pitcher	Batter	OPS	Percentage
9	LHP	LHH	0.699	7.43%
10	LHP	RHH	0.744	18.97%
11	RHP	LHH	0.759	31.81%
12	RHP	RHH	0.717	41.80%

FIGURE 12.1   Platooning Results.

Figure 12-1 tells us the following:

- We find that left-handed batters on average have a 60-point higher OPS against right-handed pitchers than against left-handed pitchers.
- Right-handed batters on average have a 27-point higher OPS against left-handed pitchers than against right-handed pitchers.
- On average, left-handed pitchers yield a 45-point larger OPS to right-handed batters than left-handed batters.
- On average, right-handed pitchers yield a 42-point larger OPS to left-handed batters than right-handed batters.
- Left-handed hitters comprise 39% of plate appearances and right-handed hitters comprise 61% of plate appearances.
- Left-handed hitters have 28% of their plate appearances against left-handed pitchers while right-hand hitters have 43% of their plate appearances against left-handed pitchers.
- 26% of plate appearances involve left-handed pitchers while 74% of plate appearances involve right-handed pitchers.

Most sabermetricians refer to these differences as platoon splits. Figure 12-1 shows that platoon splits exist.

Since most major league pitchers are right-handed, a left-handed hitter has an edge over a right-handed hitter because he will be facing mostly right-handed pitchers. This helps explain the seemingly amazing fact that 39% of plate appearances involve left-handed hitters, compared to around 10% southpaws in the US.

EXAMPLES OF HUGE PLATOON SPLITS

MLB.com (see https://www.mlb.com/news/mlb-hitters-pitchers
-with-wide-platoon-splits-c303175734) calculated for the 2017–2018
seasons the players with the largest platoon splits.

- Nolan Arenado of the Rockies had an OPS that was 166 points better against lefties.
- Eric Hosner of the Padres had a 157-point better OPS against righties.
- Kyle Schwarber had a 250-point better OPS against righties!
- Pitcher Max Scherzer of the Nationals had a 139-point better OPS against righties.
- Pitcher Julio Tehreran of the Braves had a 170-point better OPS against righties.

CHAPTER 13

WAS TONY PEREZ A GREAT CLUTCH HITTER?

Tony Perez played first base for the "Big Red Machine" during the 1960s and 1970s. He had a lifetime batting of .279, and 380 career HRs. A lifetime batting average of .279 does not often lead to a Hall of Fame selection, yet in 2000 Perez was elected to the Hall of Fame while Jim Rice, who had better statistics (.298 lifetime average and 382 HRs in seven fewer years than Tony), was not elected to the Hall of Fame until his last year of eligibility (2009). See http://bleacherreport.com /articles/30821-which-baseball-hall-of-fame-players-arent-worthy for a discussion of the Perez vs. Rice HOF issue.

One reason Perez made it to the Hall of Fame so quickly was that his manager, Sparky Anderson, said that Perez was the best clutch hitter he had ever seen. Is there an objective way to determine if Perez was a great "clutch hitter"?

Let's define a batter to be a great clutch hitter if his performance in important situations tends to be better than his overall season performance. In *Baseball Hacks* Adler defined a clutch situation to occur when a batter bats during the ninth inning or later and his team trails by one, two, or three runs. Then Adler compared the batter's OBP in these situations to his overall season OBP. If the batter did significantly better during the clutch situations, then we could say the batter exhibited clutch hitting ability. The problem with this approach is the average batter only encountered 11 clutch situations per season, which is just not enough data to reliably estimate a hitter's clutch ability.

CREATING A BENCHMARK FOR EXPECTED CLUTCH PERFORMANCE

Each plate appearance has a different level of "clutch" importance. Coming to bat with your team down by a run with two outs in the bottom of the ninth is obviously a clutch plate appearance while batting in the top of the ninth when your team is ahead by seven runs has virtually no "clutch" importance. Recall that the FanGraphs win probability added (WPA) discussed in Chapter 8 is based on the relative importance of each plate appearance toward winning or losing a game. To determine if a player exhibits significant clutch ability, we will try and determine if his WPA per plate appearance for a season is significantly higher than you would expect based on the hitter's offensive ability as defined by wOBA.

To determine a predicted WPA per plate appearance we ran a regression for the 2016 season that predicted Y = 2016 WPA per plate appearance from X = 2016 wOBA. Our work is in the file Clutch.xlsx. Our regression results are shown in Figure 13-1.

The low p-value of wOBA shows that wOBA is a significant predictor of WPA/PA. The R-squared value of .55 means wOBA explains 55% of the variation in WPA/PA. In the file Clutch.xlsx we predict WPA/PA for 2016 batters using the equation

$$WPA/PA = -0.01926 + 0.06132 * wOBA. \tag{1}$$

Using the standard error of the regression, we determine for each player a z-score, (actual WPA/PA – predicted WPA/PA)/standard error of regression.

	M	N	O	P	Q	R	S	T	U
126	SUMMARY OUTPUT								
127									
128	Regression Statistics								
129	Multiple R	0.740917606							
130	R Square	0.548958899							
131	Adjusted R Square	0.545826669							
132	Standard Error	0.001749169							
133	Observations	146							
134									
135	ANOVA								
136		df	SS	MS	F	Significance F			
137	Regression	1	0.000536228	0.000536228	175.2613704	1.1308E-26			
138	Residual	144	0.000440581	3.05959E-06					
139	Total	145	0.00097681						
140									
141		Coefficients	Standard Error	t Stat	P-value	Lower 95%	Upper 95%	Lower 95.0%	Upper 95.0%
142	Intercept	-0.019258341	0.001583132	-12.16471057	7.30109E-24	-0.02238752	-0.016129162	-0.02238752	-0.016129162
143	wOBA	0.061323864	0.004632191	13.23863174	1.1308E-26	0.052167992	0.070479736	0.052167992	0.070479736

FIGURE 13.1   Clutch Regression.

A z-score exceeding +2 would indicate the player was significantly more clutch than his overall hitting ability would indicate. A z-score less than −2 would indicate that a player was significantly less clutch than his hitting ability would indicate. During 2016 no player had a z-score exceeding +2. Bryce Harper had the highest z-score (1.97), followed by Mike Trout (1.92). As we will see later in the chapter, Mike Trout is a terrific clutch performer, even after adjusting for the fact that he is a terrific hitter.

WAS TONY PEREZ A CLUTCH HITTER?

FanGraphs does not have WPA for Tony Perez's peak years of 1967–1973. Fortunately, my dear friend and colleague Jeff Sagarin computed a version of WPA for the 1957–2006 seasons (see http://sagarin.com/mills/seasons.htm). For the years 1967–1973, Figure 13-2 shows Tony Perez's z-score for each season. To compute predicted WPA/PA we used an analog of equation (1) based on the 1980 season (the first year in which FanGraphs had WPA).

For the years 1967–1973, Perez had an average z-score of 0.48. The standard deviation of an average of 7 z-scores is $\frac{1}{\sqrt{7}} = .378$. **Therefore, Perez's clutch hitting during the years 1967–1973 was**

	D	E	F	G	H	I	J
9	Year	WPA	PA	wOBA	Actual WPA/PA	Predicted WPA/PA	Z Score
10	1967	3.5	644	0.359	0.005434783	0.003293362	0.94
11	1968	3.5	690	0.344	0.005072464	0.002002596	1.34
12	1969	5	704	0.391	0.007102273	0.006046995	0.46
13	1970	4.4	681	0.427	0.006461087	0.009144832	−1.17
14	1971	3.2	664	0.344	0.004819277	0.002002596	1.23
15	1972	3.8	576	0.373	0.006597222	0.004498076	0.92
16	1973	4.2	647	0.406	0.006491499	0.00733776	−0.37

FIGURE 13.2   Tony Perez Was a Great Clutch Hitter!

	F	G	H	I	J	K	L
3	Jim Rice Clutch?						
4				0.39371			
5	Year	WPA	PA	wOBA	Actual WPA/PA	Predicted WPA/PA	Z Score
6	1977	3.39	710	0.414	0.004774648	0.008026168	−1.4199748
7	1978	6.57	746	0.425	0.008806971	0.00897273	−0.0723889
8	1979	2.29	688	0.421	0.003328488	0.008628526	−2.3145846
9	1980	1.17	542	0.369	0.002158672	0.004153872	−0.871326
10	1981	−0.8	495	0.349	−0.001616162	0.002432851	−1.7682486
11	1982	2	638	0.383	0.003134796	0.005358586	−0.9711537
12	1983	3.71	689	0.395	0.005384615	0.006391199	−0.4395861

FIGURE 13.3   Jim Rice Was a Poor Clutch Hitter.

1.26 (0.48/0.378) standard deviations better than that of an average clutch hitter who had comparable season-long batting statistics.

The analogous results for Jim Rice's peak years of 1977–1983 are shown in Figure 13-3.

For the years 1977–1983 Jim Rice had an average z-score of −1.12. The standard deviation of the average of seven independent z-scores is again .38, so Jim Rice's clutch hitting during the years 1977–1983 was nearly three standard deviations worse than expected! Using Excel's Goal Seek command (see worksheets Tony Perez New wOBA and Jim Rice New wOBA), we found the constant that needed to be added to each player's wOBA for each season to make the mean z-score = 0. For example, in the worksheet Tony Perez New wOBA

	D	E	F	G	H	I	J	K
3								
4					wOBA .391 makes Perez Z Score = 0			
5			add					
6			0.012704					
7								
8								
9	Year	WPA	PA	wOBA	New wOBA	Actual WPA/PA	Predicted WPA/PA	Z Score
10	1967	3.5	644	0.359	0.371704069	0.005434783	0.00438656	0.46
11	1968	3.5	690	0.344	0.356704069	0.005072464	0.003095794	0.86
12	1969	5	704	0.391	0.403704069	0.007102273	0.007140193	−0.02
13	1970	4.4	681	0.427	0.439704069	0.006461087	0.01023803	−1.65
14	1971	3.2	664	0.344	0.356704069	0.004819277	0.003095794	0.75
15	1972	3.8	576	0.373	0.385704069	0.006597222	0.005591274	0.44
16	1973	4.2	647	0.406	0.418704069	0.006491499	0.008430958	−0.85

FIGURE 13.4   Perez Clutch Hitting Is Equivalent to a .391 wOBA.

(see Figure 13-4) we used Goal Seek to change cell G6 so cell G20 equals 0. In this worksheet, the predicted WPA/PA is based on Perez's original wOBA + the number in cell G6 (these numbers are listed in Column H). Perez's original average wOBA was .378, so his clutch ability was equivalent to a .391 wOBA. Applying the same analysis to Jim Rice (who had an original average .394 wOBA), we found that Rice's clutch ability was equivalent to a .364 wOBA. This analysis provides some justification for Hall of Fame voters who rated Perez as a better player than Rice.

THE 1969 AMAZING METS

The 1969 Mets exceeded all pre-season expectations. In 1968 the Mets only won 73 games and in 1969 the Mets won 102 games and the World Series. Part of the Amazing Mets' surprising success was due to the amazing clutch hitting of Art Shamsky and Ron Swoboda. Shamsky and Swoboda exhibited fantastic clutch hitting. As shown in Figure 13-5, Shamsky had a clutch z-score of 2.07 and Swoboda had a clutch z-score of 3.83!

	D	E	F	G	Actual H	Predicted I	J
9	Year	WPA	PA	wOBA	Actual WPA/PA	Predicted WPA/PA	Z Score
10	Shamsky	2.5	349	0.349	0.0071633	0.002432851	2.07
11	Swoboda	3.1	375	0.315	0.0082667	−0.000492884	3.83

FIGURE 13.5   Shamsky and Swoboda Were Amazing in the Clutch!

	E	F	G	H	I	J	K
7	Year	PA	woba	WPA	Prediction	WPA/PA	Z Score
8	2012	639	0.409	5.41	0.00582312	0.008466354	1.51114
9	2013	716	0.423	5.01	0.006681654	0.006997207	0.1804
10	2014	705	0.402	7.18	0.005393852	0.010184397	2.73875
11	2015	682	0.415	5.28	0.006191063	0.007741935	0.88663
12	2016	681	0.418	6.64	0.006375034	0.009750367	1.92968
13	2017	507	0.437	5.29	0.007540188	0.010433925	1.65435
14	2018	608	0.447	4.14	0.008153426	0.006809211	−0.76849
15	2019	600	0.436	5.17	0.007478864	0.008616667	0.65048
16							mean
17						Average Z-Score	1.09787
18						St dev Mean Z Score	0.35355
19						Trout Mean Z-score above average clutch	3.10524

FIGURE 13.6   Mike Trout Is Amazing in the Clutch.

MIKE TROUT IS BETTER THAN YOU THINK!

Recall that Mike Trout had a z-score of 1.92 in 2016. It is natural to ask whether Trout's great clutch ability (over and above his great hitting ability) extended to other years. Figure 13-6 and the worksheet Trout in the file Clutch.xlsx provide the answer.

We find that during the years 2012–2019 Trout's average clutch z-score was an amazing 3.11 standard deviations better than we would expect after adjusting for his great batting ability.

CHAPTER 14

PITCH COUNT, PITCHER EFFECTIVENESS, AND PITCHF/X DATA

In October 2003, the Red Sox were leading the Yankees 5–2 after seven innings of the seventh and deciding game of the American League Championship Series. Pedro Martinez was cruising along and had allowed only two runs. To start the eighth inning Martinez got the first batter out and then Derek Jeter hit a double. Red Sox manager Grady Little went to the mound and talked to Martinez and then left him in the game. The Yankees promptly tied the game and later won in the 11th inning on a dramatic walk-off home run by Aaron Boone. Grady Little was fired later that week. Most baseball analysts think that part of the reason that Little was fired was that he ignored the fact that after throwing 100 pitches, Martinez tended to be a much less effective pitcher. Going into the eighth inning Martinez was well over 100 pitches. According to *Baseball Hacks,* Martinez had an on-base percentage of .256 for batters faced when he had thrown fewer than 100 pitches. This means that if Pedro had thrown fewer than 100 pitches, a batter had a 26% of reaching base. **After throwing more than 100 pitches, batters facing Pedro**

had a .364 on-base percentage, or a 36% chance of reaching base.
Since average on-base percentage is 34%, this data shows that when
Hall of Famer Pedro had thrown a lot of pitches he was very in-
effective. Grady Little ignored this data and, in all likelihood, (the
Yankees had only a 10% chance of winning the game at the start of
the eighth inning) Little's decision cost the Red Sox the 2003 AL
Championship.

Teams whose front offices believe in being data-driven keep tabs
on how a pitcher's effectiveness changes as he throws more pitches.
In *The Book,* authors Tango, Lichtman, and Dolphin (2014) cleverly
analyzed how a pitcher's effectiveness varies with the number of
pitches thrown. Using every plate appearance for the 1999–2002
seasons, they analyzed how hitters performed (after adjusting for the
ability of the hitter and pitcher) each time a pitcher went through the
batting order. Their measure of hitting performance was weighted
on-base average (wOBA).

As previously discussed wOBA is a form of linear weights (see
Chapter 3) which was scaled for this analysis, so an average hitter
had a wOBA of .340 (which matched the average major league on-
base percentage).

TLD found the results shown in Table 14-1.

Therefore the first time through the order, the hitters perform
eight points worse than expected, while the third time through

TABLE 14.1

wOBA vs. Time through Batting Order

Time through Batting Order	Expected wOBA (based on hitters wOBA)	Actual wOBA	Actual-Expected wOBA
1st	.353	.345	− 8 points
2nd	.353	.354	+ 1 point
3rd	.354	.362	+ 8 points
4th	.353	.354	+ 1 point

the order, **hitters perform eight points better than expected.** The third time through the order is usually during innings five through seven. Table 14-1 shows that pitchers perform better at the beginning of a game than they do in the middle to late innings. This might be pitcher fatigue, or it might be a result of the hitters getting to know the pitchers during the game. The fourth time through the batting order the hitters perform virtually the same as expected. This may indicate that any pitcher who makes it until the seventh or eighth inning must have good stuff, and this counterbalances the fatigue effect.

In short, it seems that teams should keep results like those in Table 14-1 for each starting pitcher and on an individual basis determine if the pitcher exhibits poorer performance when his pitch count exceeds a certain level. If so, do not ignore the data; bring in the relief pitcher.

PITCH COUNT AND PITCHER INJURIES

In 1999, Kerry Wood developed a sore elbow, which broke the hearts of all Cub fans. During his rookie year Wood averaged throwing 112 pitches per game and during one start threw 137 pitches. It seems reasonable to believe that starting pitchers who throw a lot of pitches are more likely to develop sore arms. Chapter 7 of *Mind Game* (Goldman, 2005) describes the link between pitch count and the likelihood of a pitcher developing a sore arm. Keith Woolner and Rana Jazayerli defined pitcher abuse points (PAP for short) for a single start as $PAP = max(0, (\text{Number of pitches} - 100)^3)$. For example, in any start where a pitcher throws 100 or fewer pitches, $PAP = 0$. If a pitcher throws 110 pitches $PAP = 1,000$; if a pitcher throws 130 pitches $PAP = 30^3 = 27,000$. **Note that PAP is not a linear function of pitch count. See Figure 14-1 and the file PAP.xlsx.**

Woolner and Jazayerli found that pitchers with a career value of PAP/total pitches exceeding 30 are nearly twice as likely to develop a sore arm as pitchers whose career PAP/total pitches does

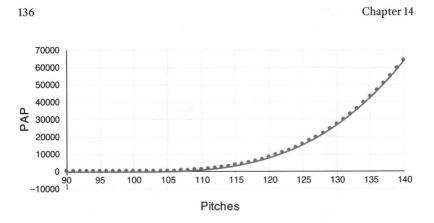

FIGURE 14.1   Pitcher Abuse Points as a Function of Game Pitch Count.

not exceed 30. For example, a pitcher with five starts throwing 100 pitches in three games and 120 and 130 pitches in his other two starts would have PAP/total pitches $= (20^3 + 30^3)/500 = 70$ and would be a candidate for a sore arm. PAP/total pitches provides a tool that managers can use to monitor the pitch counts of their starting pitchers in order to reduce the likelihood of injury.

PITCHF/X DATA

Since 2006 all major league stadiums have used two cameras to provide a cornucopia of information about each pitch thrown. To illustrate part of what's available at https://www.fangraphs.com/players/clayton-kershaw/2036/stats?position=P, Figure 14-2 (see file Kershaw .xlsx) shows Clayton Kershaw's career f/x breakdown of pitches:

We find that 58% of Kershaw's pitches are fastballs, 26% are sliders, 14% are cutters, and 2% are changeups. Kershaw throws few sinkers, and no cutters or splitters. Note that recently Kershaw is throwing fewer fastballs and more sliders.

The data shown in Figure 14-3 allows us to see which pitches make Kershaw great. This figure gives the runs above average (above average here means the pitch is more effective than an average pitch) Kershaw achieves per 100 pitches on each type of pitch. We see that sliders are Kershaw's most effective pitch, and on 100 sliders Kershaw gains 1.91 runs relative to an average pitch by a major league pitcher.

Season	Team	FA%	FC%	FS%	SI%	CH%	SL%	CU%
2008	Dodgers	0.713				0.056	0.004	0.227
2009	Dodgers	0.699	0	0	0.001	0.048	0.083	0.169
2010	Dodgers	0.715			0.001	0.013	0.203	0.069
2011	Dodgers	0.653				0.035	0.259	0.053
2012	Dodgers	0.62				0.037	0.231	0.112
2013	Dodgers	0.606	0		0.001	0.023	0.244	0.125
2014	Dodgers	0.552		0	0.002	0.011	0.291	0.143
2015	Dodgers	0.538				0.004	0.276	0.181
2016	Dodgers	0.508				0.003	0.333	0.156
2017	Dodgers	0.466			0.012	0.012	0.343	0.167
2018	Dodgers	0.409			0.002	0.004	0.419	0.164
2019	Dodgers	0.438			0.001	0.006	0.391	0.163
2020	Dodgers	0.408				0.002	0.402	0.188
Total	- - -	0.578	0	0	0.001	0.021	0.26	0.139

FIGURE 14.2   Clayton Kershaw's Pitch Breakdown.

Season	Team	wFA/C	wFC/C	wFS/C	wSI/C	wCH/C	wSL/C	wCU/C
2008	Dodgers	−0.17				2.48	−2.15	0.87
2009	Dodgers	1.47	−3.85	5.08	−2.12	−1.11	1.54	1.18
2010	Dodgers	0.69			10.91	−5.68	2.63	−0.8
2011	Dodgers	0.96				2.3	2.69	0.77
2012	Dodgers	1.14				−0.31	0.63	3.12
2013	Dodgers	1.9	15.15		0.91	−0.79	0.71	2.7
2014	Dodgers	1.36		−5.29	9.27	−6.39	2.65	2.11
2015	Dodgers	1.45				−7.73	1.76	2.81
2016	Dodgers	2.15				−3.39	3.61	1.71
2017	Dodgers	0.79			−1.33	7.21	1.85	1.55
2018	Dodgers	0.01			−30.21	0.43	1.66	1.12
2019	Dodgers	1.23			−1.36	−1.34	1.34	0.36
2020	Dodgers	0.81				−2.4	2.5	0.79
Total	- - -	1.12	5.65	−0.11	−2.31	−0.16	1.91	1.59

FIGURE 14.3   Kershaw's Pitch Effectiveness.

WILL STATCAST MAKE PITCH COUNT OBSOLETE?

The problem with pitch count is that all pitches are treated equally in the sense that two 130 pitch games are treated equally even if in one game the pitcher threw all fast balls and in the other game the pitcher threw all knuckleballs.

The Statcast pitcher statistics discussed in Chapter 5 give us the ability to measure a pitcher's in-game workload with greater accuracy. Metrics like spin rate and pitch velocity can surely help analysts develop a better statistic for pitcher abuse than our current crude PAP measure. Teams are already incorporating Statcast data into the management of their pitchers. For example (http://www.fangraphs.com/fantasy/mash-report-pitcher-spin-rates-and-injuries/), the Padres allegedly sent Ryan Butcher down to the minors after his spin rate and velocity dropped. The same blog post mentioned that Royals reliever Wade Davis's trade value suffered when teams noticed a drop of 162 in his average spin rate.

WILL STARTING PITCHERS BECOME OBSOLETE?

Throughout recent baseball history, most teams have used a five-man starting rotation. On May 19, 2018, the Tampa Bay Rays shook up baseball by starting reliever Sergio Ramos. He pitched one inning and Tampa Bay used four more relief pitchers and won the game 5–3. For the 2018 season Tampa Bay had a 46–38 record when traditional starting pitchers opened the game and a 44–34 record when relievers opened the game. The first five weeks that this strategy was used, Tampa Bay had the best ERA in baseball (prior to May 19, Tampa Bay had the eighth worst ERA in baseball). So why was this strategy a good idea?

- Tampa Bay had many pitchers on cheap contracts who could be slotted into the multiple pitcher strategy.
- As shown in Figure 14-1, during the game hitters tend to "figure out" the starting pitcher. The multiple pitcher strategy negates this edge.
- Suppose Tampa Bay starts a right-handed pitcher. The opponent might platoon and start many left-handed hitters. Then Tampa Bay brings in a left-handed pitcher, thereby negating the platoon advantage.

In 2019, Tampa Bay led the American League in ERA, and in 2020 the Rays were second in ERA, so this out-of-the-box strategy may soon become the norm. We note that the Rays were the 2020 American League Champions.

CHAPTER 15

WOULD TED WILLIAMS HIT .406 TODAY?

In 1941 Ted Williams hit .406. If Ted were in his prime in a more recent season, could he still hit around .400? In bars across the US, arguments like the following take place every day.

- Could Bill Russell dominate Shaq?
- Is LeBron better than Jordan?
- Who was better, Tom Brady or Joe Montana?

Without a time machine, we cannot actually have players from different eras compete against each other. We can, however, use mathematics to determine whether or not today's players are superior to players from an earlier time.

Let's focus on how hitters from later times (the 1940s, 1950s, . . . , 2010s) compare to the hitters in 1941. We define the level of pitching + defense in 1941 (PD1941 for short) to be the average. If, for example, PD1990 = .10, that would mean a batter hitting against PD1990 would hit .10 (or 100 points) higher than a batter hitting against PD1941. If PD1990 = −.10, that would mean that a batter hitting against PD1990 would hit .10 (or 100 points) **lower** than a batter hitting against PD1941.

Since PD1941 = 0, simple algebra shows that

$$PD2019 - PD1941 = (PD1942 - PD1941) + (PD1943 - PD1942)$$
$$+ (PD1944 - PD1943) + \cdots + (PD2018 - PD2017)$$
$$+ (PD2019 - PD\,2018).$$

How can we estimate PD1942 − PD1941? Look at all 1941 hitters who were still playing in 1942. Assume their ability has not changed from 1941 to 1942. Since young players tend to improve with added experience and older players tend to lose ability, it seems reasonable to assume that a given cohort of players will not see their ability change much from year to year. Given this assumption, suppose the 1941 players who played in 1942 had a batting average of .260 in 1941 and an average of .258 in 1942. This would indicate that PD1942 was .002 points better than PD1941 or PD1942 − PD1941 = −.002. Therefore, we conclude that PD1942 − PD1941 = BA of cohort of 1941 hitters in the 1942 − BA of cohort of 1942 players in 1941. A similar relationship holds when determining the relative merits of PD during successive years. In the file Batting.xlsx we used Excel's SUMIFS and COUNTIFS functions to determine the number of at bats and hits in years x and x + 1 for all players who played in both years, x and x + 1. Our data comes from Sean Lahman's (http://www.seanlahman.com/baseball-archive/statistics/) fantastic database.

Our predictions for Ted Williams's batting average in each year is summarized in Figure 15-1. We predicted that in 2000 Williams would have hit .355, but in 2019 our projection dropped to .306. We conjecture that this is because since 2000 there has been an increased emphasis on analytics that would hurt hitters in several ways:

- Better alignment of fielders (the shift!).
- Improved utilization of relief pitchers.
- More understanding of hitters' weaknesses.

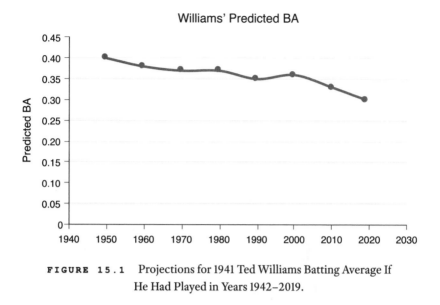

FIGURE 15.1   Projections for 1941 Ted Williams Batting Average If
He Had Played in Years 1942–2019.

We close by noting that comparing NBA and NFL players of different eras is difficult because of significant rule changes, such as
the introduction of the three-point shot (1979) and changes in NBA
(2004) and NFL (1978) hand-checking rules.

CHAPTER 16

WAS JOE DIMAGGIO'S 56-GAME HITTING STREAK THE GREATEST SPORTS RECORD OF ALL TIME?

In a beautifully written article, the late paleontologist and lifelong baseball fan Stephen Jay Gould (http://www.nybooks.com/articles /4337) argues that Joe DiMaggio's 56-game consecutive hitting streak is the greatest sports record of all time. In this chapter we use basic probability and statistics, along with Monte Carlo (MC) simulation to determine how likely it is that a 56-game hitting streak would ever occur.

In June of 1938, Johnny Vander Meer pitched consecutive no hitters. This has never been done by anyone else. Is this perhaps the greatest sports record of all time? After making some reasonable assumptions, basic probability and statistics can help us show that the occurrence of a 56-game hitting streak was less likely than the occurrence of consecutive no hitters.

While we are at it, can basic probability and statistics help us understand why between 1900 and June 2020 there were only 21 official perfect games? (See https://en.wikipedia.org/wiki/List_of _Major_League_Baseball_perfect_games.)

To address these issues, we need to study the basic mathematics of rare events.

CALCULATING THE PROBABILITIES OF RARE EVENTS: THE POISSON RANDOM VARIABLE

Let's consider random variables that always assume a value of 0, 1, 2, etc., where the value of the random variable represents the number of the occurrences of rare events. Some examples follow:

- Number of accidents a driver has in a year.
- Number of perfect games during a baseball season.
- Number of defects in a mobile phone.

Such random variables are usually modeled by the Poisson random variable. Given that a Poisson random variable has a mean λ, the probability that the random variable equals x is given by $\dfrac{\lambda^x e^{-\lambda x}}{x!}$.
Here $x! = (x)(x-1)*(x-2)\cdots*1$. Thus $3! = 6$, $4! = 24$, etc. The probability that a Poisson random variable with mean λ assumes a value x can be computed in Excel with the formula $= \text{POISSON}(x, \lambda, \text{False})$. The Excel function $= \text{POISSON}(x, \lambda, \text{True})$ gives the probability that a Poisson random variable with mean λ is less than or equal to x.

For example, if teenage drivers average 0.1 accidents per year, the probability a teen driver has exactly one accident during a year is $\dfrac{.1^1 e^{-.1}}{1!} = .0904$. This probability can also be calculated in Excel with the formula $= \text{POISSON}(1, .1, \text{False})$.

THE LAW OF RARE EVENTS

If X is a binomial random variable $\text{Bin}(n, p)$ and $\lim_{n \to \infty} n \cdot p = \lambda$, then X is asymptotically distributed as a Poisson with mean λ. Simply put, when the size of the population n is very large, and the occurrence of a certain event is rare (p small), then the binomial random variable can be approximated by a Poisson.

CALCULATING THE PROBABILITY OF
INDEPENDENT EVENTS

Two events are said to be independent if knowing that one of the events occurred tells you nothing about the likelihood of the other event occurring. For example, the event that the Cubs win the World Series during a season and the Bears win the Super Bowl during the same season are independent because if you knew that the Cubs won the World Series (as in 2016!) then you would not change your view of the probability that the Bears would win the Super Bowl. **To calculate the probability of multiple independent events all occurring, we simply multiply together the probability of the individual events.**

For example, given that the average major leaguer has an on-base percentage (OBP) of .320, what is the probability that an average pitcher would pitch a perfect game? A perfect game requires that a pitcher retire 27 consecutive batters. Assuming that the events that a batter reaches base are independent and all have probability .32, the probability that a pitcher throws a perfect game would be given by $(1 - .32)^{27} = .0000300$, or roughly one chance in 33,333. That's about one every 14 seasons or so.

WHAT IS THE PROBABILITY OF
20 PERFECT GAMES SINCE 1900?

As we previously stated, there have been 21 official perfect games since 1900. Given our previous result that the probability that a starting pitcher throws a perfect game is .00003, what is the likelihood that 20 regular season perfect games would have occurred between 1900 and October 2020?

There were 198,031 regular season games between 1900 and 2019. Each game presents two opportunities for a perfect game. Since each pitcher's start will yield either no or one perfect game, **the probability of a pitcher throwing a perfect game is equal to the**

expected number of perfect games a pitcher throws in a game. We now use the fact that for any set of random variables the expected value of the sum of the random variables is equal to the sum of the expected values of the individual random variables. This leads us to the conclusion that the expected number of perfect games since 1900 should be $2*(198,031)*(.00003) = 11.9$. Since perfect games since 1900 are rare, they should follow a Poisson random variable. Then the probability of at least 21 perfect games would be given by $1 - POISSON(20, 11.9, True) = 0.0106$. Clearly our model predicts significantly more perfect games than may have occurred. What might be wrong with our assumptions?

- We assumed every batter had an OBP of **exactly .32**. Batters like Joey Votto have OBPs much higher than .32 while batters hitting against Max Scherzer will have OBPs much less than .32. Suppose we assume that in one half of all games each hitter has an OBP of .27 and in the other half of all games each hitter has an OBP of .37. This averages out to our observed OBP of .32. Then the probability that a starting pitcher would throw a perfect game would (by the law of conditional expectation introduced in Chapter 6) be equal to

(Probability starting pitcher yields OBP of .27)$(1-.27)^{27}$ + (Probability starting pitcher yields OBP of .37) $* (1-.37)^{27} = .5 * (1-.27)^{27} + .5(1-.37)^{27} = .000051966$.

Then our expected number of perfect games since 1900 becomes $396,062 * (.000051966) = 20.6$, which is clearly more consistent with the actual number of 21 perfect games.

Our model might also be in error because we assumed that the probabilities of a batter reaching base on successive plate appearances are independent. If a pitcher has good stuff and retired the first, say, 10 batters, then you might lower your estimate that the later hitters reach base, and this would violate our independence assumption.

HOW UNUSUAL WAS THE 56-GAME
HITTING STREAK?

To determine the probability of a 56-game hitting streak occurring since 1900 we make the following assumptions:

- Only batters with at least 500 at bats in a season can have a 56-game hitting streak.
- We will not include hitting streaks that carry over between seasons (in theory these count as hitting streaks).
- Batters with over 500 at bats during a season averaged 3.5 at bats per game played so we will assume each batter had three at bats in half their games and four at bats in half their games.[1]

We can now estimate for each batter since 1900 with at least 500 at bats during a season the chance he will have a 56-game hitting streak. By summing this over all batters we obtain the expected number of 56-game hitting streaks since 1900. We will find there is roughly a 3% chance of at least one 56-game hitting streak occurring.

To illustrate the calculations let's consider a batter who played in 1900 (a 154-game season) who hit .316. We can compute the probability that the batter gets a hit in a three at bat game as $1 - $ the probability he fails to get a hit on all three at bats, or $1 - (1 - .316)^3 = .6794$. Similarly, the probability that the batter gets a hit in a four at bat game is $1 - (1 - .316)^4 = .7805$.

We now assume that each player will have 28 three at bat games and 28 four at bat games during a 56-game sequence (this assumption has little effect on our final result, even though Joe DiMaggio did have some games during his streak where he had two or five at

1. A related work by Billie, Huber, Nestler, and Costa, "Simulating rare baseball events using Monte Carlo methods in Excel and R," uses a slightly more complex assumption here—the actual number of at bats (AB) for Joe DiMaggio during the streak: 2AB—(5.4%), 3AB—11 (19.6%), 4AB—26 (46.4%), 5AB—16 (28.6%). Although the added variation does have an impact, the key results do not differ significantly from those here. See details at: http://archives.math.utk.edu/ICTCM/VOL22/S096/paper.pdf

bats). Then the probability that a batter will get a hit during each of 56 consecutive games is given by

$$\text{(Probability a batter gets hit in a three at bat game)}^{28}$$
$$\text{(Probability a batter gets hit in a four at bat game)}^{28} =$$
$$.000000019. \text{ (roughly two in 100 million).}$$

How many chances does our batter get to start a 56-game streak? He can start the streak during the first game of the season and also **during any of the first 99 games of the season for which the batter was hitless during the previous game.** We approximate the probability that the batter is hitless in a game by the average of the probability he is hitless in a three at bat game and the probability he is hitless in a four at bat game. Therefore

$$\text{Probability of hitless game} = .5 * (1 - .679) + .5 * (1 - .7805) = .27.$$

Therefore, our batter has on average $1 + 98 * (.27) = 27.46$ opportunities to start his 56-game hitting streak. This implies the expected number of 56-game hitting streaks by our batter during the season is $27.46 * (.0000000193) = .00000053$.

Summing up the expected number of 56-game hitting steaks over all batters with at least 500 at bats during the 1900–2016 seasons, we obtain an expected total of .024 56-game hitting streaks. Using the Poisson random variable, the probability of at least one 56-game hitting streak is given by $1 - \text{POISSON}(.0, .024, \text{TRUE}) = .024$.

Our calculations show that given all the opportunities that there have been for a 56-game hitting streak to occur, such a streak is highly unlikely, but certainly not impossible!

Another way to put the likelihood of a 56-game hitting streak in perspective is to determine how many years a batter of given ability would have to play before he had a 50% chance of having a 56-game hitting streak. Figure 16-1 shows the results of such calculations, assuming the batter had four at bats (a generous assumption) in each game.

	I	J
10	**Batting Average**	**Seasons**
11	0.3	134515.00
12	0.31	53294.90
13	0.32	22381.66
14	0.33	9926.43
15	0.34	4633.72
16	0.35	2269.68
17	0.36	1163.25
18	0.37	622.18
19	0.38	346.47
20	0.39	200.41
21	0.4	120.18
22	0.41	74.56
23	0.42	47.77
24	0.43	31.56
25	0.44	21.46
26	0.45	15.00
27	0.46	10.77
28	0.47	7.92
29	0.48	5.96
30	0.49	4.59
31	0.5	3.61

FIGURE 16.1   Seasons Needed to Have a
50% Chance of Getting a 56-Game Hitting Streak.

For example, we find that a .400 hitter would have to play for 120 seasons to have a 50% chance of getting a 56-game hitting streak! And how common are .400 hitters?

HOW UNUSUAL IS IT TO PITCH CONSECUTIVE NO HITTERS?

What is the probability that at last one starting pitcher (since 1900) would have pitched consecutive no hitters? To answer this question, we will make the following assumptions.

- All games are started by pitchers who start exactly 35 games during a season.

- Since 1900, .00064 of all pitchers starting a game have pitched a no hitter. We therefore assume the probability that each game started will result in a no hitter is .00064.

To determine the expected number of times a pitcher has thrown consecutive no hit games since 1900, we first determine the probability that a pitcher making 35 consecutive starts during a season has thrown consecutive no hitters. This probability is given by (probability the first of the two no hitters is game 1) + (probability the first of the two no hitters is game 2) + $\cdots$ + (probability the first of the two no hitters is game 34).

Now, the probability that a pitcher throws consecutive no hitters and the first of the two no hitters in game 1 is $(.00064)^2 = .00000041$.

The probability that a pitcher throws consecutive no hitters and the first no hitter in game i is $(1 - .00062)^{i-1}(.00062)$.

Therefore (see cell N35 of file Nohitter.xlsx) the probability that a pitcher with 35 starts will pitch consecutive no hitters during a season is approximately 0.0000138.

Now in 1900, 1,232 games were played and pitchers made $1,232 * 2 = 2,464$ starts. At 35 starts per pitcher this would imply there were around 2,464/35, or around 70 starting pitchers that year. Thus we find the expected number of times consecutive no hitters would be pitched during 1900 to be $70 * (.0000138) = .00096$.

Adding together the expected number of consecutive no hitter occurrences for the years 1900–2007 we find the expected number of consecutive no hitters during the time frame 2000–2007 to be .145. Therefore our model implies that the probability that at least one starting pitcher would throw consecutive no hitters is $1 - \text{POISSON}(0, .145, \text{True}) = 15.7\%$

Another way to put the "rareness" of consecutive no hitters in perspective is to ask how many years (with 35 starts per year) would the greatest no hit pitcher of all time (Nolan Ryan) have to have played before he had a 50% chance of throwing consecutive no hitters? Ryan pitched six no hitters in 773 starts, so the chance that any of his starts resulted in a no hitter is $6/773 = .00776$. As shown in the worksheet Ryan, this yields a .0018 chance of consecutive no hitters during a season. Ryan would have needed to pitch 384 seasons to have a 50% chance of having thrown consecutive no hitters!

In summary we find that consecutive no hitters and a 56-game hitting streak are both highly unlikely, but not beyond the realm of possibility. The 56-game hitting streak appears to be less likely to have occurred by chance than the consecutive no hitters.

A BRAINTEASER

We close this chapter with a brainteaser. A starting pitcher pitches a complete game and the game is nine innings. What is the minimum number of pitches the starting pitcher could have thrown? The answer is 25. The starting pitcher is pitching for the visiting team and allows only one runner to reach base (on a home run) and loses 1–0. The starting pitcher does not need to pitch the ninth inning, so he throws $8 * 3 + 1 = 25$ pitches!

CHAPTER 17

PROJECTING MAJOR LEAGUE PERFORMANCE

A baseball general manager does not care what a player has done for her lately, but cares about what the player is likely to do in the future. In this chapter we discuss how to predict the major league performance of a minor league player who is brought up to the major leagues. We also describe how Tom Tango's elegant Marcel projection system is used to predict the future performance of a major leaguer based on his past level of performance. The Marcel system can be easily adapted to other sports and also would prove useful to fantasy sports players trying to predict a player's next season performance.

MAJOR LEAGUE EQUIVALENTS

Major league general managers must decide every year whether a promising minor league player is ready to be brought up to the major league team. Of course, the minor player is facing inferior pitching in the minors, so you cannot expect him to duplicate his minor league statistics when he is brought up to the majors. In 1985 Bill James developed the idea of major league equivalents to help major league front office personnel determine if a minor leaguer is ready for the majors.

From Baseball-reference.com we downloaded the OBP for a set of hitters whose last minor league year was played at the AAA level, and their OBP for their first major league season. These hitters played in either the American Association (AA), International League (IL), or Pacific Coast League (PCL).

Suppose we know that a batter, Joe Hardy, had an OBP of .360 in AAA. If we bring Joe up to the major leagues, what OBP can we expect? Our data shows that in the two minor leagues for which we have a lot of data (IL and PCL), during their first year in the majors after their last year (or part of year) in the International League, the batters had an OBP that averaged 90% of their last minor league OBP. For the PCL, the batters averaged 88% of their last minor league OBP during their first year in the majors. This would indicate that the major league equivalent of an AAA minor league OBP would be roughly .89 times the minor league OBP. Therefore, we would predict Joe to achieve a "major league equivalent" OBP of $.89 * (.360) = .320$ in the major leagues.

Sabermetricians know that major league equivalents should be adjusted for the minor league park, the major league park, and the quality of pitching faced in the minor league. For example, Tucson and Albuquerque are known to be hitters' parks (see http://www.baseballamerica.com/today/features/040408parkfactors.html), so batters who played for these teams would have their major league equivalent OBPs reduced. Similarly, if a batter was being called up to a team like the Marlins, who play in a park in which it is more difficult than average to reach base, we should reduce their projected major league equivalent.

We did not collect data on slugging percentages, but let's suppose slugging percentage also drops around 11% when a player goes from AAA to the major leagues. Recall from Chapter 2 that Bill James's original runs created formula is (for all intents and purposes) computed by multiplying OBP by SLG. Then we would expect a AAA minor leaguer to retain $.89^2 = .78$ or 78% of his minor league runs creating ability and lose around 22% of his runs creating ability.

On the website https://tht.fangraphs.com/katoh-forecasting-a-hitters-major-league-performance-with-minor-league-stats/ Chris Mitchell uses minor league statistics to generate projections for WAR created for a player by age 28. Based on 2014 minor league statistics, Mitchell's top 10 prospects are shown in Figure 17-1. We also show the actual WAR each player generated through 2020. Note that two of Mitchell's top four (Mookie Betts and Kris Bryant) have won MVP awards.

	C	D	E	F
			Projected WAR	Actual WAR generated through
4	Player	Age in 2014	through Age 28	2020
5	Mookie Betts	21	21.6	40
6	Joc Pederson	22	18.3	13
7	Jose Ramirez	21	16.3	28
8	Kris Bryant	22	16	28
9	Jorge Soler	22	15.6	0
10	Gregory Polanco	22	14.6	7
11	Addison Russell	20	13.1	0
12	Arismendy Alcantara	22	12.3	−1
13	Jon Singleton	22	11.7	−1
14	Joey Gallo	20	11	10

FIGURE 17.1   Chris Mitchell Minor League–Based Projections
for Major League WAR.

THE MARCEL PROJECTION SYSTEM

Suppose we want to project Baltimore Orioles slugger Chris Davis's HRs for the 2017 season. The great Tom Tango developed the simple Marcel projection system (http://www.tangotiger.net/archives/stud0346.shtml). We begin by trying to project the fraction of Davis's 2017 plate appearances that will result in HRs. It seems logical to somehow average his HRs per plate appearance during recent (Marcel uses the last three) seasons and take a weighted average of his performance level during the three most recent seasons and the average level of American League HRs/PA for the last three seasons. Then we need a formula to predict 2017 plate appearances for Davis based on recent (Marcel uses two) seasons. Then we have a non-age-adjusted prediction for Davis's 2017 HRs.

$$(\text{Non-age-adjusted 2017 HR prediction}) =$$
$$(\text{2017 prediction for chance PA is an HR}) * (\text{2017 predicted PAs})$$

Finally, we adjust our 2017 HR prediction based on age. Since data shows the peak age for player performance is 29, Tango suggests the following age adjustment:

- Increase chance of a favorable event occurring by a factor of .006 for each year player is under 29.
- Decrease chance of a favorable event occurring by a factor of .003 for each year the player is over 29.

The file Marcel.xlsx (see Figure 17-2) illustrates the Marcel projection of Chris Davis's 2017 HRs. We now describe the required calculations.

Step 1: Enter in cells F4:G6 Davis's PA and HRs for the last three seasons, earliest year first.

Step 2: In cell range H4:H6, enter the league average (for non-pitchers) of HR/PA.

Step 3: In I4:I6 compute for each of the last three seasons Davis's HR/PA.

Step 4: In cell E8, we compute a weighted average (.0602) of Davis's HR/PA for the last three seasons. The weight for three years ago is 3 * Year-3 PA; for two years ago, 4 * Year-2 PA; for last year, 5 * Year-1 PA. Thus, the more recent the season and the more the PAs in the season, the more that season's weight. The formula in cell E8 (shown in cell F8) computes a weighted average of Davis's last three years HR/PA by dividing the weighted sum by the sum of the seasonal weights.

	A	B	C	D	E	F	G	H	I	J
2	Age	31								
3			Weights PA	Weights		PA	Statistic	League Average Rate	Player Rate	
4				3	Year-3	525	26	0.023429873	0.049524	
5			0.1	4	Year-2	670	47	0.02874408	0.070149	
6			0.5	5	Year-1	665	38	0.032105606	0.057143	
7										
8				weighted avg for player	0.060158	=SUMPRODUCT(D4:D6,F4:F6,I4:I6)/SUMPRODUCT(D4:D6,F4:F6)				
9				weighted average for league	0.029114	=SUMPRODUCT(D4:D6,H4:H6,F4:F6)/SUMPRODUCT(D4:D6,F4:F6)				
10										
11				weight for player	0.863326	=SUMPRODUCT(D4:D6,F4:F6)/(SUMPRODUCT(D4:D6,F4:F6)+1200)				
12				weight for league	0.136674	=1-E11				
13										
14			Predicted Probability Davis HR	0.055915411	=SUMPRODUCT(E11:E12,E8:E9)					
15										
16			Predicted PA	599.5	=200+SUMPRODUCT(C5:C6,F5:F6)					
17										
18						Predict 33 HRs for Chris Davis in 2017				
19			Predicted HRs	33.52128881	=D14*D16					
20			Age Adjusted Prediction	33.32016108	=D19*(1+D25)					
21			Predicted Year	2017						
22			Birth Year	1986						
23			Age	31	=D21-D22					

FIGURE 17.2   Marcel Projection for Chris Davis HRs in 2017.

Step 5: In a similar fashion, in cell E9 we compute a weighted average (.0291) of the AL HR/PA for the last three seasons.

Step 6: Now we need to calculate the weight that should be placed on E8 and E9 in a weighted average that determines our pre-age-adjusted estimate of Davis's 2017 HR probability. Tango's theory is that the more plate appearances of data you have, the more weight you give Davis's HR probability estimate. Empirically, Tango finds that giving a weight of $(5 * (\text{Year-1 PA}) + 4 * (\text{Year-2 PA}) + 3 * (\text{Year-3 PA}))/(5 * (\text{Year-1 PA}) + 4 * (\text{Year-2 PA}) + 3 * (\text{Year-3 PA}) + 1,200)$ to Davis' HR probability estimate seems to "regress" Davis's performance in the proper fashion to the league average. The formula in cell F11 computes the weight given to Davis's HR probability estimate (.8633). Then a weight of $1 - .8633 = .1367$ is given to the league average HR probability estimate.

Step 7: In cell D16 we compute (599.5) Davis's estimated 2017 plate appearances as $200 + .1 * (\text{Year-2 PA}) + .5 * (\text{Year-1 PA})$. This estimate gives more weight to Year-1 PA.

Step 8: In cell D19 we compute Davis's predicted 2017 HRs as $(.8633 * .0602 + (1 - .8633) * (.0291)) * (599.5) = 33.52$.

Step 9: Since Davis was 31 years old in 2017, we will deflate our HR estimate by $(1 - 2 * .003) = .994$. In cell D20 we obtain our age-adjusted projection for Chris Davis of 33.32 HRs in 2017. Davis hit 36 home runs in 2017.

PART II

FOOTBALL

WHAT MAKES NFL TEAMS WIN?

NFL teams want to win games. Is it more important to have a good rushing attack or a good passing attack? Is rushing defense more important than passing defense? Is it true that turnovers kill you? During the early 1960s statistician Bud Goode studied the question of what makes a team win. He found that yards per pass attempt on both offense and defense were the most important factors in predicting an NFL team's success. This is intuitively satisfying because yards per pass attempt is more of a measure of efficiency than total yards passing. Since we divide by pass attempts, yards per pass attempt recognizes that passing plays (like all plays) use up a scarce resource (a down).

Using team statistics from the 2014–2017 seasons, we built a regression model to predict each team's scoring margin, defined as (points for – points against). We used the following independent variables (see file NFLregression.xlsx):

- Team offense yards per pass attempt (PY/A). We include sacks as pass attempts, and yards lost on sacks are subtracted from yards passing.
- Team defense yards allowed per pass attempt (DPY/A). We include sacks as pass attempts, and yards lost on sacks are subtracted from yards passing.

- Team offense yards gained per rush (RY/A).
- Team defense yards allowed per rush (DRY/A).
- Turnovers committed (TO).
- Defensive turnovers (DTO).
- Penalty yards difference = Penalty yards committed by team − Penalty yards committed by opponents (PENDIF).
- Return TD difference = Return TDs − Return TDs by opponent (RET TD). It includes TDs scored on fumbles, interceptions, kickoffs, and punts.

Again, our dependent variable for each team is total regular season scoring margin, or points for − points against. After running our regression (without an intercept term) we obtain the output shown in Figure 18-1.

SUMMARY OUTPUT

Regression Statistics	
Multiple R	0.886756
R Square	0.786337
Adjusted R Square	0.76554
Standard Error	43.62831
Observations	128

ANOVA

	df	SS	MS	F	Significance F
Regression	8	840614.5193	105076.8149	55.20395801	1.70627E−36
Residual	120	228411.4807	1903.429006		
Total	128	1069026			

	Coefficients	Standard Error	t Stat	P-value	Lower 95%	Upper 95%	Lower 95.0%	Upper 95.0%
Intercept	0	#N/A	#N/A	#N/A	#N/A	#N/A	#N/A	#N/A
RET TD	9.872342	2.157991168	4.574783396	1.16958E−05	5.599669837	14.14501449	5.599669837	14.14501449
PENDIF	−0.443983	0.229640086	−1.933385498	0.055544099	−0.898654202	0.010688576	−0.898654202	0.010688576
PY/A	69.04335	5.904116666	11.69410275	1.49694E−21	57.35360686	80.73308707	57.35360686	80.73308707
RY/A	23.24033	8.870679499	2.619904014	0.009931475	5.677000772	40.80365688	5.677000772	40.80365688
TO	−4.996646	0.700715051	−7.130782236	8.14422E−11	−6.384013477	−3.609279406	−6.384013477	−3.609279406
DPY/A	−53.68362	7.309871105	−7.343990609	2.72827E−11	−68.15666081	−39.21058869	−68.15666081	−39.21058869
DRY/A	−39.19238	11.02076077	−3.556231617	0.000539424	−61.01271724	−17.37203856	−61.01271724	−17.37203856
DTO	2.014417	0.701369345	2.872120361	0.004822265	0.625754685	3.403079667	0.625754685	3.403079667

FIGURE 18.1 Results of NFL Regression.

From this output we see that for all independent variables the probability of them actually being 0 (and hence, not *important*) is small (the maximum probability is .055 for the penalty differential). The R-squared value of .79 tells us that the following equation explains 79% of the variation in team scoring margin. The standard error of 43 tells us that the actual point margin for a team will follow a normal distribution with mean equal to the point prediction for the score margin and standard deviation of 43. The point prediction equation is: Predicted Team Scoring Margin $= 9.87(\text{RETTD}) - 0.44(\text{PENDIF}) + 69.04(\text{PY/A}) + 23.24(\text{RY/A}) - 4.99(\text{TO}) - 53.68(\text{DPY/A}) - 39.19(\text{DRY/A}) + 2.01(\text{DTO})$.

From this regression we learn that for the total point margin of a season (after adjusting for the other independent variables, and given in points per season):

- An extra yard passing per attempt is associated with 69.04 points.
- An extra yard rushing per attempt is associated with 23.24 points.
- An extra turnover costs us 4.99 points.
- An extra yard allowed per passing attempt costs us 53.68 points.
- An extra yard rushing allowed per attempt costs us 39.19 points.
- An extra forced turnover aids us by 2.01 points.
- An extra yard in penalties costs us .44 points.

We find that the coefficients for offensive and defensive passing efficiency are much higher than the coefficients for offensive and defensive rushing efficiency. This is consistent with Goode's finding that passing efficiency is the key driver of success in the NFL. The standard deviation of team PY/A is 0.66 yards and the standard deviation of RY/A is .39 yards. This means that if you could move an average passing team up to the 84th percentile in PY/A (one standard deviation above average), you would improve its performance by $.66 * 69.04 = 45$ points, while if you could move an average rushing team up to the 84th percentile in RY/A, you would only improve the team's performance by $.39(23.24) = 9.06$ points. If it takes the same amount of effort to improve passing and rushing offense, you are much better off trying to improve your passing offense. These results give little credence to the belief of so-called experts that you need a good ground game to set up your passing game. In fact, the correlation between PY/A and RY/A is only .10!

Another way to show the importance of the passing game is to run a regression to predict scoring margin using only PY/A and DPY/A. These two variables by themselves explain 63% of the variation in team scoring margin. In contrast, predicting team scoring margin from RY/A and DRY/A explains only 1.7% of the variation in team scoring margin! Note: The outputs from these regression models are not shown here but are contained in the NFL-regression.xlsx file.

The offensive and defensive turnover coefficients average out to 3.13 points. This would seem to indicate that a turnover is worth 3.50 points. Note, however, that we accounted for the effect of return TDs in our RET TD term. If we drop this independent variable from our model, we find that a defensive turnover is worth 2.60 points and an offensive turnover costs us 4.79 points. This indicates that a turnover is worth around the average of 2.60 and 4.79, or 3.70 points.

The key to continued success in the NFL is to understand how the allocation of resources (primarily money spent on players) changes the values of the independent variables. For example, if it costs $10 million for an All-Pro receiver and $10 million for an All-Pro linebacker, which expenditure will improve team performance more? This is difficult to answer. The linebacker will have a big impact on DTO, DY/R, and DPY/A while our receiver will probably only affect PY/A and turnovers. The front office personnel need to determine which free agent has more impact and will be a key contributor to his team's success.

DOES A GOOD RUSHING ATTACK SET UP THE PASSING GAME?

Most fans believe that a good rushing attack helps set up the passing attack. If this is the case, we would expect to see a strong positive correlation between PY/A and RY/A. Figure 18-2 shows the correlation between our independent variables.

	RET TD	PENDIF	PY/A	RY/A	TO	DPY/A	DRY/A	DTO
RET TD	1							
PENDIF	−0.0411544	1						
PY/A	0.11048133	−0.0159983	1					
RY/A	−0.0576343	0.14013158	0.117779	1				
TO	−0.0812525	0.00623663	−0.3247161	−0.1171398	1			
DPY/A	−0.231602	−0.0861299	−0.0057858	0.04259182	0.24926699	1		
DRY/A	0.01902488	−0.220135	0.22861	−0.0454787	−0.2443873	0.23895137	1	
DTO	0.20411428	−0.0429489	0.11200158	−0.1535111	−0.0576773	−0.1431152	0.14203155	1

FIGURE 18.2   Correlation Matrix for NFL Independent Variables.

Note that PY/A and RY/A have only a .12 correlation, so a good ground game does not seem to lead to a good passing game. Of course, there are more detailed ways to explore whether there is a connection between the two (e.g., by examining the efficiency of the passing game after N number of rushing attempts, etc.). Here we should specify that the correlation looks at the relationship between the efficiency of passing and rushing. As we will see in Chapter 22, from a strategic/game-theoretic point of view rushing is still helpful.

CHAPTER 19

WHO'S BETTER

Brady or Rodgers?

Many Americans spend a nontrivial amount of time arguing about who are the best quarterbacks in the NFL. For example, is Tom Brady better than Aaron Rodgers? Brady's continued success in Super Bowl appearances makes it hard to argue that Rodgers is a better "winner" than Brady. However, one could argue based on Rodgers's consistent elite passing numbers that he is a better quarterback than Brady, as Super Bowls are a team effort. One common way in which quarterbacks can be evaluated is the use of a quarterback rating that attempts to remove team bias. In 2016, ESPN did a complete update on its calculations of NFL quarterback evaluation and created a metric called "total quarterback rating" (or Total QBR). We will get to that momentarily, but first a brief history lesson.

The traditional NFL quarterback rating system is still reliant on complex factors as described by Berri et. al in their outstanding book *Wages of Wins* (2007):

> First one takes a quarterback's completion percentage, then subtracts 0.3 from this number and divides by 0.2. You then take yards per attempts subtract 3 and divide by 4. After that, you divide

touchdowns per attempt by .05. For interceptions per attempt, you start with .095, subtract from this number interceptions per attempt, and then divide this result by .04. To get the quarterback rating, you add the values created from your first four steps, multiply this sum by 100, and divide the result by 6. Oh, and by the way, the sum from each of your first four steps cannot exceed 2.375 or be less than zero.

This formula makes quantum mechanics or Fermat's Last Theorem seem simple! We note the NCAA has its own incomprehensible system for ranking quarterbacks, which has remained unchanged in recent years.

The first thing to note is that a QB's rating is based on four statistics:

- Completion percentage (completions/passing attempts)
- Yards gained per pass attempt (yards gained by passes/passing attempts)
- Interception percentage (interceptions/passing attempts)
- TD pass percentage (TD passes/passing attempts).

Like the antiquated fielding percentage metric in baseball, this formula makes no sense. Note that all four statistics are given equal weight in the ranking formula. There is no reason why, for example, completion percentage and yards per pass attempt should count equally. In fact, since an incomplete pass gains 0 yards, completion percentage is partially accounted for by yards per pass attempt. In the TD pass percentage portion of the formula a touchdown pass of one yard is as much as a touchdown pass of 99 yards. Granted, the 99-yard pass pumps up our yards/pass attempt, but I think you see how arbitrary the system is. **Of course, any system for rating quarterbacks based on passing statistics is really rating the team's entire passing game, which is of course influenced by the quality of the team's receivers and offensive line.** With this caveat, however, we will try and develop a simpler rating system for quarterbacks.

The new "total quarterback rating" metric is largely based on Brian Burke's work. In 2007 Brian ran a regression (similar to our work in Chapter 18) to predict team games won during the 2002–2006 NFL seasons. Brian used the following independent variables:

TRUOPASS = offense (pass yds – sack yds)/pass plays

TRUDPASS = defense (pass yds – sack yds)/pass plays

ORUNAVG = offense rush yds/rush attempt

DRUNAVG = defense rush yds/rush attempt

OINTRATE = interceptions thrown/pass attempts

DINTRATE = interceptions taken/pass attempts

OFUMRATE = offensive fumbles/all plays (not fumbles lost)

DFFRATE = defensive forced fumbles/all plays

PENRATE = own team penalty yds per play

OPPPENRATE = opponent penalty yds per play

As shown in Figure 19-1, the results of his regression were as follows: the QB influences TRUOPASS (analogous to PY/PA in Chapter 18) and OINTRATE.

	D	E	F
8	**Variable**	**Coefficient**	**p-value**
9	const	5.26	0.063
10	TRUOPASS	1.54	<0.00001
11	TRUDPASS	−1.67	<0.00001
12	ORUNAVG	0.92	0.00071
13	DRUNAVG	−0.55	0.048
14	OINTRATE	−50.1	0.0012
15	DINTRATE	83.7	<0.00001
16	OFUMRATE	−63.9	0.005
17	DFFRATE	78.7	0.0001
18	PENRATE	−4.49	0.013
19	OPPPENRATE	6.58	0.0004

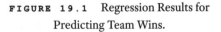

FIGURE 19.1 Regression Results for Predicting Team Wins.

Player	Our Ranking	OldQBR	TOTALQBR
Matt Ryan	1	117.1	79.6
Tom Brady	2	112.2	79.4
Dak Prescott	3	104.9	77.6
Aaron Rodgers	12	104.2	72.4
Drew Brees	5	101.7	66.8
Kirk Cousins	4	97.2	66.5
Andrew Luck	9	96.4	65.9
Matthew Stafford	13	93.3	65.2
Tyrod Taylor	18	89.7	62.4
Alex Smith	14	91.2	60.8
Ben Roethlisberger	11	95.4	60.5
Jameis Winston	21	86.1	59.5
Marcus Mariota	6	95.6	59.1
Philip Rivers	17	87.9	58.7
Russell Wilson	10	92.6	57.1
Derek Carr	7	96.7	56.1
Sam Bradford	15	99.3	53.1
Carson Palmer	20	87.2	52.9
Joe Flacco	25	83.5	52.5
Andy Dalton	8	91.8	52.3
Trevor Siemian	19	84.6	49.7
Brock Osweiler	30	72.2	49.3
Colin Kaepernick	22	90.7	49.2
Ryan Tannehill	16	93.5	48.6
Cam Newton	24	75.8	47.1
Carson Wentz	28	79.3	46.7
Eli Manning	23	86	45.7
Blake Bortles	29	78.8	43
Ryan Fitzpatrick	27	69.6	39.2
Case Keenum	26	76.4	37.5

FIGURE 19.2 NFL Quarterback Ratings, Traditional Quarterback Rating, and Total Quarterback Rating.

Since that point, ESPN has expanded this definition of total quarterback rating to account for "clutch" and "garbage time" factors to address several criticisms of the old model. Additionally, ESPN has mentioned that the next step in this calculation will be to add an additional factor to account for the defensive ability of the opponent.

While ESPN's total quarterback rating accounts for many factors, it still has the major problem of being complex, and, most importantly, not all of its details are known. In the following, we will simply rate each NFL quarterback using the coefficients from Brian's regression. This should rate quarterbacks **according to how their team's passing game favorably impacted the number of games their team wins.** Therefore our (extremely simple!) QB rating is 1.543(TRUOPASS) − 50.0957(OINTRATE).

In Figure 19-2 we show our ratings for the 2016 season (Rodgers was injured in 2017), as well as the ESPN total quarterback rating and older QB ratings.

We see that Matt Ryan, Tom Brady, and Dak Prescott are the top three quarterbacks across every method of ranking. Interestingly, the number 1 and 2 ranked quarterbacks met in Super Bowl LI, which indicates the importance of quarterback play to team success. While the old quarterback rating, as well as the revamped total quarterback rating, rate Aaron Rodgers as a top quarterback, he falls to 12 in our rating (likely because Rodgers takes a large amount of sacks, which has a strong impact in our model).

The three rating systems are remarkably consistent. The correlation between our ratings and the traditional NFL ratings is .90. The correlation between the traditional NFL ratings and the new total QBR is .87. Finally, the correlation between our ratings and total QBR is .82. Our formula does a fairly good job of providing a good evaluation of quarterback rating with just a few simple factors. While the total QBR created by ESPN seems to do an excellent job of evaluating performance more isolated from outside factors, the NFL traditional quarterback rating system seems unnecessarily complex, and less accurate than total QBR.

We will see in Chapter 20 that we can do a much better job of evaluating a team's passing attack by looking at how each play changes the expected number of points by which the team wins the game. Unfortunately, however, what is really needed is a way to decompose the constituent parts of a team's passing attack and understand what percentage is due to the QB, the receivers, and the offensive line. Since the offensive line and the QB are usually in for almost every play, this is exceedingly difficult. In the following chapters we will show that the fact that almost every NBA player sits out at least

17% of the time makes it possible to partition a team's success (or lack thereof) among their players. If only this were also possible for football. But all hope is not lost! As we will see in Chapter 27, the NFL is collecting player tracking data during every play, and this can allow us to start looking at things that can potentially lead to better player and position evaluations.

CHAPTER 20

FOOTBALL STATES AND VALUES

Recall that in Chapter 8 we discussed how in baseball knowledge of the inning, score margin, outs, and runners on bases was enough to determine the chance that a team would win a game (assuming two equal teams were playing). For example, if we are down by three runs in the top of the seventh with two outs and the bases loaded, we have a 15% chance of winning the game. We call the inning, score margin, outs, and runners on bases the **state** of the baseball game. Once we know the state of the game and have evaluated the chance of winning in each state, we can analyze strategies such as bunting, or evaluate (as shown in Chapter 8) batters and pitchers based on how they change the team's chance of winning the game.

FOOTBALL STATES

If we could define a state for football that is sufficient for us to determine the chance of winning the game (assuming two equal teams are playing), then we can analyze how plays affect a team's chance of winning. Then we can use this information to improve our evaluations of running backs, quarterbacks, wide receivers, etc. For example, we might find that when running back A carried the ball, on average he added .1 points per carry, while running back B added

.3 points per carry. This would indicate that (assuming—and this is most probably a big assumption—they ran behind offensive lines of equal quality) running back B was better. Comparing running backs based on points added per carry would be a better measure of quality than comparing running backs based on the current metric of yards per carry. We can also use football states to evaluate strategic decisions such as when to go for a two-point conversion after a TD, when to punt on fourth down or go for it, when to try a field goal on fourth down or go for it, what should be the run-pass mix on first down, etc. More on these decisions in the following chapters.

The state at any time for a football game is specified by knowledge of the following quantities:

- Yard line
- Down
- Yards to go for first down
- Score differential
- Time remaining in game.

For example, we would like to know our probability of victory in a particular situation such as: 10 minutes left in the second quarter and we have third and three yards to go on our 28-yard line and are down by seven points.

In baseball the number of states is manageable (several thousand), while in football there are millions of possible states. To simplify analysis, most analysts assume that a team's goal is to maximize, from the current time onward, the expected number of points by which it beats its opponent. We consider that the game is of infinite length. By a game of "infinite length" we mean a very long game, say 1,000 minutes long. This eliminates the need to know the time remaining in the game.[1] Of course, actual games are finite in dura-

---

1. Recent systems, such as ESPN's win probability and nflWAR (Yurko, Ventura, Horowitz, 2019), incorporate time remaining into decisions, such as whether to go for it on fourth down or go for a one- or two-point conversion.

tion and near the end of the game the goal of maximizing points scored might (will) be violated. For example, say our team is down by two points with 30 seconds left on the game clock. Then our goal is not to maximize expected points scored but simply to maximize our chance of kicking a successful field goal. Therefore, our assumption of an infinite length game will not be valid near the end of the game (or the end of the first half), but for most of the game "an expected points margin maximizer" will choose decisions that maximize the team's chance of victory. To simplify the state, we will initially assume that the state in a football game is specified by

- Down
- Yards to go for first down
- Yard line.

This simplification still leaves us (assuming we truncate yards to go for a first down at 30) with nearly $4 * 99 * 30 = 11,880$ possible states!

One way that football analysts can define the value of a state is as the margin by which a team is expected to win (from that point on) in a game of **infinite duration** given that the team has the ball on a given yard line and down and yards to go situation. These values are difficult to estimate. Carter and Machol (1971), Romer (2006), and footballoutsiders.com have all estimated these values for first and 10 situations. Cabot, Sagarin, and Winston (CSW) (1981) estimated state values for each yard line and down and yards to go situation. A sampling of the values for several first down and 10 yards to go situations is given in Table 20-1. Note the five-yard line is five yards from your goal line, while the 95-yard line is five yards from your opponent's goal line.

Carter and Machol (1971) used data from the 1969 NFL season to estimate the value of first and 10 on only the 5-, 15-, 25-, . . . , 95-yard lines. They assumed that the value to a team receiving a kickoff after a scoring play was 0. CSW (1981) analyzed the value for each down, yards to go (≤ 30) and yard line situation by inputting the probabilities from the football simulation game *Pro Quarterback*.

TABLE 20.1
State Values for NFL Football

Yard Line	Carter and Machol	Cabot, Sagarin, and Winston	Romer (read off graph, so approximate)	Football Outsiders.com (read off graph, so approximate)
5	−1.25	−1.33	−0.8	−1.2
15	−0.64	−0.58	0	−0.6
25	0.24	0.13	0.6	0.1
35	0.92	0.84	1.15	0.9
45	1.54	1.53	1.90	1.2
55	2.39	2.24	2.20	1.9
65	3.17	3.02	2.8	2.2
75	3.68	3.88	3.30	3
85	4.57	4.84	4.0	3.8
95	6.04	5.84	4.90	4.6

A unique feature of the CSW work was that they used the theory of stochastic games to solve for the offensive and defensive teams' optimal strategy mix in each situation. This model simultaneously computed the state values **and** the fraction of the time the offense and defense should choose each play in each situation. The CSW work allows us to have values for states involving second and third down. Romer used data from the NFL 2001–2004 seasons and then solved a complex system of equations to estimate the value of first and 10 for each yard line. We note that since there are around 12,000 possible states, and during a typical NFL season under 40,000 plays are run from scrimmage—with a disproportional fraction of them also being first and 10 from a team's own 20 (or 25 since 2016)—there is not enough data to estimate the value of every possible state from play-by-play data. This is most probably why the above models focused on first and 10 situations only (which appear more often), and why a simulation game was used by CSW to estimate the state values.

Romer's work caused quite a stir because his values led to the inescapable conclusion that teams should go for it on fourth down in many situations where NFL coaches either punt or try a field goal. For example, when facing fourth and five on your own 30-yard line, Romer's values imply that you should go for the first down and not punt! Few NFL coaches would go for the first down in this situation (which of course by itself does not say much about whether this decision would be right or wrong!). We will discuss this conundrum in Chapter 21.

Recently, Brian Burke of ESPN ran an analysis on NFL data to determine the value of each of these states. Burke used a slightly different definition and looked at which team is most probable to score next. He broke down the data by down and graphed it according to distance from the goal line on the x-axis and expected points on the y-axis. The results of this analysis can be found on Burke's website.[2] One particularly useful aspect of Burke's analysis is the breakdown by yards to go. This breakdown shows us that adding 10 yards to go results in approximately one less expected point. The fact that we have values for all downs, yard lines, and yards to go situations allows us to evaluate the effectiveness of every play executed by an NFL team. With values for only first down you cannot evaluate the effectiveness of second and third down plays. Since Burke's analysis includes expected points for first, second, and third downs, the impact of a singular play can be found by comparing the expected value of the starting point to the expected value of the resulting point of a play.

Most recently, Yurko, Ventura, and Horowitz (2019) (YVH) developed an open source software and model that uses a multinomial logistic regression model to identify the probability of the next scoring event. Given these probabilities and their corresponding values in terms of points, YVH can obtain the expected points from the

2. Source: http://www.advancedfootballanalytics.com/index.php/home/stats/stats-explained/expected-points-and-epa-explained

current state. They also use a generalized additive model and these values to obtain a win probability for each team. YVH has gained a lot of popularity and it has become the de facto model used by popular sports media (e.g., FiveThirtyEight.com, theathletic.com, *The Wall Street Journal*, etc.).

A SIMPLE EXAMPLE OF STATE VALUES

The actual mathematical development of the CSW model is tedious and outside the scope of this text, but the core intuition of its development is helpful, and we can provide it with a simplified example. Suppose we play football on a seven-yard field as shown in Figure 20-1. We are defending the goal line on the left and trying to reach the goal line on the right.

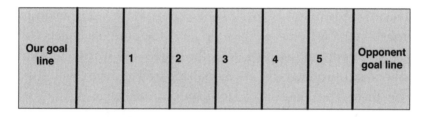

FIGURE 20.1 A Simple Example to Determine State Values.

The rules of the game are simple. We have one play to make a first down. It takes one yard to get a first down. We have a 50% chance of gaining one yard and a 50% chance of gaining 0 yards on any play. When we score, we get seven points and the other team gets the ball one yard from their goal line. What is the value of each state? We let V(i) be the expected number of points by which we should win an infinite game if we have the ball on the i yard line (based on the simplified rules discussed it will always be the only down available and one yard to go). Then we can use the following equations to solve for V(1), V(2), ..., V(5):

$$V(1) = .5V(2) - .5V(5) \qquad (1)$$

$$V(2) = .5V(3) - .5V(4) \qquad\qquad (2)$$

$$V(3) = .5V(4) - .5V(3) \qquad\qquad (3)$$

$$V(4) = .5V(5) - .5V(2) \qquad\qquad (4)$$

$$V(5) = .5(7 - V(1)) - .5V(1). \qquad\qquad (5)$$

Recall from Chapter 6 that the law of conditional expectation tells us that

Expected Value of Random Variable =

$$\sum_{\text{all outcomes}} (\text{Probability of outcome}) *$$

(Expected value of random variable given outcome)

To derive each of the equations (1)–(5), we condition on whether we gain a yard or not.

Suppose we have the ball on the one-yard line. Then with probability .5 we gain a yard (and situation is now worth $V(2)$), and with probability .5 we do not gain a yard and the other team gets ball one yard from our goal line. Now the situation is worth $-V(5)$ to us because the other team has the ball at yard 1 and the value to the other team is now equal to what the value would be to us if we had the ball at yard 5. This means that, as shown in (1), the value of having the ball on our one-yard line may be written as $.5V(2) - .5V(5)$. Equations (2)–(4) are derived in a similar fashion. To derive Equation (5) note that with probability .5 we gain a yard and score seven points. Also, the other team gets the ball on yard line 5, which has a value to them of $V(1)$ (and value to us of $-V(1)$). With probability .5 we fail to gain a yard and the other team gets the ball on its five-yard line, which has a value to it of $V(1)$ (and to us a value of $-V(1)$). Therefore, as shown in (5), the expected value of having the ball on our five-yard line is

$$.5(7 - V(1)) - .5V(1).$$

Solving these equations, we find that $V(1) = -5.25$, $V(2) = -1.75$, $V(3) = 1.75$, $V(4) = 5.25$, $V(5) = 8.75$.

Thus, each yard line we move closer to our "goal line" is worth 3.5 points = .5 touchdown. The trick in adopting our methodology to actual football is that the "transition probabilities" that indicate the chances of going from, say, first and 10 on the 20-yard line to second and four on the 26-yard line are difficult to estimate. Despite this difficulty (which CSW overcome by using the Pro Quarterback simulations), we will see in the next two chapters that the state value approach lets us gain many insights into effective football decision making and allows us to evaluate the effectiveness of different types of plays in different situations.

CHAPTER 21

FOOTBALL DECISION MAKING 101

During a football game a coach must make many crucial decisions such as

- It is fourth and four on the opponent's 30-yard line. Should we kick a field goal or go for a first down?
- It is fourth and four on our own 30-yard line. Should we go for a first down or punt?
- We gained seven yards on first down from our own 30-yard line. The defense was offside. Should we accept the penalty?
- What is the optimal pass run mix on first down and 10?

Using the concepts of states and state values discussed in Chapter 20 these decisions (and many others) are easy to make. Simply choose the decision which maximizes the expected number of points by which we win a game of infinite length (with the caveat discussed that this might be different for end-of-game or end-of-half situations).

Let's now analyze the four decision problems listed above. We will use the CSW values that are listed in the file Val2727.xlsx, but one can use any of his or her favorite expected points models.

IT IS FOURTH AND FOUR ON THE OPPONENT'S
30-YARD LINE. SHOULD WE KICK A FIELD GOAL OR
GO FOR A FIRST DOWN?

To simplify matters we assume that if we go for it and we get the first down, we gain just one yard beyond the line of gain (i.e., five yards), while if we fail, we gain exactly two yards. Let p be the probability of gaining a first down, and hence, we do not get the first down with probability $1 - p$. We define the value $V(D, YTG, YL)$ to be the number of points by which we will defeat a team of equal ability from the current point on in a game of infinite length when we have the ball YL yards from our own goal line ($YL = 20$ is our 20 and $YL = 80$ is opponent's 20), and it is down D with YTG yards to go for first down. Some examples of the CSW values follow: $V(1, 10, 75) = 3.884$, $V(4, 4, 70) = 1.732$, $V(1, 10, 28) = -0.336$.

Then after conditioning on the outcome of our fourth down play, the expected value going for the first down is:

$$pV(1, 10, 75) + (1 - p)(-V(1, 10, 28)).$$

This is because if we get the first down, we assume we will gain five yards, so we have first and 10 on yard line 75. If we do not get the first down, then the other team has the ball on its 28-yard line, which has $V(1, 10, 28)$ to it (or value $-V(1, 10, 28)$ to us). Therefore, the expected value by which we will win a game of infinite length if we go for it on fourth down is

$$3.884p + (1 - p)(-0.336).$$

EVALUATING VALUE OF A FIELD GOAL ATTEMPT

To evaluate the value of a field goal we need to know how the probability of making a field goal depends on the length of the FG attempt. Kickoffs are on average returned to the 27-yard line. We also assume that all field goal attempts are kicked from seven yards behind the line of scrimmage. For example, if the line of scrimmage is the 30-yard line, then the kick is attempted from the 37-yard line and is of length 47 yards. If the kick is missed, the defending team will get the ball on its own 37-yard line. NFL play-by-play data provide information about the success or failure of every field goal, as well as the corresponding line of scrimmage. We can use these data to determine how the probability of making a field goal depends on the length of the kick. Given that the dependent variable in this case is binary (make or miss), linear regression is not an appropriate model since several of its assumptions are violated. Luckily, we can use generalized linear models to model variables that are generated from a particular distribution in the exponential family.

GENERALIZED LINEAR MODELS (GLMs)

If the variable of interest (dependent variable) Y follows a distribution from an exponential family (e.g., Poisson, exponential, binomial, multinomial, etc.), we can model a (possibly nonlinear) function of the mean of this distribution, $g(\mu_i)$, through a linear combination of the model covariates $g(\mu_i) = x_i^T \cdot \beta$. Having this mean, we can then make predictions using the estimated distribution of Y. In our case, the variable Y follows a binomial distribution, while function g (called *link function*) is the logit function, $\text{logit}(p) = \log \dfrac{p}{1-p}$. The following table shows some popular GLMs:

Model	Probability Distribution	Link Function
Linear regression	Normal	Identity
Logistic regression	Binomial	Logit
Poisson regression	Poisson	Log
Multinomial regression	Multinomial	Generalized logit
Negative binomial regression	Negative binomial	Log

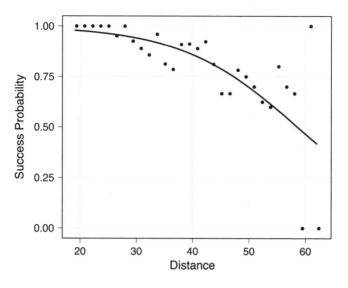

FIGURE 21.1   Field Goal Success as a Function of Its Distance.

Based on the above, we can model the probability of *success* p for this variable through the following equation for logistic regression:

$$\log\left(\frac{p}{1-p}\right) = a + b(\text{Length of kick}) \tag{1}$$

Using data from 2017, we found a = 5.697 and b = −0.097 (both with p-values <0.001). This yields the probability of making field goals shown in Figure 21-1.

Given these numbers, the probability of making a 47-yard FG is around 76%. Therefore, the value from attempting the field goal is 0.76 * (3 − V(1,10,27)) + 0.24 * (−V(1.10, 37)), which is equal to 1.84. In order to make a decision between the two we can look at what probability p of converting a fourth down is required in order for the value of a fourth down attempt to be greater than 1.84. That is,

$$3.884p + (1-p)(-0.336) > 1.84$$

Solving this inequality we find the breakeven probability of converting the fourth down in order to make going for it more valuable

TABLE 21.1
Probability of Successful 3rd or 4th Down Conversion

Yards to Go	Probability 3rd or 4th Down Play Makes the First Down
1	.67
2	.55
3	.51
4	.48
5	.45

than an FG. In this situation this breakeven point is approximately 52%. Since the probability of making a first down on third or fourth down and four yards to go does not exceed 52%, we should not go for it on fourth down—even though going for it might also not be a bad call given how close the breakeven and actual conversion probabilities are. In fact, here is where additional context can help make the final coaching decision; how do your offense and the opposing defense perform in the current game? If we fail to convert, how possible is our defense to get a three and out? How is our kicker performing? Etc.

IT IS FOURTH AND FOUR ON OUR OWN 30-YARD LINE. SHOULD WE GO FOR A FIRST DOWN OR PUNT?

Let's assume that if we get the first down we gain five yards and we now have value $V(1, 10, 35) = 0.839$. If we fail to get the first down, we assume a gain of two yards. Now the other team has the ball on yard line 68, which is worth $-V(1, 10, 68) = -3.265$ to us. Assume that if we punt we will always net 45 yards (during the 2017 NFL season the average net gain on a punt was 45.5 yards), which puts us in a situation worth to us $-V(1, 10, 25) = -0.46$ points. Therefore, we should go for it if probability p of obtaining first down satisfies $(0.839)p - 3.265(1-p) \geq -0.46$.

This inequality is satisfied for $p \geq 0.678$. Thus, we would need a 67.8% chance of success to justify going for it. Pro-football reference data indicates that in this situation we should (as NFL coaches do) punt.

As pointed out by Birnbaum on his excellent sabermetric website,[1] Romer's research indicates that you should go for it on your own 30-yard line if you have (approximately) at least a 45% chance of success. Romer's work also indicates that even on your own 10-yard line, a team should go for it on fourth and three. This does not seems reasonable. Our analysis suggests that going for it on fourth and three from your own 10-yard line requires at least a 71% chance of a successful fourth down conversion.

WE GAINED SEVEN YARDS ON FIRST DOWN FROM OUR OWN 30-YARD LINE. THE DEFENSE WAS OFFSIDE. SHOULD WE ACCEPT THE PENALTY?

After the play we have $V(2, 3, 37) = 0.956$. If we accept the penalty, we have $V(1, 5, 35) = 0.983$. Therefore, we should accept the penalty (even though based on these values, the two choices seem almost equally good). If we gained eight yards we would have $V(2, 2, 38) = 1.068$ points and we should decline the penalty. Thus gaining eight or more yards on first down is better than accepting a five-yard first down penalty.

ON FIRST AND 10 FROM ITS OWN 30-YARD LINE OUR OPPONENT RAN UP THE MIDDLE FOR NO GAIN. THE OFFENSE WAS OFFSIDE. SHOULD WE ACCEPT THE PENALTY?

After the run our opponent has $V(2, 10, 30) = 0.115$. If we accept the penalty our opponent has $V(1, 15, 25) = -0.057$, so we should accept the penalty. Since $V(2, 11, 29) = -.007$ and $V(2, 12, 28) = -0.125$, it

1. http://sabermetricresearch.blogspot.com/2007/01/are-nfl-coaches-too -conservative-on.html

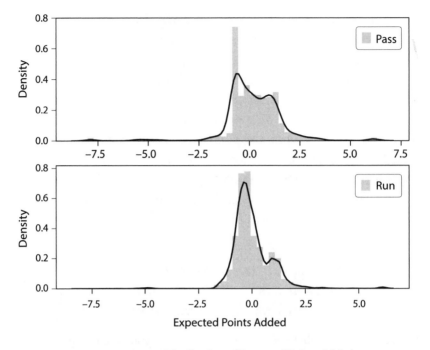

FIGURE 21.2 Distribution of Expected Points Added
on 1st and 10 (from our own 25).

appears that the defense should decline a first down five-yard penalty
if a first down play loses two or more yards.

WHAT IS THE OPTIMAL PASS RUN MIX ON
FIRST DOWN AND 10?

In order to determine this, we remade a similar analysis originally
made by Alamar in 2005. Using play-by-play data from nflscrapR, we
found that during the 2016–2018 seasons, teams passed 52.6% of the
time on first and 10 from their own 25. Traditionally, many football
data tabulation software programs (such as Pinnacle Systems) define a play on first and 10 as a success if it gains four yards or more.
However, given the notion of expected points added more recently
a different definition of a successful play has emerged. In particular,
a successful play is one that adds positive expected points. Using

the data from nflscrapR, we find that only 40% of the rushing plays are successful on first and 10, while a 54% of the passing plays are successful on first and 10 (from our own 25-yard line).[2] Given the difference in the success rates of rushing versus passing on first down, this could indicate that teams should pass more on first down. We agree with this assessment. Figure 21-2 shows the distribution of expected points added on a first and 10 for pass and run plays. As we can see, the pass plays are able to provide a higher EPA. In particular, the average EPA for a pass play on first down is 0.17, while for a run play it is −0.02. This makes it clear that running on first down is much less effective than passing. Therefore, we believe that teams should pass more often on first down. Of course, the success of a given play depends on the defensive team's setup. This means we cannot pass all the time because then the defense will anticipate the pass and always play defense for a pass attempt. This is why football teams practice the mixed strategies that are the cornerstone of the two-person zero-sum game theory discussed in the next chapter.

---

2. Also note here that we only consider plays that did not result in a penalty. Given that rushing plays result in (holding) penalties much more frequently (https:// fivethirtyeight.com/features/can-nfl-refs-do-what-analysts-never-could-get-coaches -to-pass-more/), this would make the situation even worse for rushing plays.

CHAPTER 22

IF PASSING IS BETTER THAN RUNNING, WHY DON'T TEAMS ALWAYS PASS?

In football the offense selects a play and the defense lines up in a defensive formation. Let's consider a very simple model of play selection in which the offense and defense simultaneously select their play.

- The offense may choose to run or pass.
- The defense may choose a run or pass defense.

The number of yards gained is given in Table 22-1, a payoff matrix for the game (these numbers are for illustrative purposes).

   We see that if the defense makes the right call on a run, we lose five yards and if the defense makes the wrong call, we gain five yards. On a pass, the right defensive call results in an incomplete pass while the wrong defensive call results in a 10-yard gain. Games in which two players are in total conflict (one player's gain is equivalent to the other player's loss) are called two-person zero-sum games (TPZSG). In our game, every yard gained by the offense makes the defense one yard worse off, so we have a TPZSG. The great mathematician

TABLE 22.1

Payoff Matrix for Football Game

	Run Defense	Pass Defense
Offense Runs	−5	5
Offense Passes	10	0

John von Neumann and the brilliant economist Oskar Morgenstern[1] discovered the solution concepts for TPZSG. In the game described by the payoff matrix above, the row player, i.e., the offense, wishes to maximize the payoff from the payoff matrix and the column player wants to minimize the payoff from the payoff matrix. We define the value v of the game for the row player as the maximum expected payoff of which the row player can assure himself. Suppose we choose a running play. Then the defense can choose run defense and we lose five yards. Suppose we choose a pass offense. Then the defense can choose pass defense and we gain 0 yards. Thus, by throwing a pass the offense can assure itself of gaining 0 yards. Is there a way the offense can ensure that on average it will gain more than 0 yards? A player in a TPZSG can choose to play a mixed strategy in which it chooses each of its choices (*strategies* in game theory terminology) with a given probability. Let's suppose the offense chooses run with probability q and chooses pass with probability $1-q$. On average, how does our mixed strategy do against each of the defense's "pure" strategy choices?

- If a run defense is chosen, our expected gain is
  $q(-5) + (1-q)10 = 10 - 15q$.
- If a pass defense is chosen, our expected gain is
  $q(5) + (1-q)(0) = 5q$.

1. John Von Neumann and Oskar Morgenstern, *Theory of Games and Economic Behavior.* (Princeton University Press, 1947).

For any value of q chosen by the offense the defense will choose the defense that yields minimum$(10 - 15q, 5q)$. Thus the offense should choose the value of q between 0 and 1 that maximizes minimum$(10 - 15q, 5q)$. From Figure 22-1 we see that this minimum is attained where $10 - 15q = 5q$ or $q = 1/2$.

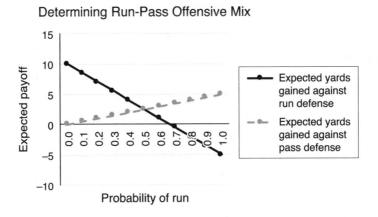

Determining Run-Pass Offensive Mix

FIGURE 22.1   Graph to Determine Offense's Optimal Mix.

Thus the optimal mixed strategy for the offense is to run 1/2 of the time and pass 1/2 of the time. This assures the offense of an expected payoff of at least 5/2 yards. By choosing $q = 1/2$ the offense is assured an expected gain of 5/2 yards! This makes the game very dull. No matter what strategy (pure or mixed) the defense chooses, we gain an average of 5/2 yards. Therefore, the value of this simple game to the row player is 5/2 yards.

Note that when the defense guesses right passing is five yards better than running and if the defense guesses wrong passing is five yards better than running. Even though passing seems like a much better option, our optimal strategy is to run 1/2 of the time. Note that TPZSGT provides a rationale for the fact that teams actually "mix up" running and passing calls. For similar reasons, players should bluff sometimes in poker when they have a poor hand. If you never bet with a poor hand, your opponents may not call your bet very often when you have a good hand. Bluffing with a bad hand will also sometimes enable you to win a hand with poor cards.

Now let's find the defensive team's optimal strategy. Let x be the probability that the defense calls a run defense and $1-x$ be the probability that the defense calls a pass defense. The defense's goal is, of course, to minimize the expected payoff to the offense. Against this defensive mixed strategy, how will the offense fare?

- If offense chooses run, then on average it gains
  $x(-5)+(1-x)(5)=5-10x$.
- If offense chooses pass, then on average it gains
  $x(10)+(1-x)(0)=10x$.

Therefore for any value x chosen by the defense, the offense will choose the play attaining maximum$(5-10x,10x)$. The defense should therefore choose the value of x that minimizes maximum$(5-10x,10x)$. From Figure 22-2 we find that choosing $x=1/4$ will attain this minimum and ensure that the offense is held to an average of $5/2$ yards per carry. Therefore the defense should play a pass defense 75% of the time. This is because the pass play is better than the run. Since the defense looks for the pass more often, the offense only passes half the time, even though the pass seems like a much more effective play.

Note that we found the offense can ensure themselves and the defense can hold the offense (on average) to the same number: $5/2$ yards. 2.5 yards is the **value of** the game. Von Neumann and Morgenstern proved that the values obtained by analyzing the row and column players' strategies in a TPZSG are **always identical.** The strategies identified above is what we call the **Nash equilibrium** of the game. Suppose Nash equilibrium strategies are chosen by each player. Then neither player will be able to improve her payoff by unilaterally changing her strategy. Simply put, if strategies σ_A and σ_B (pure or mixed) are the Nash equilibriums for players A and B, then if A decides to change her strategy away from σ_A, she will enjoy a lower payoff (as long as player B sticks with strategy σ_B). Von Neumann

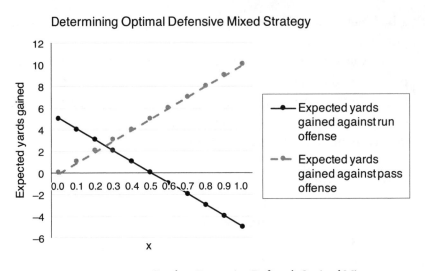

FIGURE 22.2 Graph to Determine Defense's Optimal Mix.

TABLE 22.2
Generalized Football Payoff Matrix

	Run Defense	Pass Defense
Offense Runs	r − k	r + k
Offense Passes	p + mk	p − mk

proved that every two-person zero-sum game has a Nash equilibrium if we allow for mixed strategies.

Now before you say that we condone running 50% of the time, bear in mind that these results are based on the specific payoff matrix described by Table 22-1. In general, suppose our payoff matrix looks as in Table 22-2.

If we assumed that the defense would choose a run and pass defense with equal probability, then we can interpret this matrix as saying that running plays gain r on average and passing plays gain p on average. Also the correct choice of defense appears to have m times as much effect on a passing play as it does on a running play. For example, if $m = 2$, then for a pass a correct defense makes the pass gain 4k yards worse than the incorrect defensive choice. For a run the correct choice of defense makes the run gain 2k yards worse than the incorrect defensive choice.

Note that our previous example shows that for a TPZSG with a 2×2 payoff table the optimal strategy for a player is found by equating the expected payoff against both the opponent's strategies. For the payoff matrix shown in Table 22-2 we can show the optimal strategy for the offense is to run with probability $m/(m+1)$ and pass with probability $1/(m+1)$. The optimal defensive mix can be shown to be choosing a run defense with probability $.5 + \dfrac{r-p}{2(m+k)}$ and a pass defense with probability $.5 + \dfrac{r-p}{2(m+k)}$. Note that the defense chooses to defend the better play more than 50% of the time. For $m=2$ the offense will run 2/3 of the time and pass 1/3 of the time. For $m=.5$ we run 1/3 of the time and pass 2/3 of the time. For $m=1$ we run and pass with equal probability. The idea is that if $m>1$, the defensive call has more effect on a pass than a run, and if $m<1$, the defensive call has more effect on a run than a pass. Thus we see that the offense will choose more often the play over which the defense has "less control." For $m>1$ the defense has less control over the run, so we run more than we pass. Similarly, if $m<1$ the defense has less control over the pass, so we pass more than we run. Note the optimal run-pass mix does not depend on r and p, which represent the base effectiveness levels of running and passing plays, respectively.

We can draw several other interesting insights from this simple formulation:

- Suppose we get a new quarterback who is much better. We can model this by saying that for any defense our pass play will gain, say, three more yards. Should we pass more often? We are simply replacing the current value of p by $p+3$ and leaving m unchanged, so our optimal run-pass mix remains unchanged. Even though passing is now better, the defense will play a pass defense more often, so we should play the same run-pass mix as before, according to the TPZSG solution.

- Suppose we get a new and improved running back who can gain five more yards per carry against a pass defense but no more against a run defense. How will the optimal run-pass mix change? Intuitively you would think the offense should run more often because we have a better ground game. The following example—by Mike Shor of Vanderbilt, who termed it the "*football paradox*"—illustrates that this is not the case.

TABLE 22.3
Original Payoff Matrix for Football Paradox

	Run Defense	Pass Defense
Offense Runs	-5	5
Offense Passes	5	-5

This matrix has $r = 0$, $k = 5$, $p = 0$, $m = 1$. Therefore the optimal strategy is to run $1/2 = 50\%$ of the time and pass 50% of the time. The value of this game is 0 yards. The optimal defensive mix is to choose a run defense $1/2$ of the time and a pass defense $1/2$ of the time.

The new payoff matrix is shown in Table 22-4:

TABLE 22.4
New Payoff Matrix for Football Paradox

	Run Defense	Pass Defense
Offense Runs	-5	10
Offense Passes	5	-5

The reader should be able to show that the optimal run-pass mix is to run $2/5$ of the time and pass $3/5$ of the time. The value of the game has increased to one yard and we run less often! The optimal run-pass defense mix is run defense $3/5$ and pass defense $2/5$. Since our running game has improved so much against the pass defense, the defense is terrified of the run, and plays a run defense more often. This causes us to pass more and run less, even though our running game is much better than before.

GAME THEORY AND REAL FOOTBALL!

The TPZSG examples in this chapter are admittedly simplistic and bear little relationship to actual football. Nevertheless, they provide interesting insights on how to think game-theoretic when it comes to game strategy. Is there a way that an NFL team can use TPZSG to come up with an optimal play mix? For game theory to be useful in the NFL, we believe the following information is needed for each play:

- Play called by offense.
- Defensive formation/strategy (e.g., is the defense geared toward stopping the run or the pass?).
- Down, line of scrimmage, and yards gained on play.

First we combine the down and yards to go situation into groups that might look like this:

- First and 10.
- Second and short (three yards or less for a first down).
- Second and medium (four to seven yards for a first down).
- Second and long (at least eight yards to go for a first down).
- Third and short (two yards or less for a first down).
- Third and medium (three to five yards for a first down).
- Third and long (more than five yards for a first down).

Suppose the offense has 15 plays they can call on first and 10 and the defense has 10 plays. Given the results of each play we could determine for a given down and yards-to-go situation **the average value points (the payoff does not necessarily have to be yards) gained on plays for each offensive and defensive play call combination. For** example, we might find that on first and 10, the left tackle sweeping right end for the Steelers averaged 0.4 value points against the cover 2 defense. This gives us **one of the 150 entries in the relevant payoff matrix for first and 10.** Then we can use TPZSG to solve for the optimal mixed strategies for both the offense and defense. For example, we might find on first and 10 the optimal offensive play selection mixed strategy would ensure (against any defensive mix) an expected value of 0.3 points. The numbers in the matrix could be adjusted based on the strength or weakness of the opposition. To get to this game theory "nirvana" we would need to have coaching experts break down the film of every play to tell us the offensive and defensive play calls. Currently this service exists and it is called Pro Football Focus (PFF). So maybe the day for this football game theory is not far.

CHAPTER 23

SHOULD WE GO FOR A ONE-POINT OR A TWO-POINT CONVERSION?

Since 1994, when the NFL began allowing teams to go for a two-point conversion after a touchdown, it has become important for NFL coaches to determine whether to go for one or two points after a touchdown. The success rate for a one-point conversion was over 99% until 2015, when the league changed the distance for the extra-point kick, so one could assume that there was a virtually 100% chance that a one-point conversion would have been successful. In 2015, the distance for the extra point kick moved back 15 yards, dropping the success rate to approximately 94%. The success rate for two-point conversions is around 50%.[1] On average, a one-point conversion try earns 0.94 points and a two-point conversion attempt earns $0(.5) + 2(.5) = 1$ point. So, on average, a two-point conversion earns *slightly* more points. However, while a two-point conversion maximizes the expected points earned, it does not always maximize the win probability (which is the objective of a team). For example,

---

1. Konstantinos Pelechrinis, "Decision making in American football: Evidence from 7 years of NFL data," in ECML/PKDD workshop on Machine Learning and Data Mining for Sports Analytics.

if a team scores a touchdown to tie the game with the game clock expiring, the team is better off taking the extra-point kick given its much higher success rate. Most coaches have a "chart" that tells them whether they should go for one or two points based on the score of the game. The idea of the "chart" is believed to have originated with UCLA assistant coach Dick Vermeil during the early 1970s.[2] Before talking about the chart, we will take a more detailed statistical look at the success rates and expected points for each option using play-by-play data.

THE NUMBERS

Table 23-1 presents the success rates for one and two-point conversions for the NFL seasons between 2009 and 2018 and will be the basis of this statistical analysis. These are the league average rates and based on them a two-point conversion attempt provides +0.07 points on average over an extra-point kick. Of course, the question now is whether this expected gain of 0.07 per attempt is important and whether it is *robust*.

As with any measured quantity, there is uncertainty associated with the success rates presented in Table 23-1. This uncertainty is due to various reasons. For example, the uncertainty for the two-point conversion success rate stems from a relatively small sample size as well as the selection bias for these attempts. Teams typically use the two-point conversion as situational football during the end of a game, which means that the sample is not uniformly random, but there might be biases based on which teams attempt the conversions and which defenses are defending these PATs. Considering this uncertainty, the 95% confidence interval for the expected point gain per two-point conversion attempt over an extra-point kick is [0.01, 0.15]. Given that the confidence interval does not include the value of 0, one can claim

---

2. Vermeil later became a successful NFL coach and NFL TV commentator. Sackrowitz's "Refining the point(s)-after-touchdown scenario" appears to be the first mathematical study of the one- or two-point conversion decision.

TABLE 23.1

Two-point conversion success rate is stable,
while the extra-kick success rate has significantly
reduced since the 2015 season

Year	skick	s2pt
2009	0.98	0.44
2010	0.99	0.55
2011	0.99	0.50
2012	0.99	0.55
2013	0.99	0.49
2014	0.99	0.52
2015	0.94	0.52
2016	0.94	0.55
2017	0.94	0.47
2018	0.94	0.54

that the benefit of the two-point conversion is statistically robust. While this is technically correct, in sports analytics (and of course in other domains) all results need to be placed and interpreted within the right context. In this case the context is how many times this decision appears in a game. So, while the analysis shows that there is a statistically significant gain going for two, this gain will be observed in the long run (which can span over multiple games). In an NFL game, the average number of touchdowns per team is 2.4. Let us be optimistic and assume there are three per team per game. With only three attempts there is a 12.5% probability that all three two-point conversion attempts will fail, and a 37.5% probability that two of the three attempts will fail! The risk associated with going for the two-point conversion is high when the number of attempts is small. So, for a risk-averse coach (which is what NFL coaches—or any coach for that matter—appear to be) it will be an easy choice to go for the extra-point kick, which maximizes the *risk-adjusted* expected points gain.

As a reminder, the above numbers have been obtained on a league-wide basis, where the two-point conversion attempts are observed in less than 10% of the touchdowns. Therefore, another context that needs to be considered is an individual team's ability. Some teams have better offensive units and some have better defensive units. Or the play style of a team might be more successful in two-point conversions (e.g., *mobile* QBs). These need to be considered when making in-game decisions (league-wide numbers can give you only a rough idea in these cases). Furthermore, if teams were to increase their two-point conversion attempts to a much higher fraction of their touchdowns, the defenses would adjust accordingly, and we should expect to see a reduction in the success rate of the conversion.

THE "CHART"

The coach's decision should depend on factors such as the amount of time left in the game as well as the score. To determine how the optimal strategy depends on the score of the game and time remaining, we need to use a sophisticated technique, *dynamic programming*, which allows us to work backward from the end of the game toward the beginning of the game. Dynamic programming was developed during the 1950s by Richard Bellman.[3] Using statistics since the 2009 season, we assume that on any possession there is a 19% chance of scoring a touchdown, a 13% chance of scoring a field goal, and a 68% chance of not scoring. We assume two equally matched teams are playing (the model can be adjusted to handle teams of different abilities). We ignore the possibility of a safety or that the defensive team scores on a possession. The less mathematically sophisticated reader may (without loss of continuity) now skip to the results shown in Tables 23-2 and 23-3.

We define $F_n(i)$ = the probability the team wins the game if it is i points ahead, it has just gotten the ball, and n possessions remain.

3. Richard Bellman, *Dynamic Programming* (Princeton University Press, 1957).

We also define $G_n(i)$ the probability the team wins the game if it is i points ahead, its opponent has just gotten the ball, and n possessions remain. If $n = 0$, the game has ended. If the game is tied, we assume each team has a 50% chance of winning in overtime. We will also assume that a team ahead by at least 30 points always wins (the calculations are in the file Gofortwo.xlsm). We ran our model assuming 94% of one-point conversions succeed, and for a 45% and 50% chance that a two-point conversion is successful. The interested reader is welcome to change our assumptions about the chance that a one-point or two-point conversion succeeds. Clearly

$$G_0(i) = F_0(i) = 1 \text{ for } i > 0, G_0(i) = F_0(i) = 0 \text{ for } i < 0, \text{ and}$$
$$G_0(0) = F_0(0) = .5.$$

In our subsequent explanation we assume that a one-point conversion succeeds 94% of the time and a two-point conversion succeeds 45% of the time. Now that we know what happens when the game is over, we can work backward and determine what happens with one possession left:

$$F_1(i) = MAX\{.13 * G_0(i+3) + .19 * .94 * G_0(i+7)$$
$$+ .19 * .06 * G_0(i+6) + .68 * G_0(i), .19 * .45 * G_0(i+8)$$
$$+ .19 * .55 G_0(i+6) + .68 G_0(i) + .13 * G_0(i+3)\} \qquad (1)$$

$$G_1(i) = MIN\{.13 F_0((i-3) + .19 * .94 * F_0(i-7)$$
$$+ .19 * .06 * F_0(i-6) + .68 * F_0(i), .19 * .45 * F_0(i-8)$$
$$+ .19 * .55 F_0(i-6) + .68 * F_0(i) + .13 * F_0(i-3)\} \qquad (2)$$

- Equation (1) uses the law of conditional expectation. We simply multiply the probability of each possible outcome on the game's last possession (field goal, touchdown, or no score) by the probability of winning given each possible outcome of the possession.
- With probability .13 the team kicks a field goal and is now ahead by $i + 3$ points. They win the game with probability $G_0(i + 3)$.

- If the team goes for 1, then with probability .19 ∗ .94 the team scores a touchdown and is now ahead by $i+7$ points and wins with probability $G_0(I+7)$. With probability .19 ∗ .06 the team misses the one point conversion and is ahead by six and wins with probability $G_0(6)$.
- With probability .68 the team does not score. Its probability of winning the game is now $G_0(i)$.

The part of (1) after the comma represents the chance the team with the ball wins if it goes for a two-point conversion. The team with the ball chooses the option that gives the maximum expected chance of winning the game.

Equation (2) is derived in a similar fashion. Note that the opposition chooses the strategy (one-point or two-point attempt) that minimizes the first team's chance of winning the game.

Once we have determined all $F_n(i)$ and $G_n(i)$, we can compute the probability of winning with $n+1$ possessions remaining using the following recursions:

$$F_{n+1}(i) = \text{MAX}\{.13 * Gn(i+3) + .19 * .94 * G_n(i+7) + .19 * .06 * G_n(i+6) + .68 * G_n(i), .19 * .45 * G_n(i+8) + .19 * .55G_n(i+6) + .68G_n(i) + .13 * G_n(i+3)\} \tag{3}$$

$$G_{n+1}(i) = \text{MIN}\{.13F_n(i-3) + .19 * .94 * F_n(i-7) + .19 * .06 * F_n(i-6) + .68 * F_n(i), .19 * .45 * F_n(i-8) + .19 * .55F_n(i-6) + .68 * F_n(i) + .13 * F_n(i-3)\} \tag{4}$$

Again the optimal strategy for the team is to choose a one-point conversion when it has the ball with $n+1$ possessions remaining if going for one point attains the maximum in (3). Otherwise, it chooses a two-point conversion try. We terminated our calculations with twenty-five possessions remaining the game. On average, there are 23.6 possessions per game, and thus a possession consumes 2.54 minutes on average.

Tables 23-2 and 23-3 summarize a team's optimal conversion attempt strategy based on the number of points the team is ahead for 10 or fewer possessions remaining. A highlighted cell indicates that

TABLE 23.2

Summary of Conversion Choice: Probability of Making
Two-Point Conversion = 0.5

	A	J	P	V	AB	AH	AN	AT	AZ	BF	BL
6	Margin	F1	F2	F3	F4	F5	F6	F7	F8	F9	F10
24	23	1.000	1.000	1.000	1.000	1.000	1.000	1.000	0.998	0.999	0.995
25	22	1.000	1.000	1.000	1.000	1.000	0.999	0.999	0.997	0.998	0.993
26	21	1.000	1.000	1.000	1.000	1.000	0.999	0.999	0.995	0.997	0.991
27	20	1.000	1.000	1.000	1.000	1.000	0.998	0.999	0.994	0.995	0.987
28	19	1.000	1.000	1.000	1.000	1.000	0.997	0.998	0.991	0.993	0.983
29	18	1.000	1.000	1.000	1.000	1.000	0.996	0.997	0.989	0.992	0.979
30	17	1.000	1.000	1.000	1.000	1.000	0.994	0.996	0.984	0.988	0.972
31	16	1.000	1.000	1.000	0.998	0.999	0.990	0.992	0.977	0.983	0.964
32	15	1.000	1.000	1.000	0.996	0.997	0.985	0.989	0.971	0.978	0.955
33	14	1.000	1.000	1.000	0.990	0.993	0.975	0.982	0.957	0.968	0.940
34	13	1.000	1.000	1.000	0.984	0.989	0.964	0.974	0.944	0.958	0.925
35	12	1.000	1.000	1.000	0.982	0.987	0.958	0.969	0.934	0.950	0.912
36	11	1.000	1.000	1.000	0.974	0.981	0.943	0.957	0.915	0.935	0.891
37	10	1.000	1.000	1.000	0.966	0.975	0.930	0.947	0.897	0.920	0.871
38	9	1.000	1.000	1.000	0.954	0.966	0.909	0.931	0.873	0.902	0.845
39	8	1.000	0.968	0.978	0.919	0.941	0.877	0.906	0.843	0.877	0.816
40	7	1.000	0.935	0.956	0.884	0.915	0.844	0.881	0.812	0.851	0.786
41	6	1.000	0.875	0.915	0.818	0.868	0.784	0.836	0.760	0.811	0.741
42	5	1.000	0.865	0.904	0.796	0.848	0.756	0.811	0.730	0.784	0.710
43	4	1.000	0.858	0.895	0.781	0.831	0.733	0.788	0.702	0.758	0.680
44	3	1.000	0.803	0.850	0.713	0.775	0.668	0.733	0.644	0.708	0.628
45	2	1.000	0.757	0.817	0.662	0.736	0.620	0.694	0.599	0.670	0.587
46	1	1.000	0.749	0.805	0.643	0.714	0.594	0.666	0.569	0.639	0.555
47	0	0.660	0.499	0.622	0.499	0.601	0.499	0.587	0.499	0.578	0.499
48	−1	0.320	0.251	0.440	0.357	0.489	0.405	0.509	0.430	0.518	0.444
49	−2	0.320	0.243	0.427	0.337	0.465	0.378	0.480	0.399	0.486	0.411
50	−3	0.255	0.197	0.362	0.287	0.408	0.331	0.429	0.355	0.440	0.370
51	−4	0.190	0.142	0.282	0.219	0.333	0.266	0.363	0.297	0.383	0.319
52	−5	0.190	0.135	0.271	0.203	0.311	0.243	0.335	0.269	0.351	0.289
53	−6	0.184	0.125	0.250	0.182	0.282	0.215	0.302	0.239	0.317	0.257
54	−7	0.095	0.065	0.157	0.115	0.202	0.155	0.236	0.187	0.262	0.213
55	−8	0.048	0.032	0.110	0.081	0.161	0.123	0.199	0.157	0.228	0.183
56	−9	0.000	0.000	0.062	0.045	0.119	0.090	0.161	0.126	0.193	0.154
57	−10	0.000	0.000	0.045	0.034	0.092	0.070	0.131	0.102	0.161	0.128
58	−11	0.000	0.000	0.036	0.026	0.075	0.057	0.109	0.084	0.138	0.109
59	−12	0.000	0.000	0.026	0.018	0.057	0.042	0.086	0.066	0.113	0.088
60	−13	0.000	0.000	0.023	0.016	0.049	0.035	0.074	0.055	0.096	0.074
61	−14	0.000	0.000	0.015	0.010	0.035	0.025	0.057	0.043	0.078	0.060
62	−15	0.000	0.000	0.006	0.004	0.021	0.015	0.039	0.029	0.058	0.045
63	−16	0.000	0.000	0.003	0.000	0.014	0.010	0.030	0.022	0.047	0.036
64	−17	0.000	0.000	0.000	0.000	0.008	0.006	0.021	0.016	0.037	0.028
65	−18	0.000	0.000	0.000	0.000	0.005	0.004	0.015	0.011	0.028	0.021
66	−19	0.000	0.000	0.000	0.000	0.004	0.003	0.012	0.009	0.022	0.017
67	−20	0.000	0.000	0.000	0.000	0.003	0.002	0.009	0.006	0.017	0.012
68	−21	0.000	0.000	0.000	0.000	0.002	0.001	0.006	0.004	0.013	0.009
69	−22	0.000	0.000	0.000	0.000	0.001	0.001	0.004	0.003	0.009	0.007
70	−23	0.000	0.000	0.000	0.000	0.000	0.000	0.002	0.002	0.006	0.005

TABLE 23.3

Summary of Conversion Choice: Probability of Making
Two-Point Conversion = 0.45

	A	J	P	V	AB	AH	AN	AT	AZ	BF	BL
6	Margin	F1	F2	F3	F4	F5	F6	F7	F8	F9	F10
24	23	1	1	1	1	1	1.000	1.000	0.998	0.999	0.996
25	22	1	1	1	1	1	0.999	1.000	0.997	0.998	0.994
26	21	1	1	1	1	1	0.999	0.999	0.996	0.997	0.991
27	20	1	1	1	1	1	0.998	0.999	0.994	0.995	0.988
28	19	1	1	1	1	1	0.997	0.998	0.991	0.994	0.984
29	18	1	1	1	1	1	0.996	0.997	0.989	0.992	0.980
30	17	1	1	1	1	1	0.994	0.996	0.985	0.989	0.973
31	16	1	1	1	0.998	0.999	0.990	0.993	0.978	0.984	0.965
32	15	1	1	1	0.996	0.998	0.986	0.990	0.972	0.979	0.956
33	14	1	1	1	0.991	0.994	0.976	0.983	0.959	0.970	0.942
34	13	1	1	1	0.984	0.989	0.964	0.974	0.944	0.959	0.926
35	12	1	1	1	0.982	0.987	0.958	0.969	0.935	0.951	0.913
36	11	1	1	1	0.975	0.982	0.945	0.959	0.918	0.937	0.894
37	10	1	1	1	0.967	0.975	0.930	0.947	0.898	0.921	0.872
38	9	1	1	1	0.954	0.966	0.909	0.932	0.874	0.902	0.846
39	8	1	0.971	0.980	0.923	0.943	0.880	0.909	0.846	0.879	0.819
40	7	1	0.939	0.959	0.888	0.918	0.847	0.883	0.815	0.854	0.789
41	6	1	0.875	0.915	0.818	0.868	0.784	0.836	0.760	0.811	0.741
42	5	1	0.865	0.905	0.798	0.849	0.758	0.813	0.731	0.786	0.712
43	4	1	0.859	0.896	0.783	0.833	0.736	0.790	0.704	0.760	0.682
44	3	1	0.803	0.850	0.713	0.775	0.668	0.733	0.643	0.707	0.628
45	2	1	0.757	0.817	0.662	0.736	0.620	0.695	0.600	0.670	0.588
46	1	1	0.749	0.805	0.644	0.715	0.595	0.668	0.570	0.640	0.556
47	0	0.660	0.500	0.622	0.500	0.601	0.500	0.587	0.500	0.578	0.500
48	−1	0.320	0.251	0.440	0.356	0.488	0.404	0.508	0.429	0.517	0.443
49	−2	0.320	0.243	0.427	0.336	0.465	0.377	0.480	0.398	0.485	0.410
50	−3	0.255	0.197	0.362	0.287	0.408	0.332	0.430	0.356	0.441	0.371
51	−4	0.190	0.141	0.281	0.217	0.330	0.264	0.360	0.295	0.380	0.317
52	−5	0.190	0.135	0.270	0.202	0.310	0.241	0.333	0.267	0.349	0.287
53	−6	0.184	0.125	0.250	0.182	0.282	0.215	0.302	0.239	0.317	0.258
54	−7	0.089	0.061	0.152	0.111	0.198	0.152	0.232	0.184	0.259	0.210
55	−8	0.043	0.029	0.105	0.077	0.157	0.120	0.195	0.154	0.224	0.180
56	−9	0	0	0.062	0.045	0.118	0.090	0.161	0.125	0.193	0.153
57	−10	0	0	0.045	0.033	0.091	0.070	0.130	0.101	0.160	0.128
58	−11	0	0	0.035	0.025	0.073	0.055	0.107	0.082	0.135	0.106
59	−12	0	0	0.025	0.018	0.056	0.041	0.086	0.065	0.112	0.087
60	−13	0	0	0.023	0.016	0.049	0.035	0.074	0.055	0.096	0.074
61	−14	0	0	0.014	0.009	0.033	0.024	0.054	0.041	0.075	0.058
62	−15	0	0	0.005	0.004	0.019	0.014	0.037	0.028	0.057	0.043
63	−16	0	0	0.002	0.002	0.013	0.010	0.029	0.021	0.046	0.035
64	−17	0	0	0	0.000	0.008	0.006	0.020	0.015	0.035	0.027
65	−18	0	0	0	0.000	0.005	0.004	0.015	0.011	0.027	0.020
66	−19	0	0	0	0.000	0.004	0.003	0.012	0.008	0.021	0.016
67	−20	0	0	0	0.000	0.003	0.002	0.009	0.006	0.017	0.012
68	−21	0	0	0	0.000	0.002	0.001	0.006	0.004	0.012	0.009
69	−22	0	0	0	0.000	0.001	0.001	0.004	0.003	0.009	0.006
70	−23	0	0	0	0.000	0.000	0.000	0.002	0.002	0.006	0.004

going for 2 is preferable. For the given score difference and the number of possessions remaining the number in the cell is (given our assumption) the chance the team with the ball wins. We can determine approximately how many possessions remain by dividing the time left by 2.54 (the average number of minutes in a possession). For late game situations this might need to be adjusted accordingly as teams leading in the score tend to try and run as much clock as possible when they have the ball. Thus, at the beginning of the fourth quarter we can assume $n = 5$ or 6.

INTUITION BEHIND THE CHART

Some of these results are obviously intuitive. For example, suppose it is relatively late in the game and a team is down one point and scores a touchdown. It is now up five points. Going for two points and succeeding ensures that a touchdown will not put the team behind, while if it goes for one point, a touchdown will put it behind. Some of the results are counterintuitive—based on how coaches make decisions today. Suppose a team is down fourteen points late in the game and scores a touchdown. Virtually all NFL coaches play it safe and kick a one-point conversion because they wager that they will score again, kick another one-point conversion, and win in overtime. Suppose the team has the ball with three possessions left and is down 14 points. If the team follows the one-point conversion strategy, it will win the game only by scoring a touchdown on two successive possessions, holding its opponent scoreless on a single possession and winning in overtime. The probability of winning in this fashion is $(.19)^2(.94)^2(.68)(.5) = .0108$. If the team goes for two points (as Table 23-2 indicates), after scoring with three possessions left it can win the game in one of two ways:

- The first two-point conversion try succeeds, and the team holds the opponent scoreless, scores a touchdown, and makes a one-point conversion.

- The first two-point conversion fails, and the team holds its opponent scoreless, scores a touchdown, makes a two-point conversion, and wins in overtime.

These two probabilities sum up to $0.0133 > 0.0108$, and thus teams should go for two points when they are down by 14 points and there is little time left in the game. Of course, the differences are small but in a *game of inches any edge is important*.

Suppose there are at least five possessions remaining and our team is down by four points and scores a touchdown. Most coaches reason that going for one point puts us ahead by three points, so a field goal will not beat us. If there is little time left in the game this logic is correct. If enough time remains, however, the following scenario has a reasonable chance of occurring: the opposition scores a touchdown and a one-point conversion. Then our team is down by four points and scoring a field goal will not allow us to tie the game. If our team is down four points and goes for two points, however, a successful two-point conversion ensures that if the opposition scores a touchdown and a one-point conversion, then our team can still tie the game with a field goal.

CHANGING THE QUESTION

Benjamin Morris of FiveThirtyEight.com had an interesting article[4] about the choice between one- and two-point conversions. The difference of his approach from the approaches that we described earlier is that he is asking a slightly different question to begin with: *will the first or the second point of the conversion improve more the team's chances of winning?* Morris used ESPN's win probability model to estimate the change in the win probability for different game settings. For example, Table 23-4 presents Morris's calculations for a game with 10 minutes left.

---

4. http://fivethirtyeight.com/features/when-to-go-for-2-for-real/

TABLE 23.4

Win probability changes when the size of the lead changes
(10 minutes left in the game). Calculations from
Benjamin Morris's article

Going from a Lead of	Tactical Benefit	Win Probability Change (%)
0 to 1	Take the lead	+8.4
1 to 2	A generic point	+1.8
2 to 3	Puts you up a field goal	+6.5
3 to 4	Puts you up more than a field goal	+5.0
4 to 5	A generic point	+2.9
5 to 6	Puts you up two field goals	+3.1
6 to 7	Puts you up by a touchdown	+5.2
7 to 8	Puts you up by a touchdown and a two-point conversion	+3.3
8 to 9	Puts you up two scores	+2.9
9 to 10	Puts you up by a touchdown and a field goal	+2.2
10 to 11	Puts you up by a touchdown with a two-point conversion and a field goal	+1.3
11 to 12	Puts you up more than a touchdown and a field goal	+1.1
12 to 13	Puts you up by a touchdown and two field goals	+0.4
13 to 14	Puts you up two touchdowns	+1.0
14 to 15	Puts you up two touchdowns with a two-point conversion	+0.5
15 to 16	Puts you up two touchdowns with two two-point conversions	+0.7
16 to 17	Puts you up three scores	+0.2

These changes in the win probability are all that Morris needs and uses to create his version of "the chart." The idea is simple: let us assume that we currently have a lead of x points. Using Table 23-4 (or any other win probability model that you trust) we just have to compare which win probability change is bigger, going from a lead of x + 1 or from a lead of x + 1 points to x + 2. For example, based on Morris's calculations, with 10 minutes remaining, when scoring a touchdown cuts the lead to five points, the first point (i.e., going from a five-point to a four-point deficit) will increase our win probability by 2.9%, while the second point (i.e., going from a four-point to a three-point deficit) will increase our win probability by 5%. So, in this case you should go for the two-point conversion. When the touchdown cuts the lead to seven points, then the first point (i.e., going from a seven-point to a six-point deficit) increases our win probability by 5.2%, while the second point (i.e., going from a six-point to a five-point deficit) increases our probability by a smaller margin, that is, 3.1%. So, in this case Morris's thought process suggests going for the extra-point kick. Morris further considered the quality of a team on two-point conversions—which essentially provides a range of changes in the win probability—that can eventually provide a customized chart for each team.

CHAPTER 24

TO GIVE UP THE BALL IS BETTER THAN TO RECEIVE

The Case of College Football Overtime

In college football games that go to overtime, the winner of a coin toss chooses whether to start with the ball or give the ball to its opponent. The first team with the ball begins on the opponent's 25-yard line and keeps going until it attempts a field goal, scores a touchdown, or loses possession. Then the other team gets the ball on its opponent's 25-yard line and keeps going until it attempts a field goal, scores a touchdown, or loses possession. Any team that is ahead at this point is the winner. If neither is ahead, the order of possessions is reversed, and the process is repeated. We note that after the second overtime each team must attempt a two-point conversion. Rosen and Wilson[1] tabulated the outcomes for all overtime games through 2006 and found that the team that had the ball second won

1. P. A. Rosen and R. L. Wilson, "An analysis of the defense first strategy in college football overtime games," *Journal of Quantitative Analysis in Sports*, 3(2), 2007.

54.9% of the time. This would indicate that the coach who wins the toss should elect to give the other team the ball first. The intuitive appeal of giving the ball up is that when you finally get the ball you will know what you need to do to win or keep the game going. For example, if the team with the ball first scores a TD, we know that we must go for a TD. If the team with the ball first fails to score, then we know we just need an FG to win. Can we model the "flexibility" of the second possession team in a way that is consistent with the second possession team having a 54.9% chance of winning the game?

Rosen and Wilson give the following parameter values:

- Probability team with ball first scores TD: .466
- Probability team with ball first scores FG: .299
- Probability team with ball first does not score: .235.

We will use the term "first team" to refer to the team that gets the ball first and the term "second team" for the team that gets the ball last. We will model the flexibility of the second team with the ball by estimating the following two parameters:

- EXTRAFG = This is the fraction of time that the second team would convert a possession that would have resulted in no score into an FG. The rationale here is that if the first team does not score then the second team wins with a field goal and does not need a TD. Therefore, it will play more conservatively and be less likely to commit a turnover or be sacked. This will convert some possessions that would have resulted in no score into an FG.
- PRESSURETD = Probability that second team will score a TD given that the first team has scored a TD.

The rationale here is that if the first team scores a TD, then the second team will never go for an FG and some of the possessions that would have resulted in field goals will now result in TDs. Note that if the game is tied after each team has a possession, then the first team has the ball second on the next possession and now has a .549 chance of winning. The first team's chance of winning the game may now be obtained by summing the following probabilities:

- No team scores on its first possession and the first team wins in a later sequence. The probability that the first team does not score on its first possession is .235. The second team knows it only needs an FG, so a fraction EXTRAFG of possessions that would have resulted in no score will now result in a score. **At the start of the second possession sequence the first team is now the second team so it has a .549 chance of winning.** Therefore the first team wins in this way with probability $(.235)(.235)(1 - \text{EXTRAFG})(.549)$.

- Each team scores an FG on its first possession and the first team wins on a later possession. The probability of an FG for each team on the first sequence is .299, so the first team wins in this way with probability $(.299)^2(.549) = .0491$.

- The first team scores an FG and the second team does not score on its first possession. The first team scores an FG on its first possession with probability .299. The second team fails to score with probability .235. Therefore the probability that the first team wins in this fashion is $(.299)(.235) = .070$.

- The first team scores a TD on its first possession and the second team does not score on its first possession. The probability the first team scores a TD on its first possession is .469. Since the second team knows it needs to score a TD, its chance of scoring a TD will increase because it will never try an FG (we called the probability of a TD for the second team PRESSURETD). Therefore, the probability of the first team winning in this way is $.469 * (1 - \text{PRESSURETD})$.

- Both teams score a TD on their first possession and the first team wins on a later possession. The first team scores on its first possession with probability .469 and the second team scores on its first possession with probability PRESSURETD. Then the first team has a .549 chance of

winning. Therefore the chance of the first team winning in
this way is .469(PRESSURETD)(.549).

We found that if the second team can convert 30% of no score possessions into FGs when it knows it needs only an FG to win and can
score a TD with probability .74 (instead of 46.9%) when it knows it
needs a TD, then it will win 54.9% of its games. This is fairly consistent with the observed frequency with which the second team wins
the overtime.

The college football overtime coin toss "strategy" shows the importance of "managerial flexibility." Practitioners of real option theory in finance (see Shockley, 2007) have long realized that options
such as expansion, abandonment, contraction, and postponement
of a project have real value. Here we see that the option to go for
an FG instead of a TD (or vice versa) can have real value in college
football.

CHAPTER 25

HAS THE NFL FINALLY GOTTEN THE OT RULES RIGHT?

When an NFL game goes into overtime a coin toss takes place and the team that wins the coin toss has the choice of kicking off or receiving. For many years, the OT had a sudden death format, and the team winning the coin toss invariably chose to receive so they have the first chance to score and win the game. During the 1994–2006 seasons the team that received the kickoff in overtime won 60% of the games. It seemed unfair that in NFL overtime the team winning the coin flip should have such a huge edge. To lessen the impact of the coin flip result on the game's outcome, the NFL consequently proposed moving the kickoff from the 30- to the 35-yard line. This would give the team receiving the kickoff a slightly worse field position and theoretically would decrease the chance that the team receiving the kickoff would score on the first possession. This should give the team kicking off a better chance to win the game. However, as we will see with a simple mathematical analysis (our model is a simplified version of Jones's model),[1] it will be difficult to give each

1. M. A. Jones, "Win, lose, or draw: A Markov Chain analysis of overtime in the National Football League," *College Mathematics Journal*, 35(5), 2004, 330–336.

team an equal chance to win in sudden death if the overtime begins with a kickoff. Since 2012, NFL implemented a new OT rule and the team that receives the kickoff cannot win the game with a field goal in the opening drive. If this happens, the opponent gets the ball back to either tie the game (with a field goal) and keep the OT alive triggering sudden death, or scores a touchdown and wins the game. As we will see, this change has neutralized the edge that the kickoff receiving team enjoyed under the sudden death format.

A SIMPLE MATHEMATICAL MODEL OF SUDDEN DEATH OVERTIME

Let p be the probability that an average NFL team scores (either field goal or touchdown) on a possession. If each team has a probability p of scoring on each possession, what is the probability that the team receiving the kickoff will win the game? During the regular NFL season, overtime games now last ten minutes. If no team scores during the first overtime session, the game is a tie. This happens less than 1% of the time. Our analysis is greatly simplified if we assume that the overtime can theoretically continue forever. Since less than 1% of NFL games fail to yield a winner during the overtime, our assumption differs from reality in, at most, 1% of all games, so our assumption should not cause our calculations to differ too much from reality. Let K be the probability that the team receiving the kickoff in overtime wins the game. There are two ways the receiving team can win.

- With probability p the receiving team scores on the first possession.
- The receiving team fails to score on the first possession, the kicking team fails to score on its possession, and the receiving team wins on a later possession. Assuming the outcomes of the first two possessions and later possessions are independent, the probability that the receiving team wins in this fashion is $(1-p)(1-p)K$. This follows because if overtime can go on forever, the receiving team has the same chance of

winning at the beginning of its second possession as it does at
the beginning of its first possession.

Therefore, we find that $K = p + (1-p)(1-p)K$. Solving for K we find
that:

$$K = \frac{p}{1-(1-p)^2} = \frac{1}{2-p} \tag{1}$$

The first thing to note is that since $p < 1$, K must be greater than 0.5.
As a result, if our simple model approximates reality, it is quite hard
(if not impossible) for the NFL to make a sudden death format beginning with a kickoff fair.

DOES OUR MODEL APPROXIMATE REALITY?

Our model assumes a (possibly) infinite overtime period and ignores
the possibility that the game will end on a turnover touchdown, safety,
or kick/punt return touchdown. Despite this fact, our model predicts
(correctly) that the receiving team will win 60%. In particular, as we
saw in Chapter 23, p is equal to .32, since there is a .13 chance of scoring an FG and a .19 chance of scoring a TD in a drive. Equation (1) now
implies that the probability that the receiving team wins in overtime
is given by $\frac{1}{2-0.32} = 0.595$. Since receiving teams have triumphed in
60% of sudden death overtimes, our model appears quite consistent
with reality. Of course, if the NFL moved the kickoff position a lot,
then our assumption that each team has the same chance of scoring
on each possession would be inaccurate. We will explore the idea of
changing the kickoff position later in the chapter.

IS THERE A FAIR SOLUTION TO THE OVERTIME DILEMMA?

Before examining the current format for the NFL overtime, let us
try to examine whether there are any reasonable solutions that could
give each team an equal shot at winning in overtime. Professors

Jonathan Berk and Terry Hendershott of the Haas Business School at Berkeley suggest that each team "bid" for the yard line on which its first possession starts.[2] The team that "bids" closer to its goal line wins. For example, if the Texans bid for the 20-yard line and the Steelers bid for the 10-yard line, then the Steelers begin the overtime with the ball on their own 10-yard line. If they fail to score and later lose the game, they have only themselves to blame. This approach allows teams to bid based on their strengths and weaknesses. A good offensive team (i.e., most certainly not the 2019 Steelers) is probably confident it can drive to field goal range and will bid close to its own goal line to ensure possession. A good defensive team will probably be happy to let the other team start with the ball deep in its own territory. If there is a tie on the first bid, then the teams each submit another bid.

Another fair solution to the NFL overtime problem is analogous to the solution to the famous cake-cutting problem.[3] Consider two people who want to cut a cake "fairly" into two pieces. The mathematically fair solution is to have one person cut the cake into two pieces and have the other person choose which piece he wants to eat. The cake cutter's solution to NFL overtime would be to toss the coin and let the winner choose a yard line on which the first possession will begin or let the opponent choose the yard line on which the first possession will begin. The team that does not choose the yard line gets to choose whether to take the ball. This forces the team that chooses the yard line to choose a "fair" starting situation. For example, if I choose the 50-yard line, my opponent will surely choose the ball. If I choose my 10-yard line, the opponent will probably let me start from my 10-yard line. Assume that I can choose the yard line on which the first possession starts. Assuming the two

    2. C. Bialik, "Should the outcome of a coin flip mean so much in NFL over-time?" available at http://faculty.haas

    3. See Steven J. Brams, *Mathematics and Democracy: Designing Better Voting and Fair-Division Procedures* (Princeton University Press, 2007), for a wonderful discussion of cake-cutting problems, voting methods, and other interesting social and political issues.

teams are of equal ability, I would pick the yard line x for which I feel I have a 50% chance of winning the game if I start with the ball on yard line x. If the other side gives me the ball, I have a 50% chance of winning. Assume the teams are equal. Then the other team also has a 50% chance of winning if they have the ball first and start on yard line x. Thus, my choice guarantees me a 50% chance of victory. If my opponent chooses yard line x, then I choose to start with the ball if I believe my chance of winning from yard line x is at least .5, and I choose to give the other team the ball if I believe that my chance of winning from yard line x is less than 0.5. Assuming the two teams are of equal ability, then this choice again ensures that I have at least a 50% chance of winning. In general, this solution ensures that whether or not a team wins a coin flip, it has a 50% chance of winning. Of course, given that teams are not of equal strength, our argument shows that if we win the coin flip, then we should let the opponent choose the yard line. Then our chance of winning *might* exceed 50%.

CURRENT NFL OT RULES

Since 2012[4] the NFL has adopted a modified sudden death scheme. The current rules give both teams the opportunity to possess the ball at least once in overtime unless the team that receives the overtime kickoff scores a touchdown on its first possession. In other words, the current rules force the team that receives the kickoff to attempt to score a touchdown if it wants to win in a sudden death fashion. The question then is, has this removed (partially or completely) the edge that the kickoff receiving team had under the older rules? The short answer is yes! Let us see now why.

To begin with, between 2012 and 2017 there have been 93 regular season games that went to overtime, with the team that received the kickoff having won 47 of them, i.e., 50.5% of the times. To formally

---

4. The same rules have been implemented since 2011 only for the postseason games.

examine this we build a logistic regression model that estimates the probability π that the team receiving the kickoff will eventually win the game. We start by first fitting an intercept only model, which essentially captures the correlation between receiving the ball and winning the game. Table 25-1 (Model 1) shows that receiving the kickoff does not provide a statistically significant benefit, i.e., the win probability for the receiving team is not statistically different than 50%. The obtained probability is 50.5% (exactly the ratio of W–L in the dataset), with a 95% confidence interval of [41%, 60%]. The confidence interval is very large, which basically means that the true value of the win probability for the receiving team might be as low as 41% or as high as 60%. The width of the interval is large due to the small sample size. So, even though at this point we cannot reject the hypothesis that the new rules are *fair* (i.e., $\pi = 0.5$), we will get more robust results as we collect more observations from overtime games.

We further build a model with a single covariate, namely, the closing betting point spread for the team receiving the kickoff (i.e., if the receiving team was a five-point underdog before the game, this variable will be -5, etc.). In this model (Table 25-1—Model 2), the intercept (i.e., receiving the kickoff) is still not statistically correlated with the probability of winning the game, while the betting point spread is (at the 10% level). Simply put, even with the small sample size, the team strength is a stronger predictor for the OT winner as compared to the coin flip. In particular, a one-point pre-game favorite has a 52% win probability in OT, while a five-point pre-game favorite has a 64% win probability. Figure 25-2 presents the win probabilities from Model 2 (Table 25-1) for different pre-game point spreads for the receiving team. On the same figure, we present the implied pre-game win probability for the same point spread[5] for comparison. As we can see, the overtime model implies that even though the chances for the two teams are not quite 50–50 (as many

---

5. A detailed description of obtaining win probabilities from Vegas point spreads is provided in Chapter 47.

times considered in back-of-the-envelope calculations), they are much more close to this as compared to pre-game win probabilities.

It seems that NFL has gotten it *right* this time; even though still almost every team that wins the coin toss elects to receive the kickoff, the probability of winning in OT is not correlated with receiving the kickoff, but it is correlated with the pre-game win probability! If a team elects to kick, it makes for a good week-long or more of discussions and articles—as was the case with the Patriots when they won the coin toss against the Jets in New York during the 2015 season (the Patriots eventually lost)[6] or the Steelers in their home game against the Ravens in the 2019 season.

This last situation is a good example of why a decision in this setting is not cut and dry. If we combine all the information at hand at the time of decision, choosing to kick was a sensible decision from the Steelers (casual Steelers fans still criticized Mike Tomlin for electing to kick). Their 2019 defense was strong, they had played a great game against the Ravens, while their offense was crawling. Furthermore, the Ravens have possibly the best kicker in the history of the game in Justin Tucker, which means that if the Steelers do not score in their first possession—based on their 2019 offense, their chance of doing so was much lower than .32—the Ravens can win with a 50+ yard bomb from Tucker. Factoring all these, I would have made the same decision. It is worth pointing out that even though the Steelers lost, they forced a three-and-out in the Ravens' first possession and they only lost due to a fumble from Juju Smith-Schuster in Ravens territory. It almost worked, but almost does not play well when people are used to outcome-based evaluation of decisions.

---

6. Brian Burke, using ESPN's proprietary model, estimated that New England Patriots gave up a 7.6% win probability by deferring after winning the coin toss. However, as he also mentioned under the new OT format, the team that received the opening OT kickoff had won up to that point 50.7% of the time—close to our estimates. Also note that ESPN's model gave the receiving team a 53.8% win probability, which is within the confidence interval of our simple model's prediction. http://www.espn.com/blog/statsinfo/post/_/id/112875/numbers-dont-back-up -patriots-ot-strategy

TABLE 25.1

Coefficients for the Two OT Win Probability Models.
Dependent Variable: Probability of Team Receiving Kickoff
Wins the Game (Significance Coefficients: **0.01, *0.05, .0.1)

	OT Win Probability Models	
Variable	Model 1	Model 2
Intercept	0.0215	0.0206
	(0.207)	(0.211)
Receiving Kickoff		0.0633
		(0.037)
Log-Likelihood	− 65.2	− 62.9

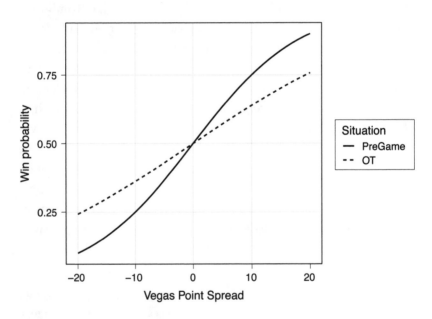

FIGURE 25.1 Win Probabilities (Pre-game and OT)
Based on Betting Point Spreads.

MASSEY AND SAGARIN WEIGH IN

Ken Massey and Jeff Sagarin have developed two of the leading computer systems for rating sports teams. In this section we share their fascinating suggestion for settling games that are tied in regulation that combines and extends some of the ideas described earlier. To illustrate their approach let's use the 2019 AFC title game between the Chiefs and the Patriots that was tied 31–31 at the end of regulation. The Patriots won the coin toss and drove down the field for a TD. Thus the Patriots won the game and KC's high powered offense never touched the ball.

After regulation ends in a tie, both teams are given three minutes to present a card to the referee with two numbers written on it: a yard line and a clock time. Say KC submits (10, 3:45) and NE submits (8, 4:00). The team that wrote the lower yard line gains possession of the football for one final drive to score. The drive starts according to the yard line and clock time written on that team's card. The other team receives one point. Now KC has the lead 32–31, and NE is given the ball at its own eight-yard line with 4:00 on the clock. The team with the ball has one drive to score in the allotted time. If it scores (FG or TD), it wins the game; otherwise it loses. Say NE does in fact drive the ball to the KC 30, and the 4:00 clock is about to expire. Then it would be forced to attempt a 47-yard FG for the win. What we have described is essentially a sealed-bid auction, whereby teams bid a yard line and clock where they would be willing to take possession, in lieu of a point. The strategic nature of this bidding would be of great interest to the fans and commentators. Neither team could blame its fate on the result of a coin toss. The teams are bidding primarily on the yard line; the clock time acts as a tiebreaker. If KC had bid (8, 3:45), it would have taken possession since it's harder to score with 3:45 on the clock than with the 4:00 that NE bids. In some games, e.g., the KC–NE game, bidding would be aggressive since both offenses were potent. In other games, e.g., the 2014 Virginia Tech–Wake Forest game that ended regulation tied 0–0, bidding would be very passive. And yes, that game could have ended with a score of 1–0 under this proposal. Following the auction, game play would start immediately with one team ahead by a point, and the other team taking possession according to its bid. The team on offense would always have four downs to move the chains, but the defensive team would need only one stop to win the game. One drive. Standard game play. Game over. Here are a few minor details to iron out:

- If both teams bid the exact same yard line *and* clock time, give possession to the designated home team. This is still equitable, because each team should bid so as to be indifferent whether it gets the ball or gets the point.
- Let the team taking possession decide which direction to face.
- Give each team one timeout for the final drive.

It is our hope that the NCAA and NFL rules committees will put this proposal on the table. Perhaps NCAA Division II would be a good testing ground. Eventually, this overtime procedure could be implemented at all levels of American football.

As mentioned above, the strategic nature of this bidding would be fascinating. Each side will be trying to anticipate how the other side will bid. Here are a few examples to illustrate different ways the KC–NE auction could turn out:

- KC bids $(10, 2:00)$ and NE bids $(8, 5:00)$. Since $8 < 10$, NE gets the ball with 5:00 on the clock. The clock time doesn't determine the auction winner, only the starting conditions of the OT drive.
- KC bids $(1, 3:10)$ and NE bids $(1, 3:15)$. Both teams desperately want the ball, but KC gets possession because it is willing to get less time.
- KC bids $(10, 2:00)$ and NE also bids $(10, 2:00)$. Possession goes to the home team, KC in this case.
- KC bids $(10, 2:00)$ and NE bids $(20, 5:00)$. KC gets the ball at the 10 with 2:00 on the clock. In this case, it may experience a "winner's remorse" knowing that it bid too aggressively. In all

cases, the auction does not really have a "loser." If a team bids too passively or too aggressively, its own strategic decision is to blame. If it bids sincerely, i.e., the starting conditions where it is truly indifferent about having possession or the point, then it should in fact be glad to "lose" the auction.

CHAPTER 26

HOW VALUABLE ARE NFL DRAFT PICKS?

The NFL in general is thought to exhibit more parity than other leagues. This means that it appears easier for a bad NFL team to improve from season to season than for a bad NBA or MLB team to do so. Most of the NFL fans believe that the major equalizer from year to year is the structure of the NFL draft. Teams draft in each round in inverse order of performance, with the worst team getting the first pick and the best team getting the last pick. Common sense tells us that an earlier pick should, on average, be a more valuable player to a team than a later pick. According to the seminal work from (Nobel Laureate) Thaler and Massey (TM), common sense may be wrong.[1]

---

1. C. Massey and R. H. Thaler, "The loser's curse: Decision making and market efficiency in the National Football League draft," *Management Science*, 59(7), 1479–1495.

ESTIMATING THE NFL IMPLIED DRAFT
POSITION VALUE CURVE

TM began by trying to estimate the relative value NFL teams associate with different picks. They collected data on all draft trades from recent years in which draft picks were dealt. For example, perhaps one team traded its ninth and twenty-fifth picks for a third pick. Letting v(n) be the relative value (v(1)=1) of the nth pick in the draft, this trade would indicate the NFL teams believe $v(9)+v(25)=v(3)$. TM found that the function

$$v(n)=e^{-a(n-1)^b} \tag{1}$$

yields an excellent fit with the observed trade data. For each trade they calculated the estimated value of both sides of equation 1. Then they chose the two constants (a and b) that minimize the sum (over all observed trades) of the squared differences between the estimated values of both sides of the trade. Thus for a trade in which pick 3 was traded for picks 9 and 25, we would try to choose α and b to minimize $(v(3)-(v(9)+v(25)))^2$. They found that the constants $a=-.148$ and $b=.7$ are the solutions to the minimization problem. The corresponding function (Weibull) is shown in Figure 26-1. TM then tried to determine how much a draft pick benefits a team. They divided a player's performance during a season into five categories that are roughly indicative of their quality/playing value:

- Not on roster
- No starts
- 1–8 starts
- 9–16 starts
- Pro Bowler

TM then identified the average salary by position for each of the five categories. For example, Table 26-1 illustrates the average salary per category for the quarterback position. For each draft pick one

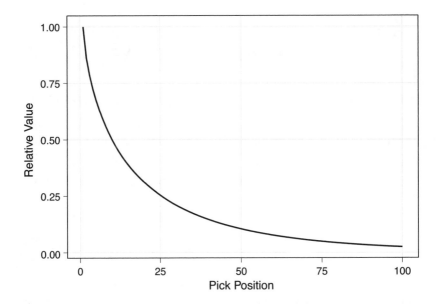

FIGURE 26.1 The NFL implied draft position value curve according to TM. Note the steepness of the curve; pick 10 is only considered half as valuable as pick 1.

TABLE 26.1

Mean Salary and Performance Level for
QBs in TM Study

Mean Salary ($)	Performance Level
—	Not on roster
$1,039,870.00	No starts
$1,129,260.00	1 to 8 starts
$4,525,227.00	> 8 starts
$9,208,248.00	Pro Bowler

can now compute the surplus value of the player picked as (player value) – (player salary). For example, a Pro Bowl QB during the TM study has a value of $9,208,248. If his salary was $5,000,000, then the team earned a $4,208,248 surplus. Using similar calculations TM found that the average surplus by position in the draft increased through pick 43, which means that later picks contributed more value than earlier picks (compared to their salary). TM viewed the fact that later picks contribute more surplus value to teams as evidence of market inefficiency. This conclusion would seem to indicate that NFL teams are not very proficient at selecting college players.

As pointed out by Birnbaum,[2] there is a major flaw in the TM analysis. TM assume that all players who play the same position and are in the same performance category are equally valuable. However, these categories are arbitrarily defined and they can include large variability in player quality. For example, a player who has a single start at a meaningless week 17 game will be in the same category with someone who has eight starts in the season. In order for TM to nail down their conclusion that the NFL draft is inefficient, they needed a better measure of player performance.

Since the TM study there are several people that have gotten interested in studying the NFL draft inefficiencies. A better approach is, instead of categorizing players in arbitrarily chosen bins, to assign a continuous value that represents their on-field value. For example, if we had a measure of wins above replacement similar to that of baseball that would be ideal. However, for football this type of metrics are hard to estimate. Nevertheless, Doug Drinen, founder of Pro Football Reference (PFR) created the Approximate Value[3] (AV) metric that attempts to describe the on-field value of a player. Using this we can create a draft curve for the average AV from each draft position.[4] Using data from Pro Football Reference—collected by Sean Taylor[5]—we can get the draft curve in Figure 26-2. For this draft curve we have used the career approximate value of a player normalized with the number of games played. Based on these average AVs we can see that the number 1 overall pick provides an average AV of 0.46 per game, while the last pick of the round provides

2. http://blog.philbirnbaum.com/2006/12/do-nfl-teams-overvalue-high-draft-picks.html [Last accessed: October, 3rd 2020]

3. https://www.pro-football-reference.com/blog/index37a8.html

4. Another possibility is to use PFF player grades, but these are not open to the public.

5. https://seanjtaylor.github.io/learning-the-draft/

FIGURE 26.2 Draft Value Curves Based on (PFR) AV/Game.

an AV per game of 0.34. This is a 26% reduction in AV, but it comes at a 70% lower price (based on 2018 salaries).

It should be obvious that the choice of performance metric is very crucial in any draft curve value analysis. For example, instead of career AV per game, one could use only the AV for the first four years of a player (rookie contract without the fifth year option). Michael Schuckers[6] and Michael Lopez[7] (currently the NFL Director of Data and Analytics) suggested that teams might not be interested in the average value they get from a draft pick but on how possible it is to find a *superstar* in a given draft position, i.e., a player's upside. What they found is that if the performance metric considers the *superstar* status of a player (e.g., based on their Pro Bowl appearances), the curves are steeper. Simply put, the probability of drafting a superstar declines much faster than the expected AV of the player. One can

6. Schuckers, "An alternative to the NFL draft pick value chart based upon player performance," *Journal of Quantitative Analysis in Sports*, 7(2), 2011.

7. https://statsbylopez.netlify.com/post/rethinking-draft-curve/

also use blended curves, which are essentially weighted averages of individual curves.

NFL MOCK DRAFTS

Leading up to the draft, pundits, media, and fans create their own mock drafts, a kind of alternative universe, where everyone can play GM and make *hot takes*. Benjamin Robinson, a graduate from the University of Pittsburgh, has created a database of NFL mock drafts[8] and has explored how well the consensus among these mock drafts matches/predicts the actual drafting order. Benjamin found that mock drafts account for about 80% of the variation in draft order. What does that mean? That most probably—on average—NFL pundits are not any better at drafting than GMs. These data currently cover only the 2018 and 2019 seasons, but as more mock draft data are collected, it will be interesting to start examining draft curves for different pundits (and know how much value to put on their opinions).

THE WINNER'S CURSE

If we accept that the NFL draft is inefficient, what might cause the observed inefficiency? Perhaps this is an instance of the well-known winner's curse. Essentially the winner's curse says that winners of an auction often pay more than the value of the object they won. The winner's curse was first observed during bidding on offshore oil leases during the 1950s. Many companies that won the rights to offshore oil sites found that the value of the site was on average less than the value they bid for the site. Suppose that the true value of a prize is $100 million (of course, nobody knows this for sure). If five people bid for the prize, they might value the prize at $60, $80, $100, $120, and $140 million.[9] The person placing the $140 million

---

8. https://benjaminrobinson.github.io

9. The average of these bids is equal to the actual value of the site, as we would expect.

value on the prize will probably bid near $140 million for the prize. The "high-value person" will win the prize but lose around $40 million. In the NFL the team that drafts a player has, in a sense, "bid the highest" for that player, so if the winner's curse is at play here, it is reasonable to expect that earlier draft picks may create less value than do later draft picks.

CHAPTER 27

PLAYER TRACKING DATA IN THE NFL

The NFL—following the example of other professional leagues—is currently collecting tracking data for the players' and the ball's location on the field. This information is collected through RFID tags that are placed in every player's shoulder pad and in the ball.[1] The data include information such as the location of the players, their direction, and their speed at a frequency of 12.5 Hz. The ball data are sampled at a higher frequency (25 Hz) and the information captured includes location, speed, and rotation. While the technology has been installed and information obtained from it has been used in broadcast since 2015, teams started obtaining these data during the 2016 season. Given that American football is essentially a *game of space,* this information has the ability to drive new insights for evaluating players and plays and understanding what wins games in the NFL. However, these data can be useful in other areas as well. For example, the NFL is aiming at using these data to adjust the rules to make the game safer![2,3] In this chapter we will show some ideas that can be analyzed with player tracking data using information from all

1. https://operations.nfl.com/the-game/technology/nfl-next-gen-stats/

2. https://www.kaggle.com/c/NFL-Punt-Analytics-Competition

3. K. Pelechrinis, R. Yurko, and S. Ventura, "Reducing concussions in the NFL: A data-driven approach," *CHANCE*, 32(4), 2019, 46–56.

the games from the first six weeks during the 2017 season, courtesy of the NFL. In fact, Mike Lopez, the Director of Data and Analytics in the NFL, has taken an innovative approach, with the league hosting data competitions for the public—appropriately called "Big Data Bowl"—using player tracking data. We would like to thank Mike for permission to use some of these data to showcase how they can be used by teams today.

QB RELEASE TIME

The time it takes for the quarterback to throw the ball (or take the sack) after the snap can reveal a variety of information. For example, a short release time could be an indicator of an offensive scheme for short passes with high completion percentage (and potentially a larger gain from yards-after-catch). A long release time could be an indicator of a good offensive line that allows the QB to stand in the pocket and wait for the play to develop, make his reads, and take the *best* decision. Of course, a long release time could also be an indicator of a mobile QB, who is on the move to avoid the pressure. The first question we are interested in examining is whether there is any relation between the release time and the air yards of the pass.[4] Using the player tracking data from the first six weeks of the 2017 NFL season, and removing scrambles and penalties, we obtain the results presented in Figure 27-1. The median release time is 2.6 seconds (mean is 2.8) after the snap is taken, while there is some right skewness due to the presence of longer release times (e.g., as mentioned earlier, a mobile QB getting out of the pocket to avoid the pressure). Figure 27-2 (left) further presents the cumulative distribution function for the release time for different offensive formations, while Figure 27-2 (right) presents the median and the corresponding confidence interval. As we can see, and as one might have expected, the median release time for *passing* formations (i.e., shotgun, empty set, and pistol) is lower compared to formations

---

4. One could do the same for play point values from Chapter 20.

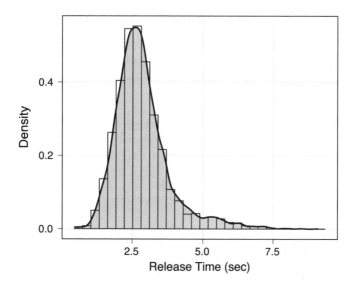

FIGURE 27.1 QB Release Time Distribution.

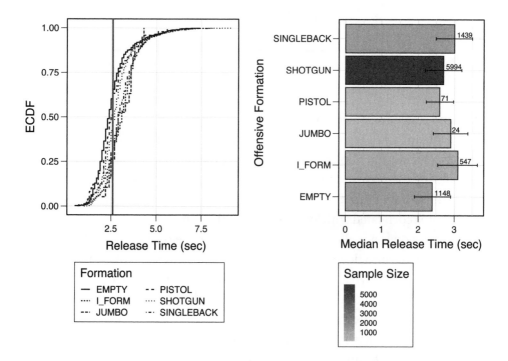

FIGURE 27.2 Release Time Based on Formation
(left: cumulative distribution, right: median).

TABLE 27.1

Pass Air Yards Regression Model

(sig: *p < 0.1, **p < 0.05, ***p < 0.001)

Variable	Pass Air Yards		
	coef	*SE*	*Sig*
Constant	−3.2	0.75	***
Release Time	3.3	0.13	***
Pass Rushers	0.7	0.14	***
Off form (I-FORM)	−1.1	0.61	*
Off form (JUMBO)	−5.9	2.23	***
Off form (PISTOL)	−2.4	1.42	*
Off form (SHOTGUN)	−0.9	0.39	**
Off form (SINGLE BACK)	−0.2	0.48	
Observations	6,019		
R2	.11		
Adjusted R2	.10		

that have the quarterback under center (e.g., jumbo, I formation, and single back formation).

Using a linear regression model we express the air yards of a pass as a linear function of the release time, the number of pass rushers during the play, and the offensive formation. Our regression results are presented in Table 27-1, where, as we can see, a faster release time is correlated with lower air yards. This is expected, since a faster release does not allow for the play to fully develop and the wide receivers to run a deep route. Furthermore, it is interesting to observe that more pass rushers are correlated with higher air yards, possibly due to weakened coverage, while an empty backfield (the reference formation for the model) is associated with the highest air yard gain. It is interesting to note that the model itself captures only 10% of the variance of the air yards of the pass, which means that other omitted variables might be more important in modeling the air yards.

TABLE 27.2

Sack Probability Model

(sig: *p < 0.1, **p < 0.05, ***p < 0.001)

Variable	Sack		
	coef	*SE*	*sig*
Constant	−8.5	0.40	***
Release Time	3.3	0.04	***
Pass Rushers	0.6	0.07	***
Off form (I-FORM)	−1.2	0.31	***
Off form (JUMBO)	−13.7	324.1	
Off form (PISTOL)	−1.4	1.1	
Off form (SHOTGUN)	−0.4	0.18	**
Off form (SINGLE BACK)	−0.7	0.22	***
Observations	6,444		
Log likelihood	−1,136.4		
AIC	2,289.3		

For example, one might expect the actual offensive (and defensive) schemes—not just the formations—to be better suited to capture the variance of a pass's air yards.

We can also examine whether and how the release time correlates with the probability of the defense getting a sack. Obviously, there are several parameters that play a role in this (similarly to the air yards from a pass), but the time that the QB has the ball in his hands is an important variable. We built a simple logistic regression model for the probability of a sack as our response variable, while our independent variables are the QB release time, the offensive formation, and the number of pass rushers. The results of the model are presented in Table 27-2, where we see that the probability of a sack increases with the number of pass rushers (as expected), while a faster release time reduces the probability of a sack (again as expected). It is also interesting to note that empty formations (the

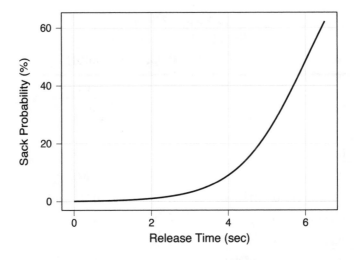

FIGURE 27.3 Sack Probability as a Function of Release
Time (Four Pass Rushers out of the Shotgun).

reference formation for our model) and the shotgun formation exhibit the highest probability of a sack. Figure 27-3 further shows the
probability of a sack for a shotgun formation with four pass rushers as a function of the release time. Sacks have traditionally been
viewed from the prism of the offensive line's quality. However, this
(simple) model shows that there are things that the QB controls that
can affect the probability and the number of sacks taken. The QB has
control over his sack rate. As Jason Lisk shows in his article "Sacks
Are a QB Stat,"[5] sack rate is among the most consistent stats for a
QB when he changes team, second only to completion percentage.

We can develop a similar model for the probability of an interception and its relationship to the time that the QB holds on
to the ball (i.e., release time), controlling for offensive formation
and number of pass rushers. Table 27-3 (first column) presents our
model, where, as we can see, the only parameter correlated with
the probability of an interception is the release time; the longer the

5. https://www.thebiglead.com/posts/sacks-are-a-quarterback-stat-01dx
qapkgvw9

TABLE 27.3

Interception Probability Model

(sig: *p < 0.1, **p < 0.05, ***p < 0.001)

Variable	Interception (no air yards)			Interception (with air yards)		
	coef	*SE*	*sig*	*coef*	*SE*	*sig*
Constant	−4.4	0.55	***	−4.3	0.56	***
Release Time	0.14	0.08	*	−0.05	0.1	
Pass Rushers	−0.01	0.1		−0.04	0.1	
Off form (I-FORM)	0.5	0.43		0.5	0.43	
Off form (JUMBO)	−11.5	333.1		−11.2	329.3	
Off form (PISTOL)	0.2	1.1		0.2	1.1	
Off form (SHOTGUN)	0.4	0.31		0.4	0.31	
Off form (SINGLE BACK)	0.2	0.37		0.2	0.37	
Air yards				0.05	0.007	***
Observations	6,020			6,019		
Log likelihood	−669.1			−649.4		
AIC	1,354.3			1,316.3		

QB holds on to the ball the higher the chance of an interception. Seems intuitive. However, is it really the longer release time that is correlated with the probability of an interception?[6] We add an additional covariate in our model, namely, the (target) air yards, and the results are presented again in Table 27-3 (second column). As we see when we add this variable, the release time is not statistically correlated with the probability of interception anymore, while the coefficient itself is closer to 0. Furthermore, the only

6. Note here that simple regression models built with observational data **cannot** reveal causal relationships, and hence, we will talk about correlations. However, in our case there is clearly no possibility for reverse causality since the events are ordered in time, i.e., an interception thrown cannot cause a longer release time.

variable that is statistically correlated with the probability of an interception is the (target) air yards. So essentially the correlation that the first model revealed between release time and the probability of interception is driven by the fact that the air yards for passes with larger release time are higher. Essentially, longer passes are more susceptible to interceptions. It might also be worth noting here that Lisk, in the same article mentioned above, found that the interception rate is one of the two least consistent stats when a QB changes team.

Variables like the air yards in our case are called **mediators** (or **mediation variables**). A mediation variable allows us to better understand the correlation between two variables X and Y (or the observed effect of an independent variable on a dependent variable in experimental research). Mediation variables is one (of the few) tools through which we can get a mechanism's underlying correlations. Formally, let us assume that the variables X, M and Y are the independent, mediation, and dependent variables, respectively. If two variables, X and Y, are correlated due to the mediator M, then X should predict M and then M should predict Y. So, with the set of linear regression equations

$$Y = A_0 + A_1 X + \varepsilon_1 \tag{1}$$

$$M = B_0 + B_1 X + \varepsilon_2 \tag{2}$$

$$Y = C_0 + C_1 X + C_2 M + \varepsilon_3 \tag{3}$$

what we are interested in is to see what happens to coefficient C_1 in equation (3). If this coefficient drops close to zero (compared to coefficient A_1 in equation (1)), then M is a mediation variable that explains the correlation observed in equation (1). Otherwise, there is no mediation. If the coefficient drops all the way to zero (i.e., C_1 is either exactly 0 or not statistically significant anymore), then this is *full mediation*. If the coefficient drops but it is still different than zero (i.e., C_1 is statistically significant), then there is *partial mediation*, i.e., variable M accounts for some of the variance. Figure 27-4 presents relationships between the variables. There are statistical tests

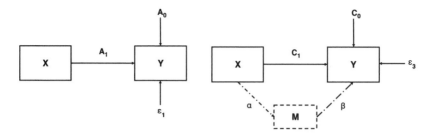

FIGURE 27.4 Mediation Variables.

to examine the presence of mediation. In particular, the Sobel test[7] examines the following hypothesis test:

$$H_0: \alpha \cdot \beta = 0$$
$$H_1: \alpha \cdot \beta \neq 0$$

Hence, if there is strong mediation from variable M, the Sobel test will reject the null hypothesis, which claims that the indirect path's effect is 0.

SCREEN AND OTHER PASSES BEHIND THE LINE OF SCRIMMAGE

Many (offensive) coaches refer to screen passes—and in general designed passes behind the line of scrimmage—as an extension of the run game. Tracking data allows us to identify the passes for which the target was behind the line of scrimmage and compare those passes with the rushing game. Figure 27-5 presents the distribution of the point value added per play for screen passes and rushing attempts, while Figure 27-6 presents the average point value added for screens and rushing attempts. For comparison, we also present the average point value added from passes that travel beyond the line of scrim-

7. Michael E. Sobel, "Asymptotic confidence intervals for indirect effects in structural equation models," *Sociological Methodology*, 13, 1982, 290–312.

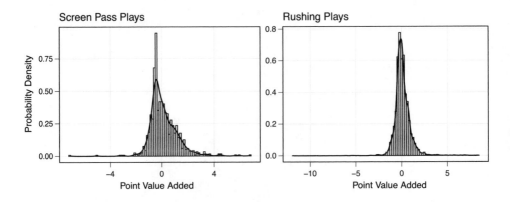

FIGURE 27.5   Play Value Distribution for Screen Passes and Rushing Plays.

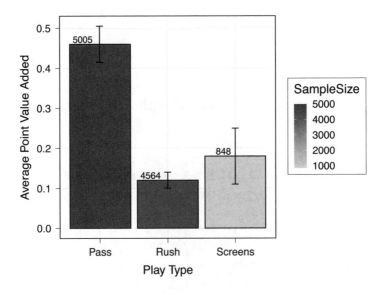

FIGURE 27.6   Point Value Added per Play Type.

mage. As we can see, the distribution for the screen passes is wider, and the corresponding variance larger, as compared to rush plays. However, their average point value added is statistically the same. Moreover, both types of plays are less efficient compared to passes beyond the line of scrimmage. The bottom line is that screen passes (more generally passes behind the line of scrimmage) indeed may be viewed as extensions of the rushing game. In fact, they are closer—in efficiency—to rushing plays than passing plays beyond the line of scrimmage. Maybe the NFL should start treating these passes as rushes and maybe incomplete passes behind the line of scrimmage should be treated as fumbles (similar to lateral passes).

OFFENSIVE LINE AND QB PROTECTION

Player tracking data allows us also to start evaluating offensive lines, a part of the game that is very important (both for passing and rushing), but there are few (if any) box score metrics to evaluate offensive lines. With information about the location of the offensive line, we can start evaluating the protection it offers to the quarterback in a passing play, as well as the gap it opens in the rushing game. For example, how much "clean" space does the offensive line provide to the quarterback during a pass play? In order to examine this we can rely on concepts from computational geometry, such as the convex hull of a set of points.

Before delving into the details of the offensive line's convex hull, let's first define the notion of a convex set. A set of points $\mathcal{L}$ is said to be convex if and only if the straight line that connects **any** pair of points within $\mathcal{L}$ is contained within the set itself. If there is a pair of points within $\mathcal{L}$ for which the straight line connecting them lays partly outside the set, then $\mathcal{L}$ is not convex. For example, the left set in Figure 27-7a is a convex set, while the one on the right is not. Now given a set of points S on the plane, the convex hull of S is the smallest convex set that contains all the points in S. For example, for the set of points shown in the left of Figure 27-7b, its convex hull is the rightmost polygon. The first polygon shown is not convex, while the circle is convex. Nevertheless, it is not the minimum one.

Now that we have a basic understanding of convex hulls, we can use them to quantify the space defined by the offensive line and the quarterback during a pass play. Figure 27-8 visualizes how we can calculate this convex hull (shaded polygon). In the play visualized, there is a running back in the back field. He is not part of the

A

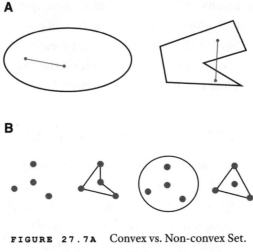

B

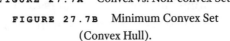

FIGURE 27.7A   Convex vs. Non-convex Set.

FIGURE 27.7B   Minimum Convex Set
(Convex Hull).

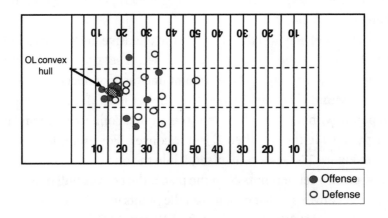

FIGURE 27.8   The Offensive Line's Convex Hull.

convex hull, even though his assignment might be to block during the play. However, we are not able to know this (a team will know the assignments), and hence, we just focus on the offensive line (the same can be true for tight ends with blocking assignments during a play).

TABLE 27.4

Pass Completion and OL Convex Hull

(sig: *p < 0.1, **p < 0.05, ***p < 0.001)

Variable	Pass Air Yards		
	coef	*SE*	*sig*
Constant	0.8	0.12	***
Target Depth	− 0.1	.004	***
OL Convex Hull Area	0.03	.009	***
Observations	4,344		
Log Likelihood	− 2,677.1		
AIC	5,340.1		

We can now begin examining whether the area of this OL's convex hull has any correlation with the passing game efficiency. For example, we build a logistic regression model for the probability of completing a pass based on the area of this polygon. This essentially captures how *comfortable* the QB feels in the pocket. However, during a play the area of this convex hull changes, as the pass rushers push the offensive line. Which area should we use? Here we choose to use the area of the convex hull at the moment of the pass release. We also remove from the data cases, where the QB is on the move to avoid the pass rush, which leads to extreme values for the area of the convex hull. Furthermore, we also include a covariate for the depth of the target since it is a variable that certainly impacts the completion probability. Table 27-4 presents the results. As we can see, the depth target is negatively correlated with the completion probability, while a larger OL convex hull area at the time of the release is positively correlated with the pass completion probability. Figure 27-9 presents the completion probability as a function of the convex hull area for different target depths. As we can see, more space (*cleaner* pocket) is associated with higher completion probability.

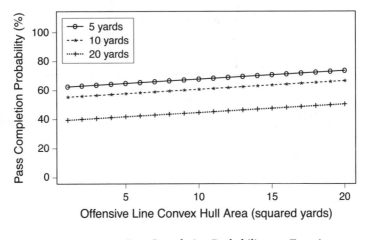

FIGURE 27.9 Pass Completion Probability as a Function of OL's Convex Hull.

Another way to look into the offensive line performance is through survival analysis, i.e., how long until the OL *collapses*. These models are also appropriate since they can deal with *censored* data, i.e., in some data points we do not observe the collapse of the OL since the quarterback threw the pass before that happened. For example, an accelerated failure time model can model the effect of various covariates on the acceleration or deceleration of the OL *lifetime*. In fact, ESPN has used similar ideas to develop an OL evaluation metric, named Pass Block Win Rate.[8]

PASS COMPLETION PERCENTAGE ABOVE EXPECTATION (CPAE)

When examined without context, pass completion percentages can be an inaccurate indicator for the performance of a QB. For example, Steeler (backup) QB Mason Rudolph had a completion percentage of 86% on 28 passes in week's 4 matchup in 2019 against the Bengals.

---

8. https://www.espn.com/nfl/story/_/id/27584726/nfl-pass-blocking-pass
-rushing-rankings-2019-pbwr-prwr-leaderboard

However, six of these passes were touch passes for jet sweeps, another six were behind the line of scrimmage, and only three traveled more than five yards beyond the line of scrimmage. So was this completion percentage that impressive when you put it in this context? Probably not, and tracking data can help put this into context. In particular, the NFL has used tracking data to build a completion probability model for a pass, adjusting for a number of covariates, including target distance, distance of the receiver from the sideline, distance of the closest defender from the QB at the time of release, etc. Estimating the completion probability for each pass a QB throws, the NFL then calculates the average expected completion percentage over all passes and compares it with the actual completion percentage of the QB, thus creating the completion percentage above expectation metric.

However, tracking data are not available to the public and hence, the public cannot rebuild similar models. To solve this problem, Sarah Mallepalle (a CMU student at the time and currently a football analyst for the Ravens) together with Ron Yurko and Sam Ventura of nflscrapR fame, and Kostas Pelechrinis, developed next-gen-scraPy[9] a computer vision software written in python that processes the pass charts from NFL NextGen Stats and obtains the raw coordinates of the passes (as well as their outcome). These raw pass locations are then used in a Bayesian framework (more about Bayes' theorem later in Chapter 52) to obtain completion percentage surfaces for different QBs and compare them with the league average. The same can be done for team pass defense to estimate the completion percentage allowed above expectation for different defensive units. Table 27-5 presents the results for the 2017 and 2018 seasons for QBs with at least 100 passes.

We also compared the estimated completion percentage above expectation using only the pass locations to the one estimated from

9. S. Mallepalle, R. Yurko, K. Pelechrinis, and S. L. Ventura, "Extracting NFL tracking data from images to evaluate quarterbacks and pass defenses," *Journal of Quantitative Analysis in Sports*, 16(2), 2019, 95–120.

TABLE 27.5

Completion Percentage above Expectation for
2017 and 2018 Seasons

QB	CPAE17	npasses_2017	CPAE18	npasses_2018
Drew Brees	4.21	439	6.14	473
Ryan Fitzpatrick	0.47	112	3.42	157
Nick Foles	−3.64	152	3.42	229
Russell Wilson	5.77	309	3.39	295
Matthew Ryan	2.77	524	3.22	552
Carson Wentz	0.07	333	3.08	313
Derek Carr	0.27	300	2.96	429
Kirk Cousins	−0.15	394	2.53	467
Derrick Watson	2.53	110	2.43	492
Cameron Newton	−1.18	352	2.14	392
Marcus Mariota	0.95	495	1.75	275
Jared Goff	−0.41	428	1.7	553
Ben Roethlisberger	2.46	394	1.29	518
Patrick Mahomes			1.27	445
Philip Rivers	0.34	416	1.15	560
Rayne Prescott	−0.14	408	1.11	434
Jameis Winston	2.55	268	0.44	295
Andrew Luck			0.33	559
Mitchell Trubisky	−1.36	262	0.27	323
Ryan Tannehill			0.08	191
Brock Osweiler			0.06	163
John Stafford	3.14	384	−0.04	480
Aaron Rodgers			−0.15	573
Baker Mayfield			−0.38	269
Alexander Smith	4.31	418	−0.88	254
Tom Brady	3.23	524	−0.89	519
Elisha Manning	−2.22	369	−1	536
Sam Darnold			−1.05	289

TABLE 27.5 (*continued*)

QB	CPAE17	npasses_2017	CPAE18	npasses_2018
Casey Keenum	0.33	382	− 1.38	509
Joseph Flacco	− 0.23	438	− 1.67	367
Nicholas Mullens			− 1.87	118
Andrew Dalton	− 1.25	307	− 1.89	195
Lamar Jackson			− 2.07	112
Joshua Allen			− 3.44	237
Casey Beathard	− 4.94	185	− 4.37	168
Joshua Rosen			− 4.54	260
Jeffrey Driskel			− 4.83	110
Robby Bortles	− 1.9	399	− 5.04	336

NextGen Stats using several different attributes from the tracking data. The correlation is pretty high, $\rho = .83$, which means that the majority of the information for the completion percentage is included in the actual pass locations (target distance). Furthermore, there is a moderate correlation ($\rho = .41$, p-value <0.05) between the CPAE values of a player across the two different seasons examined.

In this chapter, we have only scratched the surface of what is possible with the player tracking data in the NFL. We expect to see a lot more research coming out using these types of data and gaining new insights for the game. For example, deep learning models have been developed to estimate the distribution of point values play in real time[10], which can further be used to identify specific parts of the play that were crucial. Tracking data can also lead to an analyst looking at traditional questions from a different lens. For example, Mike

10. R., Yurko, F., Matano, L., Richardson, N., Granered, T., Pospisil, K., Pelechrinis, and S. Ventura, "Going deep: Models for continuous-time within-play valuation of game outcomes in American football with tracking data." *Journal of Quantitative Analysis in Sports,* 2020, doi:10.1515/jqas-2019-0056

Lopez analyzed the actual yards needed" for a first down in fourth and short (1–2 yards) situations and identified that not all fourth downs with the same yardage to go in the play-by-play data are created equal. Fourth and one have anywhere between inches and 1.8 yards to cover. Insights from tracking data will be tied not only to winning games, but also to making the game safer for athletes, as mentioned earlier. For example, is there any correlation between punt formations and concussion incidents? Can we use these data to gain a better understanding of what happens on the field when there is a concussion? What actionable insights can we obtain? A new era in football analytics has certainly started.

11. M. Lopez, "Analyzing the National Football League is challenging, but player tracking data is here to help," 2019 (preprint).

PART III
BASKETBALL

CHAPTER 28

BASKETBALL STATISTICS 101

The Four Factor Model

Like enthusiasts of other sports, basketball fans and professionals are interested in what factors correlate with winning. Obviously, scoring more points than the opponent (as in every sport) helps, but what other aspects correlate with winning. Dean Oliver in his seminal book *Basketball on Paper* (2004) identified the so-called "four factors" that strongly correlate with winning. These metrics capture the different aspects of basketball: scoring, crashing, protecting, and attacking. In reality, these four factors are eight metrics, since they are defined for both the offense and the defense of a team. Let's take a look at them in what follows.

SCORING: EFFECTIVE FIELD GOAL PERCENTAGE

Many coaches, players, fans, and media commentators evaluate shooting performance by looking at field goal percentage. For example, suppose in a Mavericks–Knicks game the Mavericks make 45 out of 100 field goals and shoot 45%. The Knicks make 50 out of

100 field goals and shoot 50%. At first glance the Knicks shot better than the Mavericks. Suppose, however, the Mavericks shot 15 for 20 on three pointers and the Knicks shot one for five on three pointers. Then on the same number of shots the Mavericks scored 105 points and the Knicks scored 101 points. This indicates that the Mavericks actually shot better—i.e., used their shots more efficiently—from the field than the Knicks. To capture this, we can use the **effective field goal (EFG) percentage** defined as follows:

$$EFG = \frac{\text{All field goals made} + .5 * (\text{Three-point field goals made})}{\text{All field goal attempts}}$$

In essence, EFG gives 50% more credit for making a three pointer because a three pointer is worth 50% more points than a two-point field goal. For our example, the Mavericks EFG = $(45 + .5 * 15)/100 = 52.5\%$ and Knicks EFG = $(50 + .5 * 1)/100 = 50.5\%$. This example shows that EFG better captures the quality of a team's shooting than the traditional metric of field goal percentage.

CRASHING: REBOUNDING PERCENTAGE

Raw rebounds for a team or player can be misleading. Imagine a team with no offensive rebounds. It either is very bad at rebounding or is shooting perfectly from the field, and hence there are no rebounds to grab to begin with. Hence, what really matters is the percentage of available rebounds a team gets. Based on whether the team is on offense or defense we have the offensive rebounding percentage and the defensive rebounding percentage, respectively.

PROTECTING: TURNOVER PERCENTAGE

How often does the team turn the ball over in a game? Again, raw numbers might be misleading. Are 15 turnovers a lot? Depends on over how many possessions or plays these turnovers were committed. Hence, the third factor is the turnover percentage, i.e., the number of turnovers a team makes per 100 plays.

ATTACKING: FREE THROW RATE

There is a reason the free throw line is called the "charity line." Theoretically, it is the easiest point to score. How do you get to the free throw line? By attacking and getting fouled. Hence, the fourth factor that Dean Oliver examined was the fraction of free throws over the total number of field goal attempts.

Figure 28-1 contains the values of the four factors for teams during the 2015–2016 NBA season. Figure 28-2 gives the ranking of each team for each of the four factors as well as the number of games won by each team.

For example, the Warriors had an EFG of 56.3% and held their opponents to an EFG of 47.9%. They also committed 14.9 turnovers per 100 possessions and *forced* 13.8 turnovers per 100 possessions

	A	B	C	D	E	F	G	H	I	J
1		Mean	0.502467	0.502467	0.143567	0.1436	0.2383	0.2380667	0.2092667	0.2092
2	Team	Wins	Offensive Shooting	Defensive Shooting	Offensive Turnovers	Defensive Turnovers	Offensive Rebounding	Defensive Rebounding	Offensive Free Throws	Defensive Free Throws
3	Atlanta Hawks	48	0.516	0.48	0.149	0.159	0.191	0.254	0.185	0.193
4	Brooklyn Nets	21	0.491	0.534	0.15	0.142	0.241	0.243	0.186	0.176
5	Boston Celtics	48	0.488	0.487	0.136	0.16	0.251	0.255	0.208	0.231
6	Charlotte Hornets	48	0.502	0.496	0.124	0.137	0.2	0.202	0.222	0.191
7	Chicago Bulls	42	0.487	0.485	0.138	0.117	0.245	0.251	0.189	0.182
8	Cleveland Cavaliers	57	0.524	0.496	0.138	0.137	0.251	0.215	0.194	0.205
9	Dallas Mavericks	42	0.502	0.504	0.129	0.14	0.207	0.238	0.211	0.198
10	Denver Nuggets	33	0.489	0.515	0.148	0.137	0.258	0.228	0.216	0.216
11	Detroit Pistons	44	0.491	0.504	0.136	0.135	0.27	0.207	0.197	0.196
12	Golden State Warriors	73	0.563	0.479	0.149	0.138	0.235	0.24	0.191	0.208
13	Houston Rockets	41	0.516	0.516	0.156	0.161	0.257	0.272	0.244	0.219
14	Indiana Pacers	45	0.497	0.489	0.147	0.156	0.234	0.24	0.205	0.205
15	Los Angeles Clippers	53	0.524	0.48	0.129	0.153	0.202	0.262	0.22	0.222
16	Los Angeles Lakers	17	0.46	0.523	0.136	0.127	0.232	0.253	0.228	0.202
17	Memphis Grizzlies	42	0.477	0.518	0.136	0.165	0.253	0.25	0.231	0.251
18	Miami Heat	48	0.508	0.485	0.141	0.131	0.238	0.222	0.21	0.196
19	Milwaukee Bucks	33	0.499	0.51	0.155	0.156	0.25	0.269	0.206	0.221
20	Minnesota Timberwolves	29	0.498	0.524	0.15	0.148	0.243	0.254	0.263	0.2
21	New Orleans Pelicans	30	0.498	0.523	0.134	0.136	0.212	0.212	0.201	0.225
22	New York Knicks	32	0.483	0.487	0.135	0.115	0.237	0.242	0.205	0.204
23	Oklahoma City Thunder	55	0.524	0.484	0.159	0.129	0.311	0.241	0.228	0.205
24	Orlando Magic	35	0.5	0.513	0.141	0.15	0.231	0.235	0.175	0.215
25	Philadelphia 76ers	10	0.487	0.51	0.159	0.148	0.207	0.26	0.186	0.24
26	Phoenix Suns	23	0.487	0.523	0.169	0.146	0.254	0.23	0.204	0.237
27	Portland Trailblazers	44	0.511	0.503	0.146	0.133	0.259	0.238	0.202	0.225
28	Sacramento Kings	33	0.51	0.521	0.155	0.156	0.239	0.251	0.214	0.202
29	San Antonio Spurs	67	0.526	0.477	0.134	0.153	0.23	0.209	0.197	0.182
30	Toronto Raptors	56	0.504	0.498	0.131	0.139	0.246	0.223	0.255	0.201
31	Utah Jazz	40	0.501	0.495	0.155	0.145	0.259	0.223	0.213	0.21
32	Washington Wizards	41	0.511	0.515	0.142	0.159	0.206	0.223	0.192	0.218

FIGURE 28.1   Four Factors for NBA 2015–2016 Season.

	A	B	C	D	E	F	G	H	I	J
			Offensive	Defensive	Offensive	Defensive	Offensive	Defensive	Offensive Free	Defensive Free
1	Team	Wins	Shooting	Shooting	Turnovers	Turnovers	Rebounding	Rebounding	Throws	Throws
2	Atlanta Hawks	48	6	3	20	4	30	24	29	5
3	Brooklyn Nets	21	24	30	22	16	15	19	27	1
4	Boston Celtics	48	21	8	8	3	9	26	14	27
5	Charlotte Hornets	48	13	12	1	20	29	1	7	4
6	Chicago Bulls	42	25	6	12	29	13	21	26	2
7	Cleveland Cavaliers	57	3	13	13	21	10	5	23	14
8	Dallas Mavericks	42	14	16	2	17	25	13	12	8
9	Denver Nuggets	33	23	21	19	22	5	10	9	20
10	Detroit Pistons	44	22	17	9	24	2	2	21	6
11	Golden State Warriors	73	1	2	21	19	19	15	25	17
12	Houston Rockets	41	7	23	27	2	6	30	3	22
13	Indiana Pacers	45	20	10	18	6	20	16	16	15
14	Los Angeles Clippers	53	4	4	3	9	28	28	8	24
15	Los Angeles Lakers	17	30	26	10	28	21	23	5	11
16	Memphis Grizzlies	42	29	24	11	1	8	20	4	30
17	Miami Heat	48	11	7	14	26	17	6	13	7
18	Milwaukee Bucks	33	17	18	24	7	11	29	15	23
19	Minnesota Timberwolves	29	18	29	23	12	14	25	1	9
20	New Orleans Pelicans	30	19	27	5	23	24	4	20	25
21	New York Knicks	32	28	9	7	30	18	18	17	13
22	Oklahoma City Thunder	55	5	5	28	27	1	17	6	16
23	Orlando Magic	35	16	20	15	11	22	12	30	19
24	Philadelphia 76ers	10	26	19	29	13	26	27	28	29
25	Phoenix Suns	23	27	28	30	14	7	11	18	28
26	Portland Trailblazers	44	8	15	17	25	3	14	19	26
27	Sacramento Kings	33	10	25	25	8	16	22	10	12
28	San Antonio Spurs	67	2	1	6	10	23	3	22	3
29	Toronto Raptors	56	12	14	4	18	12	7	2	10
30	Utah Jazz	40	15	11	26	15	4	8	11	18
31	Washington Wizards	41	9	22	16	5	27	9	24	21

FIGURE 28.2   NBA Team Rankings for Four Factors 2015–2016.

of their opponents. Furthermore, they rebounded 23.5% of their missed shots and 76% of their opponents' missed shots (or equivalently their opponents rebounded 24% of their missed shots). Finally, they made 19.1 FTs per 100 FGAs while their opponents made 20.8 FTs per 100 FGAs. This shows the Warriors bested their opponents only in shooting, but by such a significant margin that this factor alone helps explain their significant success in the 2015–2016 season.

THE FOUR FACTORS ARE VIRTUALLY UNCORRELATED!

The interesting thing about the four factors is that there is little correlation between them (see Chapter 5 for an explanation of correlation). Recall that correlations are always between –1 and +1

with a correlation near +1 for two quantities x and y indicating that
when x is big y tends to be big, and a correlation near –1 indicating
that when x is big y tends to be small. Using the Correlation option
from Excel's Analysis ToolPak we found the correlations shown in
Figure 28-3.

	A	B	C	D	E	F	G	H	I
1		Offensive Shooting	Defensive Shooting	Offensive Turnovers	Defensive Turnovers	Offensive Rebounding	Defensive Rebounding	Offensive FTs	Defensive FTs
2	Offensive Shooting	1							
3	Defensive Shooting	–0.11	1						
4	Offensive Turnovers	–0.27	0.12	1					
5	Defensive Turnovers	–0.1	0.05	–0.02	1				
6	Offensive Rebounding	–0.47	–0.04	0.46	0.05	1			
7	Defensive Rebounding	–0.0006	–0.67	0.003	–0.39	0.06	1		
8	Offensive FTs	–0.25	–0.24	–0.34	–0.05	0.25	0.06	1	
9	Defensive FTs	–0.31	0.04	0.41	0.22	0.44	0.05	0.36	1

FIGURE 28.3 Correlations between the Four Factors.

Notice that most of the correlations are near 0. For example, the
correlation between offensive shooting and defensive turnovers
is –.10. This indicates that if a team is better than average on offensive shooting, our best guess is it will be slightly worse than average
in causing defensive turnovers. Let's examine the three largest (in
absolute value) correlations shown in Figure 28-3.

- There is a –.67 correlation between defensive shooting percentage and defensive rebounding. This means that teams that
  give up a high shooting percentage tend to be poor defensive
  rebounding teams. This is reasonable because if you fail to rebound your opponents' missed shots, they will likely get many
  easy inside shots or dunks on follow-up shots.
- There is a –.47 correlation between offensive shooting and offensive rebounding. This means that good shooting teams tend
  to be poor offensive rebounding teams. The 2015–2016 Spurs
  (second in offensive shooting and 23rd in offensive rebounding)
  are an illustration of this phenomena. Perhaps teams loaded
  with good shooters do not choose to crash the offensive boards
  as frequently, preferring to turn to defense.

- There is a .46 correlation between offensive rebounding and offensive turnover percentages. This means teams that are good at offensive rebounding also tend to turn the ball over a lot. This could be explained from the fact that good offensive rebounders tend to also be poor ball handlers.[1]

DIFFERENT PATHS TO TEAM SUCCESS (OR FAILURE!)

We can use Figure 28-2 to quickly zero in on the keys to success (or failure) for an NBA team. For example,

- The 2016 champion Cavaliers were great overall shooters, were average on holding opponents to poor shooting, had good offensive rebounding, and had great defensive rebounding.
- The 2016 Spurs were successful because they had great shooting, had the best defense on opposing shooters, forced many turnovers, had excellent defensive rebounding, and performed well on defensive free throws.
- The 2016 76ers were the worst team in the league. They performed poorly in every category.
- The 2016 Thunder performed well with 55 wins in the regular season. Its season might have been even more successful if it committed fewer turnovers and forced more turnovers.

HOW IMPORTANT ARE THE FOUR FACTORS?

Can we estimate the relative importance of the four factors? Recall that in Chapter 18 we used regression to predict NFL team performance from measures of passing and rushing efficiency and turnover frequency. In a similar fashion we can use regression to evaluate the

---

1. Here it would be interesting to examine the chance of an offensive rebound leading directly to a turnover.

importance of the four factors. We ran a regression using the data in Figure 28-1 to predict a team's number of wins from the following four independent variables:

- EFG–OEFG
- TPP–DTPP
- ORP–DRP
- FTR–OFTR.

The results of the regression are shown in Figure 28-4.

	A	B	C	D	E	F
1	Regression Statistics					
2	R	0.91696				
3	R-square	0.84081				
4	Adjusted R-square	0.81534				
5	Standard Error	5.96607				
6	Observations	30				
7						
8	ANOVA					
9		d.f.	SS	MS	F	p-level
10	Regression	4	4700.15004	1175.03751	33.01224	1.21589E–09
11	Residual	25	889.84996	35.594		
12	Total	29	5590			
13						
14		Coefficient	Standard Error	t Stat	p-level	
15	Intercept	59.04678	20.30285	2.9083	0.00752	
16	Shooting Dev.	383.31318	36.11647	10.61325	9.518E-11	
17	Turnover Dev.	–244.36862	80.41288	–3.03892	0.0055	
18	Rebound Dev.	34.49081	38.71857	0.89081	0.38153	
19	Free Throw Dev.	84.26961	42.89746	1.96444	0.06069	

FIGURE 28.4   Four Factor Regression.

Therefore, we predict

- Games Won = 59.05 + 383.31(EFG–OEFG) – 244.36(TPP–DTPP) + 34.49(ORP–DRP) + 84.27(FTR–OFTR)

We found that these four independent variables explain 84% of the variation in the number of games won, with the standard error being 5.97 wins.

To measure the impact of our four factors on wins we can look at the correlations between each of the four factors and wins.

- EFG–OEFG has a .87 correlation with wins and by itself explains 76% of variation in wins.
- TPP–DTPP has a –.28 correlation with wins and by itself explains 8% of variation in wins.
- ORP–DRP has a –.11 correlation with wins and by itself explains 4% of variation in wins.
- FTR–OFTR has a .19 correlation with wins and by itself explains 4% of variation in wins.

This analysis indicates that a team's differential on shooting percentage is by far the most important factor in NBA team success.

The relative importance of the four factors is summarized as follows:

- A .01 improvement in EFG–OEFG is worth 3.8 wins. That is, any of the following improvements

1. Improve our EFG 1% (say from 47% to 48%),
2. Reduce our opponent's EFG by 1%, and
3. Improve our EFG by .5% and cut our opponent's EFG by .5%

could on average translate to 3.8 additional wins.

- A .01 improvement in TPP–DPPP is worth 2.4 additional wins. Thus, either of the following improvements

1. Committing one less turnover per 100 possessions or
2. Committing one less turnover per 200 possessions and forcing one more turnover per 200 defensive possessions

could translate to 2.4 more wins.

- An increase of .01 in ORP–DRP would lead on average to 0.3 more wins per season. Thus, any of the following combinations would be expected to translate to 0.3 more wins.

1. One more offensive rebound per 100 missed shots.
2. One more defensive rebound per 100 shots missed by an opponent.
3. One more offensive rebound per 200 missed shots and one more offensive rebound per 200 shots missed by an opponent.

- An increase of .01 in FTR–OFTR would be expected to lead to .84 wins. Therefore, any of the following three combinations *could* lead to .84 additional wins.

1. One more FT made per 100 FG attempts.
2. One less FT given up per 100 FG attempts by an opponent.
3. One more FT made per 200 FG attempts and one less FT given up per 200 FG attempts.

In summary, Dean Oliver's decomposition of a team's ability into four factors provides a quick and effective way to diagnose a team's strengths and weaknesses. We have showcased the four factors by regressing a team's total wins at the end of the season to its season-long four factors. This type of model is what we call a descriptive model, that is, we try to identify factors that correlate and can *describe* the dependent variable observed (in this case the total wins for a team). We could use the same four factors of a team to build predictive models. For instance, during the season, we track the running value of the four factors for a team, and we can use them to predict the final total wins of a team or the chance of it winning an upcoming matchup against another team. These models can be built and back-tested using historical data from past seasons. The error of these models is expected to be higher than that of the descriptive model we built in this chapter, since in order to make inferences, say, for an upcoming matchup, we will not know the exact values of the independent variables (i.e., the four factors that the teams will post in the matchup), but, rather, we will have to use some estimate of them (typically their running values). Simply put, we will have an error from the model itself, but also an error due to uncertain estimates in the input of the model.

CHAPTER 29

LINEAR WEIGHTS FOR EVALUATING NBA PLAYERS

Recall that in Chapter 3 we discussed the use of linear weights to evaluate Major League Baseball hitters. We found that by determining appropriate weights for singles, walks, doubles, triples, home runs, outs, stolen bases, and caught stealings we can do a pretty good job of estimating the runs created by a hitter.

Given the wealth of information in an NBA box score, many people have tried to come up with linear weights formulas that multiply each box score statistic by a weight and equate the weighted sum of a player's statistics as a measure of the player's ability. In this chapter we will discuss several linear weighting schemes used to rate NBA players:

- The NBA Efficiency Statistic
- John Hollinger's PER and game score ratings
- Berri, Schmidt, and Brook's (BSB) win scores (explained in *Wages of Wins* (2007)).

NBA EFFICIENCY RATING

Let's begin by discussing the NBA efficiency rating which is computed as efficiency per game = (points per game) + (rebounds per game) + (assists per game) + (steals per game) − (turnovers per game) − (missed FGs per game) − (missed FTs per game).

This simplistic system essentially says that all good statistics are worth +1 and all bad statistics are worth −1. This does not make sense. For example, a player who shoots 26.67% on three-point field goals (which is a poor shooting percentage, almost 10% below league average) would raise his efficiency by taking more three pointers. For example, suppose a player shot five for 18 on three pointers. He scored 15 points and missed 13 shots. This player's three-point field goal attempts would yield $15 - 13 = 2$ efficiency points. If he shot 10 for 36, his three-point field goal attempts would yield $30 - 26 = 4$ efficiency points. Any player who shot this badly would be told not to shoot! Similarly consider a player who shoots 36.4% (four for 11) on two pointers. If he takes 11 shots, he scores eight points and misses seven shots. These shots add $8 - 7 = 1$ point to his efficiency rating. If he shot eight for 22 (which is almost 10% below league average), his shots would add two points to his efficiency rating[1].

The 2015–2016 leaders in efficiency rating are listed in Table 29-1.

In Chapter 30 we will see that certain players on this list are vastly overrated by the NBA efficiency statistic. Other players are vastly underrated by the NBA efficiency statistic.

---

1. The reader should be able to show that a player who shoots above 25% on three pointers or above 33.33% on two pointers would improve his efficiency rating by taking more shots. For example, if a player makes x three pointers and misses $(100 - x)$ three pointers, he would contribute $3x - (100 - x)$ to the efficiency rating. This is positive as long as $4x - 100 > 0$ or $x > 25$. Thus a player shooting > 25% on three pointers increases his efficiency by taking more three pointers. This is a disquieting property of the efficiency rating.

TABLE 29.1

NBA Efficiency Leaders 2015–2016 Season Ranked
Efficiency/Game

Player, Team	GP	MPG	EFF
Stephen Curry, GSW	79	34.2	31.5
Kevin Durant, OKC	72	35.8	28.2
Russell Westbrook, OKC	80	34.4	27.6
LeBron James, CLE	76	35.6	27.5
Chris Paul, LAC	74	32.7	26.2
Kawhi Leonard, SAS	72	33.1	26.0
Hassan Whiteside, MIA	73	29.1	25.7
James Harden, HOU	82	38.1	25.3
Anthony Davis, NOP	61	35.5	25.0
Enes Kanter, OKC	82	21.0	24.0
DeMarcus Cousins, SAC	65	34.6	23.6
Jonas Valanciunas, TOR	60	26.0	22.6
Karl-Anthony Towns, MIN	82	32.0	22.5
LaMarcus Aldridge, SAS	74	30.6	22.4
Damian Lillard, POR	75	35.7	22.2
Kyle Lowry, TOR	77	37.0	22.2
Greg Monroe, MIL	79	29.3	21.8
Brook Lopez, BRK	73	33.7	21.7
Pau Gasol, CHI	72	31.8	21.7
Derrick Favors, UTA	62	32.0	21.6

PER RATING SYSTEM

Now let's turn our attention to John Hollinger's well-known PER (Player Efficiency) and Game Score ratings. Hollinger's ratings can be found on ESPN.com. An average NBA player has a PER score of 15. Hollinger's PER rating formula is complex, but we used data from the great site Basketball-Reference.com to reverse-engineer the formula. For all players in the 2018–2019 season who played at least 1,000 minutes we ran a regression (see file PER.xlsx) to predict the player's PER rating from the following independent variables (all measured per minute):

- Field Goals
- Field Goal Attempts
- Free Throws
- Free Throw Attempts
- Offensive Rebounds

- Defensive Rebounds
- Assists
- Blocks
- Turnovers
- Steals
- Personal Fouls
- Points.

The following equation yielded an R^2 of 0.99 and standard error or 0.47. This means that for 95% of all players the formula predicts a player's PER rating within 0.94.

$$\text{Predicted PER} = 0.68 + 26.77\text{FG} - 35.58\text{FGA} + 26.32\text{FT} - 25.47\text{FTA} + 42.74\text{OR} + 11.45\text{DR} + 29.7\text{AST} + 47.3\text{STEAL} + 38\text{BLK} - 47.9\text{TO} - 18.8\text{PF} + 41.1\text{PTS}.$$

The sign of each coefficient is reasonable, but the coefficients involving shooting are unreasonable. Let's look at a player who shoots 33 for 100 in 300 minutes on two-pointers. He is a horrible shooter, yet his shooting contributes $(1/300) * (26.77 * 33 - 100 * (35.58) + 41.1 * 66) = 2.3$. Thus being a horrible two-point shooter increases your PER!

WIN SCORES AND WINS PRODUCED

Now we turn to BSB's win scores and wins produced. Although the win score formula published on its blog[2] is only an approximation of its more complex formula, BSB states that the following formula closely approximates its more complicated method for ranking players.

$$\text{Player Win Score} = \text{Points} + \text{Rebounds} + \text{Steals} + \tfrac{1}{2}\text{Assists} + \tfrac{1}{2}\text{Blocked Shots} - \text{Field Goal Attempts} - \text{Turnovers} - \tfrac{1}{2}\text{Free Throw Attempts} - \tfrac{1}{2}\text{Personal Fouls}.$$

2. http://dberri.wordpress.com/2006/05/21/simple-models-of-player-performance/

Unlike the NBA efficiency metric or Hollinger's PER rating, the linear weights in the player win score metric pass the "smell test." For example, to raise his rating by shooting more, a player needs to shoot over 50% on two-point field goals or over 33.33% on three-point field goals. This is very reasonable. It also seems reasonable to give equal weight to a turnover and rebound since a rebound gets you a possession and a turnover gives up possession. By a complex method, BSB converts win scores into wins produced. The nice thing about wins produced is that the sum of wins produced by all of a team's players will come very close to the team's total number of wins.

We note, however, that the fact that for most teams the wins produced by each player on a team add up to near the team's total wins does not imply that the team's "wins" are accurately partitioned among the team's players. The main reason for this is the fact that the components of the win score formula do not include information with regard to the quality and impact of a player's defense, apart from blocks and steals. Basketball is half offense and half defense, so the shortage of defensive measures in the box score causes BSB to shortchange great defenders. This means that not enough of a player's defensive ability can be picked up by BSB.

For most players (other than some defensive stalwarts), we believe that BSB win scores do an excellent job of determining the weighting of box score statistics that is best for player evaluation. The only problem is that BSB only rates players based on **the statistics tabulated in a box score.** During at least 80% of a game **nothing is occurring** that is tabulated in a box score. For example, the box score does not give credit for the following:

- Taking a charge.
- Deflecting a pass.
- Boxing out so my teammate gets the rebound.
- Passing before the pass that earns an assist, or passing to a player who misses an open look.

- Helping out on defense when my teammate is beaten by a quick guard.
- Setting a screen that leads to an open Joe Harris three-pointer!

The NBA is currently providing information about several *advanced* statistics not present in the box score that can possibly be integrated on similar metrics. For example, the NBA tracks deflections, charges drawn, screen assists, and several other nontraditional metrics that quantify events that help the team, but are not measured in the box score. In the next chapter we turn our attention to rating players by adjusted +/−. Versions of this approach have been used by the Dallas Mavericks to rate NBA players since the 2000–2001 season and variations of adjusted +/− are currently used by virtually all NBA teams. Adjusted +/− takes the view that a good player helps his team. We do not care about a player's box score contribution; we simply care about how the player's team does when he is on and off the court. In the next chapter we give a thorough discussion of adjusted +/−.

CHAPTER 30

ADJUSTED +/− PLAYER RATINGS

Basketball is a team game. The definition of a great player should be somebody who makes his team better, not a player who scores 40 points per game. Terry Pluto, in his excellent book *Tall Tales* (1992), talks about what defines a great player. The late, great Celtics coach Red Auerbach said that whenever the Celtics practiced, the late K. C. Jones's team always won. This must mean he was a good or great player. Yet during his peak years his PER rating was around 10, indicating he was a poor player. It must have been that K. C. did some important things that did not show up in the box score.

PURE +/− RATINGS AND THEIR FLAWS

The first statistic that tied a player rating to team performance was hockey's +/− statistic (hockey analytics are discussed in Chapter 40). A player's +/− statistic is simply the number of goals by which a player's team outscores its opponents when the player is on the ice (power plays, when one team has more players on the ice, are excluded from the calculation). Since 1968 the NHL has kept track of +/− for each player (more about hockey analytics in Chapter 40). The highest recorded +/− was Bobby Orr, who during the 1970–1971 season earned a +124 goal +/−.

For an NBA (WNBA) player his (her) pure +/− per game was thought to be a valid measure of how he (she) helped his (her) team. Raw +/− is now shown in NBA box scores and is **updated while the game is being played.** The problem with pure +/− statistics is that a player's pure +/− statistic depends on the quality of the players you play with and against. Suppose Player A had a pure +/− of 0 and played for the 2016–2017 Brooklyn Nets, the league's worst team. Suppose Player B also had a pure +/− of 0 and played for the 2016–2017 NBA champion Golden State Warriors. It should be clear that Player A is the much better player. Player A made a terrible team average. Player B made a great team average. To prove that unadjusted (or pure) +/− statistics are not that useful Figure 30-1 shows the players (who played at least 20 minutes a game) with the 10 best pure +/− during the 2016–2017 season.

	F	G	H	I
7	Rank	Player	Team	Plus Minus / 48 mins
8	1	Stephen Curry	GS	18.6
9	2	Draymond Green	GS	16.4
10	3	Kevin Durant	GS	16.4
11	4	Klay Thompson	GS	14.7
12	5	Chris Paul	LAC	12.9
13	6	Andre Iguodala	GS	12.6
14	7	Blake Griffin	LAC	9.9
15	8	David Lee	SAS	9.9
16	9	Kevin Love	CLE	9.4
17	10	LeBron James	CLE	9.2

FIGURE 30.1   Pure +/− 2016–2017 NBA (Includes Playoff Games).

All these players played for teams with great records (Warriors, Spurs, Cavs, and Clippers). Even David Lee's agent would not say he was among the 10 best players in the league. Lee had a great +/− ratings because he was often on the court with great players like Kawhi Leonard and Manu Ginobili. By the way, the player with the best raw +/− who played for a losing team was Nikola Jokic, with a raw +/− of 7.00. Warrior center Zaza Pachulia has a raw +/− of 16.25, but nobody would claim he was a better player than Jokic.

ADJUSTED +/−: A BETTER APPROACH

The trick to making sense of pure +/− is to **adjust each player's rating based on the ability of the players he is on the court with and the players he plays against.** The data we use consists of (by the end of the season) over 38,000 rows of play-by-play data. Each row represents a segment of time (stint or shift) where the players on the court remain unchanged. For example, a sample stint might look like this:

GOLDEN STATE HOME AGAINST CLEVELAND

Curry, Durant, Thompson, Green, and Pachulia on the court for Golden State

LeBron, Love, Kyrie, Thompson, J. R. Smith on the court for Cleveland

Length of time segment: 3 minutes

Score during time segment: Golden State ahead 9–7.

We adjust the score during each time segment based on a home edge of 3.2 points per 48 minutes (given that the home field advantage has been in the decline, in Chapter 46 we will see how one can get a better estimate for its value). Thus our adjusted score would be Golden State ahead by $9 − (3/48) * .5(3.2)$ to $7 + (3/48) * .5(3.2)$ or Golden State ahead by 8.9 to 7.1.

We have a variable for each player that represents his adjusted +/− rating per 48 minutes. An average NBA player will have an adjusted +/− of 0. Thus an adjusted +/− rating of 5 means that if the player replaced an average NBA player for 48 minutes, his team would improve by an average of five points per game. For any time segment we predict the home team's margin of victory per minute to be $(3.2/48) + ((\text{sum of home team player ratings for players on court}) − (\text{sum of away team player ratings for players on court}))/48$.

The 3.2/48 represents the average number of points per minute by which the home team defeats the away team. Wayne Winston's colleague, Jeff Sagarin of *USA Today* fame, wrote a program (called WINVAL, short for winning value) to solve for the set of player ratings that best match the scores of the game. Essentially, we use a trial set of ratings to predict the score margin for each time segment and adjust the ratings until the set of ratings makes the most accurate predictions possible (aggregated over our approximately 38,000 rows of data!).

To illustrate how adjusted +/− differs from pure +/− let's look at a simplified example. Team 1 has played 20 full length (48-minute) games against Team 2. Team 1 consists of players 1–9 and Team 2 players 10–18. The results of the games (from the standpoint of Team 1) are shown in Figure 30-2. See file Newadjustedplusminusex.xls.

	G	H	I	J	K	L	M	N	O	P	Q
3	Result	P1	P2	P3	P4	P5	P6	P7	P8	P9	P10
4	−13	4	1	7	5	2	15	16	10	17	14
5	19	1	6	2	5	4	11	17	14	15	18
6	−4	1	9	2	8	4	15	14	10	17	13
7	29	1	6	5	3	2	16	17	18	14	11
8	−3	9	7	1	5	6	17	15	12	18	10
9	12	7	2	5	1	4	17	11	15	16	18
10	−5	6	5	8	9	1	13	16	12	15	10
11	−32	4	2	9	5	3	17	12	10	18	15
12	18	8	3	9	1	7	17	16	15	14	11
13	17	1	2	9	6	4	13	16	10	11	18
14	−11	7	3	2	5	6	14	17	15	12	15
15	−14	7	8	4	6	3	18	11	12	17	15
16	29	4	5	9	2	6	11	13	14	17	18
17	17	1	8	4	2	7	13	12	14	17	18
18	−4	6	9	8	7	1	15	12	10	17	14
19	−7	6	3	2	1	8	17	18	16	14	10
20	9	3	2	5	6	7	13	16	14	10	11
21	24	1	7	6	7	4	18	13	18	15	11
22	18	1	2	5	8	6	14	13	12	15	18
23	−24	2	4	3	8	5	11	18	16	17	10

FIGURE 30.2 Sample Data for
Adjusted +/− Calculations.

	A	B	C
3		1.45661E−13	
4	Player	Adjusted +/−	Pure +/−
5	1	12.42497415	8.428571
6	2	2.425071482	4.214286
7	3	−6.575047563	−4
8	4	−10.57506265	2.818182
9	5	−0.575004084	2.333333
10	6	0.424985887	7.769231
11	7	1.425023952	5.363636
12	8	−6.574961831	−0.55556
13	9	5.425007418	2
14	10	16.42495706	6.6
15	11	−13.57506636	−11.9
16	12	0.425010347	4.25
17	13	−9.574964372	−13.125
18	14	−6.574989062	−8.33333
19	15	7.424986222	0.428571
20	16	1.425046331	−4
21	17	−0.574971597	−0.8
22	18	−1.575004236	−7.78571

FIGURE 30.3   Pure and Adjusted +/− for Our Example.

Thus in the first game, players 4, 1, 7, 5, and 2 lost by 13 points to players 15, 16, 10, 17, and 14. The pure +/− and adjusted +/− for this data are shown in Figure 30-3. To see that the adjusted +/− are right for this example, note that the predicted score for each game is the

(Sum of player ratings for Team 1 players in game) −
(Sum of player ratings for Team 2 players in game).   (1)

For each game (1) perfectly predicts that game's score, so these adjusted +/− ratings must be correct. For example, in Game 1,

Team 1 Total ratings = −10.57 + 12.42 + 1.43 − .57 + 2.45 = 5.13
Team 2 Total Ratings = 7.42 + 1.43 + 16.43 − .58 − 6.57 = 18.13

Thus, we would predict Team 1 to win by 5.13 − 18.13 = −13 points and this is exactly what happened.

How did we find the adjusted +/−? We used the Excel Solver to choose each player's adjusted +/− rating to minimize

$$\sum_{i=1}^{14}(\text{Sum of Team 1 player ratings for game i} -$$

Sum of Team 2 player ratings for game i −

Margin of victory for Team 1 for game i$)^2$

(See the chapter appendix for an explanation of the Excel Solver.) This sum is simply the sum of our squared prediction errors over all games. The set of adjusted +/− in Figure 30-2 makes this sum of errors equal to 0. In most cases it is impossible, of course, to get a zero sum of squared errors, but the Solver can easily minimize the sum of squared errors.

Note that Player 4's adjusted +/− is much worse than his pure +/− while for Player 10 his adjusted +/− is much better that his pure +/−. Let's explain the first anomaly. When Player 4 was on court his team won by an average of 2.81 points. Averaging the ability of his teammates, we find he played with players that on average totaled to 8.33 points better than average. Player 4 played against opponents' lineups that averaged 5.06 points worse than average. Thus, ignoring Player 4, we would have predicted Team 1 to win by 8.33 + 5.05 = 13.38 points per game. Since Team 1 won by an average of only 2.81 points we would estimate Player 4's rating as 2.81 − 13.38 = −10.57.

In our simple example, **every game was predicted perfectly by the simple model:**

Points Team 1 wins by = (sum of Team 1 player ratings for players on court) − (sum of Team 2 player ratings on court).

However, we would like to note here a few things. First, these predictions are, as we call them, *in-sample*, i.e., we are evaluating how well the identified ratings *describe* the stints used to learn these data. So these ratings are purely descriptive (i.e., assign credit on what happened during the games played). Second, for real data, there is lots of variability and no set of ratings will come close to matching the margins for our over 38,000 time segments. Still, there is a unique set of ratings that best fits the scores, and that is what is found by the WINVAL program. Finally, in the above definition of adjusted +/– we have expressed the rating as a per 48 minutes measure. Other versions of the metric can further use as *time reference unit* the number of possessions involved in a stint (and hence be expressed in terms of per possession or per 100 possessions). Given the play-by-play data of a stint there are several approaches that one can use to obtain an approximate value of the number of possessions involved. Nylon Calculus had an overview[1] of the different calculations that are used and how little they differ. For example, ESPN uses totals from both teams and estimates the number of possessions as $(FGA + 0.44 * FTA − ORB + TOV)/2$.

2016–2017 ADJUSTED +/– RATINGS

Figure 30-4 shows our 10 best rated players for the 2016–2017 season (among players averaging at least 20 minutes per game).

	F	G	H	I	J	K
7	Rank	Player	Team	Points	Offense	Defense
8	1	LeBron James	CLE	18	11	−7
9	2	Stephen Curry	GS	18	22	4
10	3	Kawhi Leonard	SAS	18	17	−1
11	4	Russell Westbrook	OKC	15	26	11
12	5	Kyle Lowry	TOR	12	6	−6
13	6	Paul George	IND	11	9	−2
14	7	Robert Covington	PHI	11	3	−8
15	8	Chris Paul	LAC	11	5	−6
16	9	Mike Conley	MEM	11	13	2
17	10	Devin Booker	PHX	10	9	−1

FIGURE 30.4 WINVAL Top 10 Players Points Rating 2016–2017.

Note that we have rated each player's offensive ability and defensive ability. More on this in a minute. LeBron James's point rating of

1. https://fansided.com/2015/12/21/nylon-calculus-101-possessions/

18 means that, after adjusting for whom he played with and against, we believe LeBron playing a whole game in place of an average NBA player would improve the team's performance by 18 points per 48 minutes. LeBron has an offense rating of +11. This means that after adjusting for who LeBron played with and against, we believe that replacing an average NBA player by LeBron would result in the team scoring 11 more points per game. A good offense rating can be created by doing many things: scoring points, getting rebounds, throwing good passes, setting screens, reducing turnovers, etc. **We do not know how LeBron creates the points (we leave this to the coaches or to the tracking data** discussed in Chapter 37) **but we can tell you how much he helps the offense.** LeBron has a defense rating of −7 points. While a positive offense rating is good, a negative defense rating is good. Thus when LeBron replaces an average NBA defensive player for a game, his team gives up seven **fewer** points. Again, a good defensive rating can be created by doing many things: blocking shots, stopping the pick and roll, reducing turnovers, causing turnovers, rebounding, etc. Note that for each player, points rating = offense rating − defense rating. The beautiful thing about a WINVAL player Points rating is that it is based **half on offensive ability and half on defensive ability.** Clearly, offense and defense play an equal role in basketball, so this is reasonable. Since NBA box scores calculate many more offensive statistics than defensive statistics, NBA Efficiency, PER, and other ratings based on linear box score weights cannot help but be biased toward great offensive players, as we mentioned in the previous chapter. In an effort to include a larger defensive component, Win Score is adjusted based on team defense statistics. Unfortunately, such an adjustment fails to understand that Green, Pachulia, West, and Iguodala (not Curry and Durant) were the keys to the Warriors defense.

As a byproduct of player ratings, WINVAL rates teams. The beauty of the WINVAL team rating system is that since we know who is on the court, we can incorporate knowledge of strengths of opponents faced (rather than just team strength) to rate teams. **WINVAL player ratings weighted by the number of minutes played**

always average out to the WINVAL team rating. Any valid rating system should have this property.

Note that there are several possible paths to basketball excellence.

- You can (like LeBron, CP3, and Kyle Lowry) be great on both ends of the court.
- You can be an offensive stud but nothing special on the defensive end (like Russell Westbrook and Devin Booker).
- You can be a defensive stud but nothing special on the offensive end. In 2016–2017 Rudy Gobert of the Utah Jazz (not Draymond Green) should have been defensive player of the year. Gobert's offensive rating was −5 and defensive rating was −15! This means that putting Gobert on the court with four average NBA defenders would result in his team's giving up 15 fewer points per game than an average NBA defensive team. By the way, Green's defensive rating was −7.

To convince you of the validity of adjusted +/− as a measure of a player's ability try to name your top five players for the 2000–2009 and 2010–2019 decades. By adjusted +/− our top five players of that decade were Kevin Garnett, Tim Duncan, LeBron James, Dirk Nowitzki, and the late Kobe Bryant. If you disagree with our list, who would you take off the list? For 2010–2019 our five best players were LeBron, Curry, CP3, KD, and Lillard. You might want to replace Lillard or CP3 with James Harden, but Harden was not a top player during his years with OKC.

RIDGE REGRESSION

One of the problems that one can face when calculating adjusted +/− is that there can be cases with severe collinearities, that is, there are specific pairs (or triplets or quadruplets or in the extreme case full lineups) of players who share most of their on-court time. In this case the variance of the coefficients that are calculated through the ordinary least squares is large, and hence the coefficients themselves can be very far from their true value. To deal with this problem, one can add a penalty factor in the objective function of the regression that will essentially shrink the values of the coefficients, and reduce their variance. This technique is called regularization. As we will further discuss in the next chapter, this technique can also improve the predictive performance of a model on new unseen data.

For our previous example (Figure 30-2), we can obtain a regularized solution by solving the following minimization problem:

$$\left[\sum_{i=1}^{14}(\text{Sum of Team 1 player ratings for game i} - \right.$$

$$\text{Sum of Team 2 player ratings for game i} -$$

$$\left.\text{Points team 1 wins game i by})^2\right] + \lambda \sum_{j=1}^{18}(\text{player j rating})^2$$

The only difference is the term $\lambda \sum_{j=1}^{18}(\text{player j rating})^2$, which essentially forces us to shrink the players' ratings if we want to minimize the objective function. The benefit of this is that it can prevent *overfitting* to outliers. For example, if there is a player who appeared in one lineup (of mainly average players), and this lineup performed extremely well, the nonregularized objective function will try to assign a high rating to this player in order to *fit* this lineup's performance. The regularization term will *force* the optimizer to reduce this player's rating. When the regularization term is the sum of the squares of the coefficients (as above), the corresponding regression is called **ridge regression.**

Solving the above optimization (see the worksheet Regularization of the workbook Newadjustedplusminusexample.xlsx) provides the values shown in Figure 30-5 for the regularized adjusted +/− of the players in our fictitious example ($\lambda = 2$).

As we can observe, the extreme coefficients have shrunk. One of the issues with (ridge) regression is the amount of regularization that we should do, i.e., the choice of λ. Small values for λ (e.g., $\lambda \to 0$) will provide us with the OLS solution, while large values for λ will

	A	B
4	Player	Adjusted +/−
5	1	7.453124323
6	2	−0.165542274
7	3	−5.675792386
8	4	−4.659767867
9	5	−0.45937132
10	6	2.991972037
11	7	1.039209585
12	8	−3.890266596
13	9	2.745415332
14	10	11.78720674
15	11	−7.820003539
16	12	3.193179291
17	13	−6.676403858
18	14	−7.256651397
19	15	4.542154588
20	16	−0.530887608
21	17	1.479942183
22	18	−3.061821406

FIGURE 30.5   Regularized Adjusted +/−.

provide us with an intercept-only model. In a practical setting, one will use a validation set to choose the value of λ. More specifically, we should split the dataset into three parts: a training set, a validation set, and a test set. Using the training set, we can learn the adjusted +/− coefficients of the players for different values of λ. Consequently, using the validation set we can choose the value of λ that provides the coefficients with the best performance on the validation set. Finally, if we want to obtain a measure of the predictive performance of the selected model on new (or, as it is called, *out-of-sample*) data—which is always a good idea to do—we can use the test set.

THE GREATNESS OF LEBRON AND DEVIN BOOKER AND THE SAGA OF KEVIN MARTIN

As Red Auerbach said, a great player makes his teammates better. Figure 30-6 shows the greatness of LeBron during the 2016–2017 season.

	BH	BI	BJ	BK	BL	BM	BN	BO	BP	BQ
7		Shumpert	Smith	Thompson	D. Williams	Korver	R. Jefferson	Love	Irving	James
8	James and other in	7 1227m	8 919m	6 1747m	7 256m	-1 544m	10 948m	11 1467m	10 1954m	8 2795m
9	James out and other in	-9 691m	-15 254m	-9 582m	-16 216m	-6 314m	-5 649m	1 417m	-7 570m	DNP
10	James in and other out	10 1568m	8 1875m	11 1047m	8 2538m	11 2250m	8 1847m	6 1327m	4 840m	DNP
11	James out and other out	-10 459m	-8 896m	-10 568m	-8 934m	-11 836m	-16 501m	-15 733m	-11 580m	-9 1150m

FIGURE 30.6   The Greatness of LeBron.

Cell BQ8 shows that with LeBron in, the Cavs played eight PPG better than average and with LeBron out, the Cavs played nine PPG worse than average. In virtually every situation listed in the figure, Lebron's presence made at least a 15-PPG difference. For instance, with Irving and LeBron in, cell BP8 tells us that the Cavs played 10 PPG better than average, but from cell BP9 we see that with Kyrie in and LeBron out, the Cavs played seven PPG worse than average. With LeBron in and Kyrie out, cell BP10 tells us the Cavs did fine (four PPG better than average), but with Kyrie and LeBron both out, cell BP11 tells us the Cavs played 11 PPG worse than average.

Let's examine the 2016–2017 performance of Phoenix Sun guard Devin Booker. According to the widely used PER rating, he had a below-average (14.64) PER during the 2016–2017 season and was the 21st best shooting guard in the NBA. Devin Booker scored 70 points against the Celtics (10th highest single game point total in NBA history) and he is below average. Figure 30-7 shows why we have Devin Booker as a great player.

	BI	BJ	BK	BL	BM	BN	BO	BP	BQ	BR	BS
7		Chriss	Jones	Len	Ulis	Warren	Bender	Booker	Williams	Dudley	Chandler
8	Booker and other in	-8 1428m	7 220m	-7 977m	-8 553m	-2 1574m	4 192m	-3 2731m	-4 346m	1 636m	2 1174m
9	Booker out and other in	-20 316m	-5 325m	-23 583m	-12 570m	-18 474m	-15 382m	DNP	-7 362m	-7 727m	-24 124m
10	Booker in and other out	3 1303m	-4 2510m	0 1754m	-1 2178m	-3 1157m	-3 2539m	DNP	-3 2385m	-4 2095m	-6 1557m
11	Booker out and other in	-12 920m	-17 910m	-6 653m	-17 666m	-12 761m	-14 854m	-14 1236m	-17 874m	-25 509m	-13 1112m

FIGURE 30.7   Devin Booker's 2016–2017 On-Off Statistics.

From column BP we see that with Booker in, the Suns played three PPG worse than average, and with Booker out, the Suns played 14 PPG worse than average. In many other situations the figure shows that the absence of Booker made the Suns at least 10 PPG worse. This figure should convince you that Booker was a great player during the 2016–2017 season.

Finally, for the years 2006–2015 Kevin Martin had an average PER rating of 19, well above average. The supposedly analytic savvy Rockets and Thunder traded for Martin. Adjusted +/− tells us that during the 2006–2015 time period Martin was an excellent offensive player (average +5 PPG rating), a poor defender (average +8 PPG rating!), and a below-average player (average −3 PPG rating.)

IMPACT RATINGS

We have spent the chapter singing the praises of WINVAL. So what are its shortcomings? First, there is a lot of "noise in the system." It takes many minutes or possessions to get an accurate player rating. For example, we do not have a great deal of confidence in WINVAL ratings for a player who plays less than 500 minutes in a season. Another problem with WINVAL is that you can accumulate your "good or bad" statistics at times when the game is decided. For example, suppose a player enters the game with three minutes to go and our team is down by 20 points. The game outcome is certainly decided at this point. Yet if this player's team cuts the opponent's lead to three points by the end of the game, this player will pick up a lot of credit toward our WINVAL point rating. To remedy this problem, WINVAL has created the WINVAL impact rating. The WINVAL impact rating is similar to baseball's WPA rating described in Chapter 8. We have looked at thousands of NBA game scores as the game progresses and determined the probability (for two teams of equal ability) of a team winning given the length of time left in the game and the score of the game. To create the WINVAL impact rating we create an alternate scoreboard where the score at any point in the game is the team's chance of winning the game. Then a player gains credit for the change in the team's chance of winning, instead of the change in the score of the game. For example, at the start of the game our alternate scoreboard has a score of 50–50. If I play for five minutes and the score after five minutes is 14–5 in favor of my team, the regular scoreboard says we have picked up nine points (which feeds into my point rating), while the impact scoreboard says that my chance of winning is now 72–28, so my team has picked up 22 impact points. Suppose we are down by five with two minutes to go. The alternate scoreboard says we are down 11–89. If we win the game, my team has picked up 89 impact points. We then use the change in the alternate scoreboard as the input to our impact rating analysis.

We interpret a player's impact rating as follows. A player with an impact rating (as a decimal) of x would win a fraction .5 + x of his games if he played with four average NBA players against a team of five average NBA players. For example, during the 2016–2017 season Kyle Lowry had a 40% impact rating. This means that in a 48-minute game Kyle Lowry with four average NBA players would beat five average NBA players 90% of the time.

Another way to measure a player's clutch ability is to look at his fourth quarter rating. We break down each player's rating by quarter. Figure 30-8 gives the top 10 fourth quarter performers for the 2016–2017 season. Russell Westbrook's fourth quarter rating of +23, for example, means the Thunder played 23 points (per 48 minutes) better during the fourth quarter than they would have if an average NBA player had replaced Westbrook.

	F	G	H	I
6	Rank	Player	Team	Adusted 4th quarter +/–
7	1	Russell Westbrook	OKC	23
8	2	Kawhi Leonard	SAS	21
9	3	Demarcus Cousins	SAC	17
10	4	Khris Middleton	MIL	15
11	5	Paul Millsap	ATL	15
12	6	George Hill	UTA	14
13	7	Stephen Curry	GS	14
14	8	Kyle Lowry	TOR	13
15	9	Jae Crowder	BOS	12
16	10	Bradley Beal	WAS	12

FIGURE 30.8   Top 10 4th Quarter
Adjusted +/– for 2016–2017 Season.

NBA ANALYTICS COMES OF AGE: THE KHRIS MIDDLETON CONTRACT

In April 2017, the Bleacher Report[2] called Khris Middleton the NBA's most overlooked star. We have known for years that Khris Middleton is an advanced analytics superstar. During 2013–2014 NBA season, his PER was a below-average 13. During the 2014–2015 season, his PER was a slightly above-average 16. **During the 2014–2015 season, however, Middleton had an adjusted +/− of +10 and an impact rating of +33. He ranked 10th on impact and 15th on points. This is all-star level production.** The Bucks rewarded Middleton with a five-year $70 million contract. No way this happens without advanced analytics. Props to the Bucks and, of course, Khris Middleton.

CHAPTER 30 APPENDIX USING THE EXCEL SOLVER TO FIND ADJUSTED +/− RATINGS

To install the Excel Solver in Excel go the File tab, and after selecting Options choose Add-ins. After clicking Go, check the Solver Add-in and hit OK. You have now installed the Solver Add-In. You may now activate the Excel Solver by selecting the Solver from the right-hand portion of the data tab. When you activate the Solver, you will see a dialog box that contains the three important parts of a Solver or Excel optimization model.

- **Target Cell:** This is an objective that you wish to maximize or minimize. In our example, our objective or target cell is to minimize the sum of the squared prediction errors for each game.
- **Changing cells:** These are the cells in the spreadsheet that Solver is allowed to change to optimize the target cell. In our example the changing cells are the adjusted +/− ratings for each player.

2. http://bleacherreport.com/articles/2700851-milwaukee-bucks-khris-middleton-is-the-nbas-most-overlooked-star

- **Constraints:** Constraints are restrictions on the changing cells. In our example, the only constraint we included was to make the average player rating equal to 0. This ensures that an above-average player will have a positive rating and a below-average player will have a negative rating.

Below we describe the setup in the worksheet Adjusted Plus Minus of the workbook Newadjustedplusminusexample.xlsx for determining Adjusted +/− Ratings:

Step 1: In Column B we begin by entering any trial set of adjusted +/− ratings.

Step 2: These ratings are averaged in cell B3.

Step 3: In Column S we use lookup functions to determine the sum of the adjusted +/− for our team during each game.

Step 4: In Column T we use lookup functions to determine the sum of the adjusted +/− for our opponents during each game.

Step 5: In Column E we created our "prediction" for each game by taking [sum of our player rating] (from Column S) − [sum of opponent player ratings] (from Column T).

Step 6: In Column D we compute the squared error for each of our game predictions by computing (Column G − Column E)^2.

Step 7: In cell D2 we compute the sum of squared prediction errors for each game.

Step 8: The Solver window inputs needed to compute our Adjusted +/− ratings is shown in Figure 30-9. Our target cell is D2 (minimize sum of squared errors). The changing cells are the adjusted +/− ratings (B5:B22). We constrain the average rating (in cell B3) to equal 0. This makes a better-than-average player have a positive rating and a below-average player have a negative rating.

Note that if the games are not of equal length, then for each "game" you should predict the point differential per minute and then compute the squared prediction error for each game on a per minute basis. **Then to compute the target cell you weight each game's per minute squared error by the number of minutes in the game.**

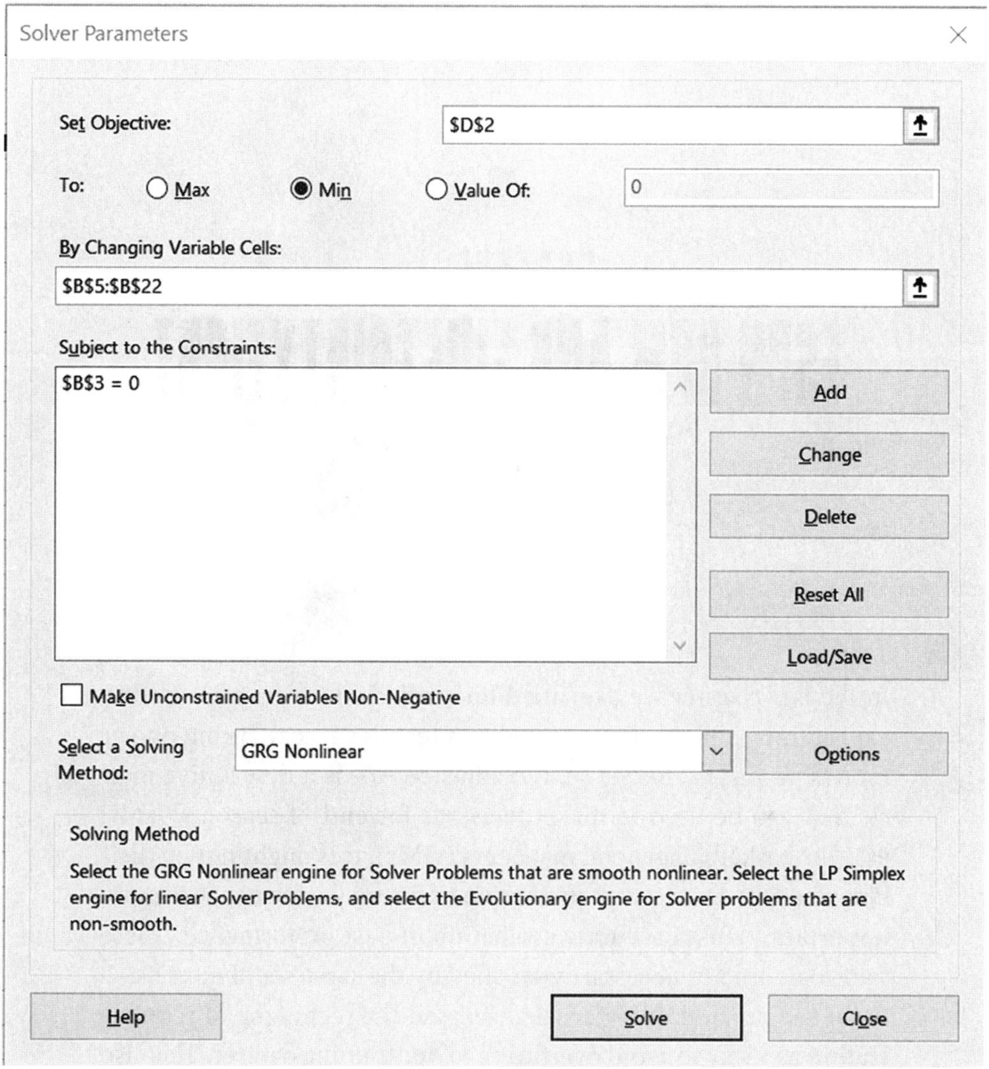

FIGURE 30.9 Solver Window for Adjusted +/−.

CHAPTER 31

ESPN RPM AND FIVETHIRTYEIGHT RAPTOR RATINGS

In the last chapter we explained how adjusted +/− can be used to explain how a player contributed to a team's success during one or more seasons. In this sense, the adjusted +/− is a descriptive metric that can be used to make decisions for end-of-season awards, etc. For basketball general managers (GMs), this might not be their first priority. An accurate prediction of future performance is more important than an accurate evaluation of past performance. These two goals are not necessarily satisfied by the same set of metrics.

In the previous chapter we discussed the technique of **regularization** as a way to avoid overfitting to our training dataset. This also leads to learning models that are what we call **generalizable**, i.e., models that have not just memorized the data they saw but rather identified actual patterns in them. A generalizable model will be able to make more accurate predictions in previously unseen data, and hence it can enable evaluation of past performance to be transformed into more accurate estimates of future performance. Basically, exactly what GMs are dreaming of. Joe Sill (now employed by the Washington Wizards) first introduced applied regularization to NBA player ratings at the 2010 MIT Sloan Sports Analytics Conference.[1]

Together Jeremias Engelmann, currently a senior analyst for the Dallas Mavericks, and Steve Ilardi, a professor of clinical psychology at the University of Kansas, refined Sill's work and developed RPM (Real Plus-Minus). Engelmann and Ilardi essentially used a modified ridge regression that instead of shrinking the coefficients or ratings to 0, shrinks them toward the box plus-minus of the player, which is based on more stable individual measures from the box score.

We have summarized the RPM leaders for the 2018–2019 season in the file RPM201819. Figure 31-1 shows the 2018–2019 RPM leaders.

	A	B	C	D	E	F	G	H	I	J	K	L	M
1	RANK	NAME	TEAM	GP	MPG	ORPM	DRPM	RPM	WINS	RAPM	PIMP	RAPTOR	PREDATOR
2	1	Paul George, SG	OKC	77	36.9	4.55	3.08	7.63	19.86	6.08	6.33	9.53	9.96
3	2	James Harden, SG	HOU	78	36.8	7.4	0.02	7.42	18.54	3.24	5.58	10.72	12.57
4	3	Stephen Curry, PG	GS	69	33.8	5.99	0.85	6.84	15.24	5.43	7.34	7.79	8.7
5	4	Giannis Antetokounmpo, PF	MIL	72	32.8	3.16	3.53	6.69	15.22	4.31	7.66	6.81	7.08
6	5	Nikola Jokic, C	DEN	80	31.3	3.89	2.59	6.48	14.91	2.14	4.41	8.7	9.02
7	6	Joel Embiid, C	PHI	64	33.7	2.68	3.72	6.4	12.9	4.11	6.62	7.53	7.15
8	7	Anthony Davis, PF	NO	56	33	2.54	3.2	5.74	10.81	2.18	5.83	7.38	7.32
9	8	Nikola Vucevic, C	ORL	80	31.4	1.9	3.59	5.49	13.67	2.76	4.2	4.8	4.93
10	9	LeBron James, SF	LAL	55	35.2	3.61	1.83	5.44	11.04	3.32	4.39	5.58	4.3
11	10	Damian Lillard, PG	POR	80	35.5	5.67	-0.59	5.08	14.52	3.38	4.9	5.6	6.12
12	11	Kevin Durant, SF	GS	78	34.6	4.36	0.6	4.96	14.1	5.78	5.79	5.06	4.97
13	12	Chris Paul, PG	HOU	58	32	2.49	2.27	4.76	9.17	2.27	3.23	4.63	6.44
14	13	Danny Green, SG	TOR	80	27.7	2.66	2.07	4.73	11.23	5.84	3.97	4.06	4.91
15	14	Jimmy Butler, SF	MIN/PHI	65	33.6	2.55	2.16	4.71	10.97	2.11	2.45	5.62	5.56
16	15	Kyle Lowry, PG	TOR	65	34	2.83	1.82	4.65	10.97	3.49	3.84	5.6	6.95
17	16	Rudy Gobert, C	UTAH	81	31.8	0.29	4.35	4.64	13.1	2.42	4.54	5.43	6.15
18	17	Jrue Holiday, SG	NO	67	35.9	3.38	1.23	4.61	12.28	4.76	3.96	5.57	5.46
19	18	Al Horford, PF	BOS	68	29	1.77	2.66	4.43	9.51	3.07	3.05	3.77	4.17
20	19	Paul Millsap, PF	DEN	70	27.1	1.57	2.74	4.31	8.77	4.36	2.9	3.87	3.93
21	20	Kyrie Irving, PG	BOS	67	33	3.77	0.46	4.23	10.34	2.66	2.97	5.69	6.96

FIGURE 31.1 2018–2019 RPM Ratings.

We use Giannis's numbers to interpret the given information. Unlike WINVAL's adjusted +/–, all RPM numbers are per 100 possessions.

1. J. Sill, "Improved NBA adjusted +/– using regularization and out-of-sample testing," in MIT SSAC, 2010.

1. **GP**—Giannis played in 72 regular season games.
2. **MPG**—Giannis played an average of 32.8 minutes per game.
3. **RPM**—Per 100 possessions, Giannis added 6.69 points more to a team's ability than an average NBA player.
4. **ORPM**—Per 100 possessions, Giannis added 3.16 more points scored to a team's ability than an average NBA offensive player.
5. **DRPM**—Per 100 possessions, Giannis caused his team to give up 3.53 fewer points than an average NBA defender. Note that, unlike WINVAL, a positive defensive rating is better. It is always true that that a player's RPM = ORPM + DRPM.
6. **Wins**—This column captures the number of wins contributed by a player to his team's wins total above those of a replacement player. This considers the number of possessions a player played for the team as well as his RPM.

LINKING RPM TO WINS

The file RPM16.xlsx contains ESPN's RPM ratings and wins for the 2015–2016 season. Clearly, if you are an above average player, then playing more minutes should generate more wins. We found that the formula

$$\text{WINS} = 0.00064 * (\text{RPM} - (-3.15)) * \text{minutes played}$$

predicts wins and is off by an average of only 0.016 wins. This model indicates that ESPN believes an RPM of −3.15 points per possession defines a replacement player.

ALTERNATIVES TO RPM

ESPN has recently updated its RPM calculations utilizing tracking data (we will further discuss tracking data in Chapter 37). In particular, the advanced box score stats derived from tracking data (e.g., who guards whom) are incorporated in RPMs prior. Kevin Pelton wrote a detailed article[2] on ESPN describing the updates of RPM 2.0.

FiveThirtyEight also introduced its own player rating metric before the 2019–2020 season, named RAPTOR. RAPTOR is again expressed as points relative to a league-average player over 100 possessions. Nate Silver provides a detailed description of what goes into the metric,[3] but overall it combines publicly available player statistics with modern tracking data. There are two versions of the statistic provided: (a) a descriptive version that aims at describing who contributed to a team's winning, and (b) a predictive version that aims at predicting future matchups (termed PREDATOR). The latter uses depth charts to project the minutes played by each player.

In general, there are various versions of similar adjusted +/− metrics. Ryan Davis has created a luck-adjusted version[4] of regularized adjusted +/−, where instead of considering the outcome of a three-point shot or a free throw (made or missed) its expected value is considered in the calculations. Similarly, Player Impact Plus-Minus[5] is another luck-adjusted version of adjusted +/−.

These different versions will most probably provide different *views* for the same player (in terms of the actual plus-minus value). Figure 31-2 shows the relationship between the different versions of adjusted +/− (which are expressed at per 100 possessions) for the top 10 players as per RPM for the 2018–2019 season. As we can see from Figures 31-2 and 31-3, there is some correlation between them, but clearly there are differences. Casual fans might be confused by this since they are used to unambiguous, "factual statistics"—e.g.,

2. https://www.espn.com/nba/story/_/id/28309836/how-real-plus-minus-reveal-hidden-nba-stars

3. https://fivethirtyeight.com/features/how-our-raptor-metric-works/

4. http://nbashotcharts.com/rapm?id=1109440799

5. https://www.bball-index.com/

points scored. However, for an analyst the different results are an
artifact of potentially different objectives (e.g., descriptive versus
predictive) and/or different information used and weighted in the
development of the metric. When using these metrics we should
understand what each one attempts to capture and what is the purpose of using it. Darryl, a casual fan from Houston, might pick one
metric over the other simply because it supports his and his friends'
view that Harden was snubbed for the MVP vote (e.g., choose RAP-
TOR instead of PIMP because Harden ranks higher than Giannis in
RAPTOR). However, an analyst is not looking into data and metrics
for confirming her prior beliefs, but rather coming to an objective
conclusion—even if it does not match her previously held views.

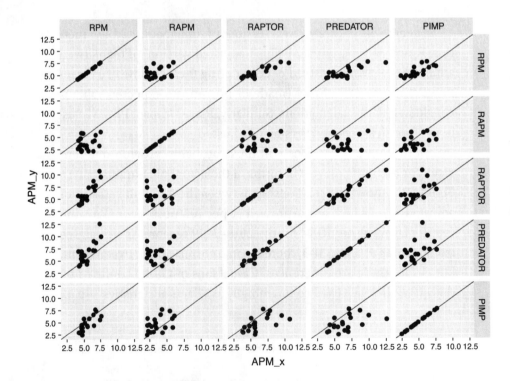

FIGURE 31.2 Relationship for the Various APM Player Metrics (Top 20
Players in the 2018–19 Season According to RPM).

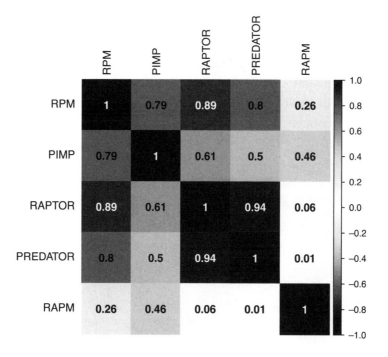

FIGURE 31.3 Correlation between the Different APM Metrics.

IS MICHAEL JORDAN THE G.O.A.T?

ESPN, CBS Sports, Fox Sports, *SLAM* magazine, and Bleacher Report all rank Michael Jordan as the G.O.A.T. **If you pick your G.O.A.T purely on RPM, you cannot choose MJ as the G.O.A.T.** In the file RPM.xlsx we have summarized Engelmann's RPM ratings for the years 1990–2014. This data excludes Jordan's great early seasons (1985–1990), but still includes his second three-peat and many of his great seasons. For the 1990–1998 period MJ has an average RPM of 4.3, **the same as his HOF teammate Scottie Pippen.** During these years, teammate Dennis Rodman had an average RPM of 2.7. Shaq's average RPM was 6.6. The Admiral, David Robinson of the Spurs, began playing in 1989, so this dataset omits only his rookie year. Robinson had an average RPM of 7.3.

During the years 1990–1998, John Stockton had an average RPM of 5.3. Through the 2017 season LeBron had an average RPM of 8.

My (WW) personal G.O.A.T. is Wilt Chamberlain, but no +/− data is available for those days, and during the 1961–1962 season Wilt averaged 48.5 minutes per game, so he would not have been out of the game enough to estimate an accurate adjusted +/−!

CHAPTER 32

NBA LINEUP ANALYSIS

In Chapter 30 we described our methodology for creating adjusted +/− ratings. These can facilitate decisions such as trades and salaries. During the season, however, few players are traded and a team's major concern is how to win more games with its current roster. One of the regular and crucial decisions coaches make during the season is which lineups to play when. For example, would a team have been better off trying to go big or small against the Warriors' "Death Lineup" of Durant, Curry, Thompson, Green, and Iguodala?

During a year the typical team plays more than 500 different lineups. Is there any rhyme or reason to coaches' lineup choices? Good lineup decisions can help win more games, and lineup ratings could facilitate this. Play your better lineups more and worse lineups less.

Once we have player ratings, it is "*easy*" to develop lineup ratings. Suppose we want to rate the aforementioned "Death Lineup." Let's call this lineup Warriors 2A. This lineup played the second most minutes of any lineup during the 2016–2017 season. In particular, Warriors 2A played 288 minutes and outscored its opponents by 162 points. This means that Warriors 2A played $288/48 = 6$ games and has a pure +/− of $162/6 = 27$ points per 48 minutes. Then we look at each minute the Warriors lineup 2A was on the court and average the total abilities of the opponents (adjusting for the 3.2-point league home edge). We find this lineup was on the court against opponents averaging +2.67 points in ability per 48 minutes. This means that the Warriors' lineup 2A should have an adjusted +/− rating of $27 + 2.67 = 29.67$ points. In short, this means that our best estimate is that the Warriors' lineup 2A played 29.67 points per game better than an average NBA lineup (a lineup where the sum of the five player ratings is 0).

The amazing thing is that many teams play inferior lineups far more minutes than their better lineups. To illustrate this situation, we note that for the 2016–2017 Atlanta Hawks the most played lineup was Bazemore, Howard, Millsap, Sefelosha, and Schroder. This lineup played 426 minutes and played 2.4 points per game worse than an average lineup. The lineup that simply replaced Bazemore by Tim Hardaway Jr. played only 126 minutes but played 28 points per game better than average. It's hard to understand why this lineup did not play more than the below-average lineup!

During the years 2000–2007 the Dallas Mavericks factored WINVAL lineup ratings into their coaching decisions. Through those seven years the Mavs won more games than any team except for the Spurs. We cannot prove the link between our lineup ratings and the Mavs performance, but playing good lineups more and worse lineups less makes perfect sense.

During the 2012–2013 season under the leadership of outstanding New York Knicks General Manager Glen Grunwald, WINVAL was hired to analyze players and lineups. During March 2013, the Knicks had lost four games in a row. Our lineup analysis showed that the Prigioni-Felton backcourt had played an amazing 41 points per game better than average. The Knicks began starting this backcourt duo and promptly won 13 games in a row! The 2012–2013 Knicks won 54 games, the most since 1997. The Knicks were perhaps one Roy Hibbert block away (it was goaltending!) from making the Eastern Conference Finals. We would be remiss if we did not note that under Grunwald's successor (you may have heard of Phil Jackson), the Knicks won only 34% of their games!

Critics might say that lineups do not play many minutes, so there must be a lot of variability in the lineup ratings. And this is true. For instance, in the example above, the Hardaway Jr., Howard, Millsap, Sefelosha, and Schroder lineup played 30% of the time that the Bazemore, Howard, Millsap, Sefelosha, and Schroder lineup played. How much of this difference can be *attributed* to the sample size? What is the probability that a given lineup is better than another lineup? The spreadsheet Lineupssuperiority.xlsx (see Figure 32-1) answers this question. Enter the minutes played and lineup rating for the inferior lineup in row 4 and the same data for the better lineup in row 5. Then cell E8 gives the chance that the lineup with the higher rating is truly better. We find there is over a 99% chance that the Hawks lineup with Tim Hardaway Jr. is superior to the lineup with Bazemore. This should convince the Hawks that the Tim Hardaway Jr. lineup should have played more often!

	B	C	D	E	F
1	Howard, Schroder, Millsap, Sefelosha				
2					
3	5th	Lineup	Rating	Minutes	Games
4	Bazemore	1	−2.4	426	8.875
5	Hardaway Jr.	2	28	126	2.625
6					
7	Difference	Variance difference	Sigma difference	Probability Lineup 2 is better	
8	−30.4	71.082495	8.431043528	0.9999289	

FIGURE 32.1 Lineup Superiority Calculator.

How did we determine the probability of Lineup 2 being better than Lineup 1? Data indicates that the actual performance of a lineup over 48 minutes is normally distributed with a mean equal to its lineup rating and a standard deviation of 12 points. The standard deviation of a lineup's rating can then be shown to equal

$$\frac{12}{\sqrt{\text{Games played}}}.$$

For example, for a lineup playing 192 minutes our rating has a standard deviation of $12/2 = 6$ points. A basic theorem in statistics tells us that the variance of the difference of independent random variables is the sum of the random variable variances. Since the standard deviation of a random variable is simply the square root of the random variable's variance, we find that the standard deviation of the difference in the rating of two lineups is given by

$$\sigma = \sqrt{\frac{144}{\text{Games Lineup 1 played}} + \frac{144}{\text{Games Lineup 2 Played}}}.$$

Here, games played = minutes played/48.

Then we find the probability that Lineup 2 is better equals the probability that a random variable with mean Lineup 2 rating − Lineup 1 rating and standard deviation σ is > 0. This probability may be found with the Excel formula

$$1 - \text{NORMDIST}(0, \text{Lineup 1 rating} - \text{Lineup 2 rating}, \sigma, \text{True}).$$

Figure 32-1 shows that there is over a 99% chance that the Tim Hardaway Jr. lineup is better than the Bazemore lineup.

LINEUP CHEMISTRY 101

Coaches often say that a given lineup has great "on-court chemistry" when the lineup really clicks. When a lineup plays poorly, coaches say these guys play "like they have never seen each other before." We can potentially identify lineups with good or bad chemistry by utilizing lineup and individual player ratings. We define the lineup's chemistry value = (lineup rating) − (sum of individual player ratings). A positive chemistry rating indicates that a team played better than expected and exhibits positive chemistry, while a negative chemistry rating indicates that a team played worse than expected and exhibits negative synergy.

See Figure 32-2 for some examples of team chemistry from the 2016–2017 season. For example, the previously discussed Tim

Hardaway Jr. lineup played 27 points per game better than the sum of the individual players' adjusted +/− ratings. On the other hand, the Celtics lineup of Bradley, Crowder, Smart, Thomas, and Olynyk played 14 points per game worse than the sum of the individual players' adjusted +/− ratings.

	B	C	D	E	F	G	H	I
3	Team	Player 1	Player 2	Player 3	Player 4	Player 5	Minutes	Chemistry
4	ATL	Hardaway	Howard	Millsap	Schroeder	Sefelosha	120	27
5	BOS	Bradley	Crowder	Smart	Thomas	Olynyk	114	−14
6	CHI	Butler	Gibson	Grant	Lopez	Wade	133	16
7	CLE	Smith	James	Frye	Love	Irving	83	18
8	GSW	Stephen Curry	Green	Iguodola	Mcgee	Thompson	94	−11
9	DAL	Seth Curry	Barnes	Harris	Matthews	Mejri	57	−14
10	HOU	Beverly	Dekker	Gordon	Capella	Brewer	91	−15
11	LAC	Crawford	Griffin	Jordan	Paul	Mbah_a_Moute	80	19
12	LAL	Ingram	Mozgov	Randle	Russell	Young	63	−23

FIGURE 32.2   Examples of Good and Bad Lineup Chemistry.

ADJUSTING LINEUP RATINGS FOR PLAY TIME

As mentioned earlier, one of the criticisms with lineup ratings has to do with the observation sample. For example, are 50 minutes of play time enough to have a good picture of how the lineup performs? Is +28 the best estimate for the Tim Hardaway Jr. lineup above? We can adjust these estimates using a Bayesian approach (we provide more details on Bayesian inference in Chapter 52). In particular, we can use the Bayesian average to account for the observation sample size. Let us consider a lineup that has played 60 minutes and has a +/− of 12.5 per 48 minutes. If we believe that an average lineup will have a +/− of 0 after 100 minutes of play, we can update our +/− estimate for the lineup using the following equation:

$$\widehat{\text{Lineup}}_{+/-} = \frac{(12.5 \cdot 60) + (0 \cdot 100)}{(60 + 100)} = 4.69$$

As we can see, the new estimate for the lineup's +/− per 48 minutes has shrunk closer to 0. The 0 +/− over 100 minutes is what we call our *prior.* This prior can be either an "arbitrary" choice, or informed by data (e.g., the average time a regular rotation lineup plays and its average +/−). One thing to observe is that the prior in the Bayesian average essentially acts the same way that the regularization term in a ridge regression does. The choice in the example above shrinks the lineup rating closer to 0, while if we pick another value for the prior, this will be similar to, say, the RPM that uses box score statistics to regress the players' ratings toward a (potentially) non-zero prior. It should be evident that as we accumulate more "evidence" (i.e., playing time) for a lineup, the Bayesian average will get closer to the observed +/− of the lineup. On the contrary, when we have little evidence, the Bayesian average will be closer to the prior (in this case 0). Figure 32-3 shows the pure +/− for the Atlanta Hawks lineups during the 2016–2017 season and their Bayesian adjustment (using the same prior as above). The size of the circles corresponds to the minutes played by each lineup. As we see for lineups with small playing time, their Bayesian +/− is close to 0, as expected from the above discussion. Note the difference in scales between the two axes. The y-axis, which represents the Bayesian average, has a much smaller

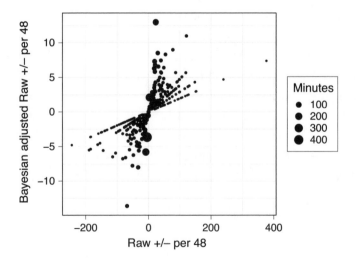

FIGURE 32.3 Atlanta Hawks Lineups (2016– 2017).

range, compared to the observed +/−. For example, the lineup Howard, Humphries, Sefolosha, Schroder, and Hardaway Jr. played one minute the whole season and outscored its opponents six points. This leads to a +/− of 288 per 48 minutes. It is obvious that this is not an accurate estimate. After the Bayesian adjustment, the +/− is (the more realistic) +2.85 per 48 minutes.

CHAPTER 33

ANALYZING TEAM AND INDIVIDUAL MATCHUPS

There are many qualities needed to make a successful coach. A great coach needs to be a master psychologist who can motivate players to play for the team rather than themselves. Remember that there is "no 'I' in 'team'," and "the whole is greater than the sum of its parts." Great coaches must have a sound offensive and defensive strategy and get players to buy into executing the coach's strategic concepts. Great coaches also have an instinct about whom to put in the game at a given time to best match up with the opponent's lineup. As we have stated in Chapter 30, we have adjusted +/− ratings for each NBA player. We also break down each player's adjusted +/− by opponent. This breakdown aids coaches in determining how to match up with opponents.

SPURS–MAVERICKS 2006 WESTERN CONFERENCE SEMIFINAL

A great example that shows the usefulness of the team-by-team breakdown of adjusted +/− comes from the Dallas Mavericks' surprising journey to the NBA finals during the 2006 playoffs. Few analysts gave the Mavs any chance against the World Champion Spurs. A key coaching move during this series was the insertion of guard Devin Harris into the starting lineup in place of Adrian Griffin. This surprise decision enabled the Mavs to "steal" game 2 at San Antonio, and the Mavs went on to win the series in seven games by winning an overtime thriller game 7 in San Antonio. Our team-by-team adjusted +/− ratings were an important input into the coaching decision. Devin Harris's overall 2005–2006 rating was −2.1 points, and his impact rating was −15%. **Against the Spurs, Harris's rating was +9.4 points and his impact was +8%. Against the Spurs, Griffin had a −5 point rating and a −18% impact. More importantly, Griffin had a −18 points offense rating.** This indicates that against the Spurs, Griffin killed the Mavs' offense. Given this data, it seems clear that Harris should have started in lieu of Griffin. So how did Harris do? During the playoffs, we do a two-way lineup calculation that shows how the Mavs do with any combination of Mavs players on or off court against any combination of opponent players on or off court. For the 2005–2006 regular season, the data indicated that Harris could outplay future Hall of Famer Tony Parker. So what happened during the first six games of the Spurs–Mavs series? With Harris on the court against Parker, the Mavs (excluding game 2, in which the Mavs took the Spurs by surprise by starting Harris) beat the Spurs by an average of 102–100 per 48 minutes with Harris in against Parker. With Parker on the floor and Harris out, the Mavs **lost by an average of 96–81 per 48 minutes!** As the series progressed, we learned other interesting things. When Marquis Daniels was on the court against Manu Ginobili, the Mavs lost by an average score of 132–81 per 48 minutes!! When Ginobili was on the court and Daniels was out, the Mavs won by an average of 94–91 per 48 minutes. By the way, Daniels did not play in game 7 of the series.

Next the Mavs played the Suns in the 2006 Western Conference finals. Here we knew Devin Harris would not be as effective because he had trouble with Steve Nash. For this series, when the Harris and Jason Terry backcourt played against Nash, the Mavs were down by 113–90 per 48 minutes. On the contrary, **the Terry and Jerry Stackhouse**

TABLE 33.1

Spurs–Mavericks 2006 Playoff Lineup Analysis

Dallas Mavericks

						Rating	Minutes	Simple	Raw+/−	Lineup Code#
1	Diop	Harris	Howard	Nowitzki	Terry	16.36	41.04	1.18	1	8816_DAL_2006
2	Dampier	Howard	Nowitzki	Stackhouse	Terry	29.24	31.06	15.45	10	12868_DAL_2006**
3	Dampier	Harris	Stackhouse	Nowitzki	Terry	60.55	21.34	49.47	22	12836_DAL_2006**
4	Dampier	Harris	Howard	Nowitzki	Terry	−1.78	15.89	−18.12	−6	8804_DAL_2006***
5	Dampier	Harris	Howard	Nowitzki	Stackhouse	−20.83	14.19	−33.80	−10	4708_DAL_2006***
6	Diop	Harris	Nowitzki	Stackhouse	Terry	15.54	11.48	4.18	1	12848_DAL_2006
7	Dampier	Daniels	Howard	Stackhouse	Terry	−65.28	8.92	−75.26	−14	12364_DAL_2006***
8	Diop	Howard	Nowitzki	Stackhouse	Terry	−26.13	8.79	−43.62	−8	12880_DAL_2006***
9	Diop	Harris	Howard	Nowitzki	Stackhouse	81.91	8.02	65.92	11	4720_DAL_2006**
10	Daniels	Harris	Howard	Nowitzki	Stackhouse	78.79	7.82	67.55	11	4712_DAL_2006**
11	Diop	Howard	Nowitzki	Terry	Griffin	−9.51	6.92	−34.70	−5	41552_DAL_2006***
12	Harris	Howard	Stackhouse	Nowitzki	Terry	44.11	6.78	35.43	5	12896_DAL_2006**
13	Dampier	Daniels	Nowitzki	Stackhouse	Terry	−42.30	6.37	−52.81	−7	12812_DAL_2006***

San Antonio Spurs

1	Bowen	Duncan	Finley	Ginobili	Parker	21.08	45.83	6.29	6	1054_SAS_2006**
2	Barry	Bowen	Duncan	Finley	Parker	37.90	33.42	20.12	14	1039_SAS_2006**
3	Bowen	Duncan	Ginobili	Horry	Parker	17.94	28.23	5.10	3	1078_SAS_2006
4	Barry	Bowen	Duncan	Horry	Parker	−26.88	10.05	−47.75	−10	1063_SAS_2006***
5	Barry	Duncan	Finley	Ginobili	Parker	71.11	9.52	60.37	12	1053_SAS_2006**
6	Barry	Bowen	Duncan	Ginobili	Parker	−25.71	8.48	−45.28	−8	1047_SAS_2006***
7	Barry	Duncan	Finley	Ginobili	Van Exel	46.74	7.70	37.35	6	4125_SAS_2006**
8	Bowen	Duncan	Finley	Ginobili	Van Exel	32.55	7.67	18.79	3	4126_SAS_2006
9	Duncan	Finley	Ginobili	Horry	Van Exel	−22.46	5.84	−24.68	−3	4156_SAS_2006
10	Barry	Bowen	Duncan	Finley	Van Exel	−53.95	5.66	−59.24	−7	4111_SAS_2006***
11	Barry	Finley	Ginobili	Horry	Parker	−14.32	5.53	−26.06	−3	1081_SAS_2006
12	Bowen	Duncan	Ginobili	Horry	Van Exel	46.51	4.13	34.84	3	4150_SAS_2006
13	Barry	Bowen	Finley	Horry	Parker	−179.83	4.11	−186.96	−16	1067_SAS_2006***

Note: *denotes the player's team; ** denotes a good lineup; *** denotes a bad lineup. Simple = Raw +/− rating per 48 minutes.

backcourt (with Harris out) was up 116–96 per 48 minutes with Nash in. The Mavs used this information to adjust their rotation to play the more effective combination more often.

During a playoff series we also track how well each lineup plays. Through five games of the Spurs–Mavs playoff series, Table 33-1 shows how the most frequently used lineups performed. For example, lineup 3 for the Mavs (Dampier, Harris, Nowitzki, Terry, and Stackhouse) played 21.34 minutes and was up by 22 points. This lineup played at a +60.55 level. Lineup 7 for the Mavs (Dampier, Daniels, Howard, Terry, and Stackhouse) played very poorly. This lineup lost by 14 points in 8.92 minutes and played at a −65.28 level.

THE NON-TRANSITIVITY OF NBA MATCHUPS

Mathematics classes often cover the transitivity property. In basketball terms, transitivity would indicate that if player A outperforms player B during a matchup and player B outperforms player C in another matchup, then player A will outperform player C during their matchup. The following discussion shows that transitivity does not hold for basketball player matchups!

We have seen that Devin Harris can outplay Tony Parker and Steve Nash can outplay Devin Harris. During the 2006–2007 season, Tony Parker got the better of the matchup against Steve Nash. For the season Nash had an impact rating of +28%, and against the Spurs he had a 0% impact. For the season Parker had an impact of −3%, **but against the Suns Parker had an impact of +35%.** We have seen that Harris beats Parker and Parker beats Nash and Nash beats Harris. This example shows that basketball matchups can exhibit a lack of transitivity! The great coaches probably intuitively understand matchups, but our WINVAL analysis allows coaches to be more data-driven when they make crucial decisions about how their lineups should be selected to best perform against their opponents' on-court lineup.

ANALYTICS HELPS THE MAVS BECOME 2011 WORLD CHAMPIONS!

At the start of the 2011 NBA playoffs the Miami Heat, with its Big 3 of LeBron James, Dwayne Wade, and Chris Bosh, was the overwhelming favorite to capture the NBA title. The Dallas Mavericks upset the Heat, and lineup analysis played a huge role. During the regular season, the Mavericks best lineup was Jason Terry, Jason Kidd, Dirk Nowitzki, Shawn Marion, and Tyson Chandler. During the regular season, this lineup played an average of only two minutes per game, but during the regular season this lineup played 28 points per game better than average. Recognizing the greatness of this lineup, the Mavs increased the lineup's playing time to eight minutes per game during the playoffs, and this lineup played an incredible 46 points per game better than average!

One of the most important decisions a coach must make is determining the optimal lineup to play when its star player (in this case Dirk Nowitzki) is out. During the three Western Conference playoff series, the Mavericks rested Dirk by inserting Peja Stojakovic. In 126 minutes, Peja, in for Dirk, played three points per game better than average. If you can rest your star and still play above average, you are doing great!

In the first three games of the NBA finals against the Heat, the Mavs continued to insert Peja for Dirk. Unfortunately, in these games the Mavs lost by 24 points in 18 minutes. All these games were close, so something needed to be done. In Game 4 the Mavs benched Peja and began to use Brian Cardinal to replace Dirk. In games 4–6 of the finals, the Mavs lost by only three points in 24 minutes with Cardinal replacing Dirk. This lineup change improved the Mavs' performance by seven points per game.

In the first three games of the series the Mavs starting lineup of Kidd, Dirk, Chandler, Marion, and Stevenson was −7 points in 28 minutes while with guard J. J. Barea for the Mavs, they were −14 points in 51 minutes. During the first three games of the series, Barea came off the bench and was usually matched up against Mario

Chalmers. This matchup did not go well for the Mavs. During the first three games, Chalmers owned Barea (Mavs −31 points in 21 minutes with both in). On the other hand, during the first three games, Barea owned Heat starter Mike Bibby (Mavs +17 points in 33 minutes with both in). Based on this information, Carlisle inserted Barea into the starting lineup, replacing Stevenson. This new lineup was +8 points in 28 minutes in games 4 and 5. Starting Barea placed Barea more often against Bibby than against Chalmers. The Heat finally figured out that starting Barea was hurting them, and in game 6 they started Chalmers. This rendered the Mavs new starting lineup ineffective, but the damage had been done and the Mavs were World Champions!

CHAPTER 34

NBA SALARIES AND THE VALUE OF A DRAFT PICK

In Chapter 9 we determined salaries for baseball players based on how many wins a player generated over and above the number of wins that would be achieved with a team of "replacement players." Using the ESPN's RPM Wins, we may use the same approach to come up with an estimate of a fair salary for an NBA player.

During the 2017–2018 season, the average team payroll was $93 million. The minimum player salary depends on how many years a player has played in the NBA and varies between around $500,000 and $1,500,000. Let us assume average minimum salary is $1 million. Based on our analysis of ESPN RPM WINS we define the RPM value of a "replacement player" as −3.1. This is the point value per 100 possessions for a 2016–2017 player in the bottom 10% of the league point values. A team made up of replacement players would lose by $5*(−3.11) = 15.55$ points per 100 possessions to an average NBA team, or 14.88 per game (the average pace in 2017–2018 was 96 possessions/game). After noting that an average NBA team scored 105.6 points per game we find that our team of replacement players would have a scoring ratio of 0.86. Using the basketball Pythagorean

Theorem of Chapter 1 we find that our team of replacement players would be expected to win a fraction $.86^{14}/(.86^{14}+1)=10.7\%$ of their games. This means our team of replacement players should win $0.107*82=8.7$ games during an 82-game season.

CALCULATING A FAIR PLAYER SALARY BASED ON ESPN RPM WINS

Assuming that a player's 2017–2018 wins will equal his 2016–2017 wins (and this might be a big assumption of course, but let's go with it), we can estimate a "fair" salary based on his projected wins. Assuming a 12-player roster we have $93 − $12 = $81 million to spend to bring a team of replacement players from approximately nine wins to 41 wins. Therefore, 32 wins is worth $81 million. Let' s just say $80 million = 32 wins, which implies each win above replacement is worth $2.5 million.

As an example of a "fair value" calculation, let's compute a fair salary for a player who generated 20 wins. Our back-of-the-envelope calculation implies that this player should earn $(20)*2.5=\$50$ million. One of the things to keep in mind is again the difference between descriptive and predictive metrics. In the above calculations we have assumed that a player's RPM WINS will be the same the following year. We might want to distinguish between identifying players that provided surplus value for our team this year compared to what we expect them to do the following years. FiveThirtyEight used similar calculations to estimate the players in the 2018–2019 season that provided the most value for money.[1] Pascal Siakam topped this list, generating a surplus value of $35.8 million!! Paul George (surplus $30 million), James Harden (surplus $27 million), and Giannis (surplus $22 million) are all in the top 10 of this list, supporting the view that NBA superstars are underpaid in several cases based on their on-court production.

---

1. https://fivethirtyeight.com/features/forget-giannis-pascal-siakam-is-our-mvp/

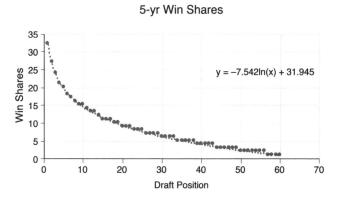

FIGURE 34.1   5-Year Win Shares vs. Draft Position.

IS THE NBA DRAFT EFFICIENT?

Recall that in Chapter 26 we reviewed the Thaler–Massey (TM for short) study of the NFL draft. TM found that the NFL draft was inefficient in that later draft picks created more value than earlier picks. TM results drew a lot of criticism as being flawed, however, because they lacked an accurate measure for player value.

In the file Winshares.xlsx, we have summarized an excellent study of the NBA draft.[2] The player value is based on Justin Kubatko's win shares metric.[3] A player's win share purports to measure the number of wins a player generates. Usually the sum of the win shares for players on a team closely matches the team's actual wins. Figure 34-1 shows the average win shares generated by the draft picks during the 1985–2010 seasons for each position in the NBA draft (1–60) for the first five years (rookie contract) in the league. As shown in Figure 34-1, the curve five-year win shares $= -7.5542 * \ln(\text{draft position}) + 31.945$ does a great job of fitting actual draft value. The chart can be used when evaluating trades. For example, it implies that 2 #8 picks will, on average, generate the same five-year value

2. https://www.reddit.com/r/nba/comments/36wv9m/trying_to_create_an
_nba_draft_trade_value_chart/

3. https://www.basketball-reference.com/about/ws.html

as a single #1 pick.[4] Of course, salary cap considerations matter and surplus value—as in the TM study—needs to be considered. Also, the NBA is clearly a star-driven league, and if a GM traded a LeBron James for 2 #8 picks, he would quickly have become an ex-GM!

4. Pick protections can further complicate such analysis. Foster and Binns recently had an interesting study on the value of protections on NBA draft picks ("Analytics for the front office: Valuing protections on NBA draft picks," MIT SSAC, 2019).

CHAPTER 35

ARE NBA OFFICIALS PREJUDICED?

The sports pages of the May 2, 2007, *New York Times*[1] contained the headline: "Study of N.B.A. Sees Racial Bias in Calling Fouls." The article was based on a study by Wharton professor Justin Wolfers and Cornell Professor Joseph Price (2007). Wolfers and Price (WP) claim that "more fouls are awarded against players when officiated by an opposite-race officiating crew than when officiated by an own-race officiating crew." In this chapter we discuss their insightful analysis of the referee bias question.

WHAT'S THE BEST DATA TO USE TO TEST FOR REFEREE BIAS?

An NBA officiating crew consists of three officials. The ideal way to determine whether the racial composition of the officiating crew influences the rate at which fouls are called against players of different race would be to look at a set of NBA games and determine the rate at which Black officials and White officials call fouls on White and Black players, respectively. The data might look something like that of the 1,000 games of (synthetic) data excerpted in Figure 35-1

1. http://www.nytimes.com/2007/05/02/sports/basketball/02refs.html?ex=1335844800&en=747ca51bedc1548d&ei=5124)

	C	D	E	F	G	H	I	J	K	L	M
1		total	978	768	288		total	9204	11351	19667	5453
2			1	2	3		ref	black ref	white ref	white ref	black ref
3	Game	Whites	ref 1	ref 2	ref 3	black min	white min	bl pl	wh pl	bl pl	wh pl
4	1	1	1	0	0	396.85	83.15	35	1	6	10
5	2	2	1	1	0	283.98	196.02	14	14	20	8
6	3	2	1	1	0	274.56	205.44	6	14	14	9
7	4	3	1	1	1	369.24	110.76	0	9	38	0
8	5	3	1	1	1	387.83	92.17	0	8	44	0
9	6	2	1	1	0	350.36	129.64	12	6	18	6
10	7	3	1	1	1	342.29	137.71	0	19	35	0
11	8	2	1	1	0	315.09	164.91	9	9	26	5
12	9	2	1	1	0	337.87	142.13	10	11	24	9

FIGURE 35.1   Ideal Dataset to Test for Referee Bias.

(see file Refsim.xls.). Note that for each game we can classify each foul into one of four groups:

- A Black official calling a foul on a Black player
- A White official calling a foul on a White player
- A White official calling a foul on a Black player
- A Black official calling a foul on a White player.

A "1" in columns E through G denotes a White official while a "0" in columns E through G denotes a Black official. For example, in game 1 there was one White and two Black officials. Black players played around 397 minutes during the game and White players around 83 minutes. Thirty-five fouls were called by Black officials against White players, one foul was called by the White official against a White player, six fouls were called by the White official against Black players, and 10 fouls were called by a Black official on a White player.

Combining this data over all 1,000 games we can determine the rate at which each of the four types of fouls occurs. We would find that:

- White referees call 1.454 fouls per 48 minutes against Black players while Black referees call 1.423 fouls per 48 minutes against Black players.

- Black referees call 1.708 fouls per 48 minutes against White players while White officials call 1.665 fouls per 48 minutes against White players.

Note that this data shows a clear (but small) bias in that an official of a different race calls more fouls on either White or Black players than an official of the same race.

As an example of these calculations we found that 9,204 total fouls were called by Black officials against Black players. Black player minutes occurring when a Black official is on the court totaled 310,413 minutes (note that during a game in which Black players play, for example, 200 minutes and there are, say, two Black officials we count this as 400 Black player Black official minutes). Therefore Black officials call $48 * 9{,}204/310{,}413 = 1.423$ fouls per minute on Black players.

The problem is that creating the data shown in Figure 35-1 requires that we know which official called each foul. This data is kept by the NBA but is not publicly available. The NBA says its study of this proprietary data does not show any evidence of prejudice. WP's analysis, however, tells a different story.

WP'S APPROACH: REGRESSION ANALYSIS WITH INTERACTION

Since WP do not know which official made each call, they worked with box score data. For each player in each game, they have a data point containing the following: fouls per 48 minutes committed by player, race of player, percentage of game officials who are White. WP refer to the first variable as foul rate, the second variable as Blackplayer, and the third variable as %Whiteref.

For example, if a Black player played 32 minutes and committed three fouls and there were two White officials, then the data point would be: $(3 * 48/32, 1, 2/3) = (4.5, 1, 2/3)$. WP use $1 =$ Black player and $0 =$ White player. Next, WP run a regression (see Chapter 3 for an explanation of regression) to predict player fouls per 48 minutes from the following independent variables:

- %Whiteref * Blackplayer
- %Whiteref
- Blackplayer.

WP weighted each data point by the number of minutes the player played in the game. They found the following equation best fits the data:

$$\text{Foul Rate} = 5.10 + .182\text{Blackplayer} * (\%\text{Whiteref}) - \tag{1}$$
$$.763(\text{Blackplayer}) - .204(\%\text{Whiteref}).$$

WP found that each of their independent variables has a p-value less than .001. This means that each independent variable is significant at the .001 level. **In other words, for each of the three independent variables there is less than one chance in a 1,000 that the independent variable is not a useful variable for predicting foul rate.**

So what can we conclude from (1)? Table 35-1 shows, for all possible values of %Whiteref $(0, 1/3, 2/3, 1)$ and all possible values for Blackplayer (1 or 0), the predicted foul rate per 48 minutes.

Table 35-1 shows that when there are three Black officials, Blacks are called for 0.76 fewer fouls per 48 minutes than Whites, while when there is an all-White officiating crew, Blacks are **only** called for 0.58 fewer fouls per 48 minutes than Whites. Table 35-1 implies that for any composition of the officiating crew, Whites commit more fouls per 48 minutes than Blacks. **The discrepancy between the rate at which Blacks and Whites foul shrinks by 23% (.18/.76) as the officiating crew shifts from all Black to all White.**

WP included other independent variables, such as player height, player weight, indicators of whether the player was an all-star, and player position. Inclusion of these extra independent variables did not change the conclusion that the racial makeup of the officiating crew has a small (but statistically significant effect) on the frequency with which Black and White players commit fouls.

TABLE 35.1
Predicted Foul Rate per 48 Minutes

%WhiteRef	Blackplayer = 1	Blackplayer = 0	Black-White Rate Difference
0	$5.10 + .182(0)(1) - .763(1) -$ $.204(0) = 4.337$	$5.10 - .204(0)$ $= 5.1$	-0.763
1/3	$5.10 + .182(1/3)(1) - .763(1) -$ $.204(1/3) = 4.329$	$5.10 - .204(1/3)$ $= 5.032$	-0.702
2/3	$5.10 + .182(2/3)(1) - .763(1) -$ $.204(2/3) = 4.323$	$5.10 - .204(2/3)$ $= 4.964$	-0.642
1	$5.10 + .182(1)(1) - .763(1) -$ $.204(1) = 4.315$	$5.10 - .204(1)$ $= 4.896$	-0.581

THE MEANING OF A SIGNIFICANT INTERACTION

Interaction terms in a regression can help our understanding of
the relationships among the variables in the model. The presence
of a (significant) interaction between two independent variables
indicates that the effect of one predictor variable on the response is
different for different values of the other predictor variable. In particular, WP found that when predicting foul rates there is a significant **interaction** between a player's race and the racial makeup of
the officiating crew. The presence of an interaction term changes
drastically the interpretation of the coefficients. If there were no
interaction term in equation (1), the coefficient −.763 would be the
unique effect of being Black on the rate of being called for a foul.
However, in the WP model there is an interaction term, and so the
unique effect of being a Black player on the rate of being called for a
foul is not limited to the single variable coefficient (−.763), but also
depends on the percentage of White referees (multiplied by .182).

The way WP spotted the player race official race interaction
was by including the new independent variable Blackplayer ∗
(%Whiteref), which multiplied two independent variables. In
general, to see if two independent variables interact, include
the product of the two independent variables as an independent
variable in the regression. If the coefficient of the product term
is not zero, then the two independent variables interact. If one
cannot reject the null hypothesis that the product term is equal
to zero, the independent variables do not exhibit a significant
interaction.

IS THERE STILL RACIAL BIAS?

In 2010, Devin Pope, Joseph Price, and Justin Wolfers (PPW) revis-
ited[2] the original WP study and found interesting results. A *Washington Post* article[3] summarized and compared the results. PPW
analyzed NBA data from 2003 to 2006 before the original WP study
was revealed and from 2007 to 2010 after the original study was revealed.[4] Using Wolfer and Price's model, PPW found that before the
study was revealed (2003–2006), there still was significant racial
bias in the NBA regarding fouls called. Using the same model for
2007 to 2010, PPW found that there was no longer significant racial
bias from officials in the NBA. This is a valuable finding, since it
provides evidence that knowledge of (possibly unconscious) biases
can help in reducing and eliminating them.

2. D. Pope, J. Price, and J. Wolfers, "Awareness reduces racial bias," *Management Science*, 64(11), 2018, 4988–4995.
3. https://www.washingtonpost.com/news/wonk/wp/2014/02/25/what-the
-nba-can-teach-us-about-eliminating-racial-bias/
4. Note that the original WP study analyzed data from 1991–1992 to 2003–2004
seasons, and hence there is no overlap between the data used in the two studies.

CHAPTER 36

PICK-N-ROLLING TO WIN, THE DEATH OF POST UPS AND ISOS

One of the newest types of data that the NBA provides through its website (stats.nba.com) is play types (powered by Synergy). We can now get information about pick-n-roll plays involving specific players, isolation plays, post ups, off screens, put backs, cuts, etc. This allows us to start understanding the style of play for a team or individual players. It also allows us to see play trends over time, with all the caveats that come with manual charting (which is how Synergy compiles the data) and changes in the corresponding protocols over the years.

PICK-AND-ROLL: RETURN TO THE FUNDAMENTALS

This is particularly important to one of the authors of the book (KP), and we will start with a personal story. August 31, 2006: this was the day that I moved from Greece to the US to start my Ph.D. studies. It was also the day that the Greek national basketball team achieved its biggest win ever against Team USA—which included LeBron, Wade, Carmelo, Bosh, Howard, and Paul, among other NBA stars—in the semifinals of the 2006 World Cup in Japan. Why is this relevant here? Because Greece did so by repeatedly running the same play almost every possession (unfortunately we do not have the actual data here, but you have to take our word for it—or ESPN's).[1] What was that play? Pick-n-roll! Even though Stockton and Malone mastered the play in the 1990s, the NBA relied heavily on isolations and one-on-one game during that era (while in other parts of the world, pick-n-roll was one of the fundamental plays being taught in youth basketball). As the skillset of big men started changing and being more versatile, this play started taking the league by storm. According to Synergy Sports, in 2006 approximately 20% of the possessions in the NBA finished using pick-and-roll (either from the ball handler or the roll man), while this number in 2018–2019 jumped a bit over 30% (a 34% increase in the play frequency)! Figure 36-1 presents the data for the 2018–2019 season for possessions[2] that ended up with a pick-n-roll play. The data are also further broken down by who took the final action (the ball handler or the screener). As we can see, almost two out of the three PNRs end with an action from the ball handler, despite the fact that they are less efficient, as captured by the points per possession (PPP). Overall, a PNR generates .93 PPP (when the PNR ends with an action from the ball handler or the screener). Portland generated the best PNR offense in the season with 1.023 PPP, largely driven by the almost 1 PPP from the ball handler PNRs (Damian Lillard had 10.9 possessions per game in PNR situations as a ball handler who finished with a whooping 1.08 PPP!). Dallas also had the best roll man PNR efficiency, with two players, at the 1.39 and 1.35 PPP (Kleber and Powell, respectively), exploiting the playmaking abilities of 2018–2019 ROY Luka Doncic. Now one thing to understand here is what exactly the data

---

1. https://www.espn.com/olympics/wbc2006/news/story?id=2568543

2. A possession according to Synergy is essentially a chance. For instance, an offensive rebound renews the possession and provides a new chance for the offense to score. However, Synergy considers this as a new possession. So, keep this in mind whenever you compare points per possession from Synergy to points per possession for other sources.

on the NBA stats site includes. The PNR offense that ends with a ball handler action includes only shots taken (or turnovers) by the ball handler. It does not include other plays after the screen (e.g., passes out of the screen). The latter are highly efficient and generate good offense—however, these data are not part of the public data feed provided by the league (even though Synergy tracks this as well and provides teams with information on what they call PNR derived offense).

But why do teams run PNR with ball handlers more than with roll men if they are less efficient? Well, here is where an analyst needs to understand what the data being analyzed capture and how they were logged. A ball handler will be "charged" with the termination of a chance with 0 points if he tries to hit the roll man but instead turns the ball over. Simply put, the ball handler PNR category does not include only shots taken from the ball handler out of a PNR. On ball events are more probable to generate turnovers, creating several of the differences in the efficiency numbers.[3] While inconvenient for the analyst, proper understanding of the situation will not lead to misinterpretations of the data.

Tracking data provided from Second Spectrum (more details about it in the next chapter) include additional metadata such as the type of defense the ball handler and the screener defenders chose (e.g., over, show, blitz, etc.) and the actual action from the screener (e.g., pop, roll, or slip). Given that Second Spectrum provides detailed information for the players involved in each PNR, we could imagine a PNR-specific adjusted +/− metric depending on the role that the players have in the play. The last point is interesting since traditionally the screener is a "big," while the ball handler is a guard. However, a recent article[4] by FiveThirtyEight discussed

---

3. Seth Partnow of The Athletic (and previous head of data analytics for the Milwaukee Bucks) wrote an interesting article on this: https://theathletic.com/1733785/2020/04/10/nba-offensive-styles-analysis-part-ii-variety-is-the-spice-of-life/

4. https://fivethirtyeight.com/features/want-to-confuse-an-nba-defense-have-a-guard-set-a-ball-screen/

TEAM	TOT_FREQ	BHANDLER_POS	BHANDLER_PPP	SCREENER_PPP	SCREENER_POS	PPP
ATL	26.1	23	0.8	1.18	7.8	0.90
BOS	21.3	16.3	0.91	1.18	7.5	1.00
BKN	27.2	23.7	0.89	1.05	7.5	0.93
CHA	29.2	25	0.93	1.04	7.4	0.96
CHI	27.7	22	0.85	1.1	8.4	0.92
CLE	26.1	23.3	0.86	1.06	5.2	0.90
DAL	24.6	21.3	0.86	1.21	6.1	0.94
DEN	20.2	16	0.88	1.03	6.6	0.92
DET	21.9	19	0.88	1.05	5.4	0.92
GSW	14.5	12.1	0.99	0.93	4.1	0.97
HOU	21	16.2	0.9	1.11	7	0.96
IND	25.8	17.2	0.81	1.03	11	0.90
LAC	29.7	25.8	0.9	1.21	7.9	0.97
LAL	22.9	19.7	0.83	1.13	6.8	0.91
MEM	23.7	17.6	0.85	1.03	7.9	0.91
MIA	25	21	0.75	1.16	6.9	0.85
MIL	17.1	13.9	0.91	1.13	5.6	0.97
MIN	23.3	19.4	0.84	1.18	7.4	0.93
NO	18.4	15	0.81	1.14	6.5	0.91
NYK	26.5	24	0.88	1.1	5.8	0.92
OKC	24	21.1	0.85	1.02	7.2	0.89
ORL	24.4	20.3	0.83	1.11	6.5	0.90
PHI	15	13.1	0.91	1.1	4	0.95
PHX	26.5	23.4	0.83	1.09	6.3	0.89
POR	24.5	21.5	0.98	1.17	6.3	1.02
SAC	22.6	19	0.84	1.08	7.1	0.91
SAS	24.1	20.2	0.92	1.06	6.2	0.95
TOR	22.3	16.8	0.87	1.07	8.1	0.94
UTA	27.9	24.1	0.86	1.22	7.2	0.94
WAS	21.1	17.2	0.86	1.08	6.8	0.92

FIGURE 36.1   Pick-n-Roll Play Type Data.

the increasing trend of PNRs where the screener is a guard. These PNRs add +.09 PPP, compared to PNRs where the screen is set by a "big." Similar metadata can also help with another limitation from play type data that is important to understand. Play type data include situations where the chance ended with this play type (e.g., PNR). However, many times the defense forces the offense to reset after running a PNR (e.g., a good hedge from the roll man defender, forcing the ball handler to hold on to the ball and reset the offense instead of hitting the roll man or finishing the chance). From play type data, we will not be able to credit the good defense in these situations, because they are not being recorded. Overall, the *take away* from this discussion is that data are great to have, but understanding how they were collected, what they are logging, etc., is just as important.

THE DEATH OF POST UPS

If you grew up in the 1980s, 1990s, or even the 2000s, you are accustomed to teams looking for a big guy that can push his way to the low block and dunk all over the defense. You might be even missing it a little bit (even though that would be a bit strange since the league introduced the "5 seconds back to the basket" violation for players posting up more than five seconds, because that play was "destroying" the league). It certainly is not a big part of today's offense—at least as direct offense. And there is a good reason for this. Post up shots are not as efficient (anymore). In particular, possessions that end up with a post up move (shot or turnover) are worth .89 PPP. Figure 36-2 shows the post up value for each team for the 2018–2019 season. There are few teams that are able to generate high value from post up shots. Philadelphia is one of them, having Embiid, one of the best—if not the best—post up center today in the league. However, long gone are the days where a player like Shaq would play one-on-one any defender and get his way to the basket. Kirk Goldsberry, in his excellent recent book *SprawlBall,* provides longitudinal data on the fraction of post ups per 100 possessions that show that post up activity has declined by 45% between 2013–2014 and 2018–1019! Now, again, this does not mean that there is no value in post ups. In particular, offense **generated** from the low post can be particularly efficient. One of the possible reasons that post ups shots are not very efficient in today's NBA might be that players have become faster and defenses manage to quickly double-team the post. This, in conjunction with rules introduced ("5 seconds back to the basket"), can make these plays inefficient. However, if the player posting up is an excellent passer, then this post action can generate some high-quality shots from elsewhere on the floor. Data from the DribbleHandoff

TEAM	POSSESSIONS	PPP
ATL	2.8	0.92
BOS	4.6	1
BKN	0.4	0.81
CHA	4.8	1
CHI	2.6	0.83
CLE	5.9	0.87
DAL	3.7	0.8
DEN	8.2	0.94
DET	6.3	0.91
GSW	5	0.86
HOU	2.8	0.86
IND	5.3	0.92
LAC	7.4	0.89
LAL	8.3	0.95
MEM	6.2	0.89
MIA	3.2	0.97
MIL	7.2	0.98
MIN	5	0.89
NO	2.9	0.9
NYK	7.8	0.88
OKC	4.3	0.96
ORL	6.1	0.73
PHI	12.6	1.01
PHX	5.6	0.93
POR	6.2	0.95
SAC	5.3	0.84
SAS	8.7	0.98
TOR	5.3	0.82
UTA	2.3	0.95
WAS	1.9	0.87

FIGURE 36.2 Post Up Play Type Data.

.com show that shots coming from post passes are worth approximately 1.14 points. This is +.25 points compared to shots off the post. In fact many NCAA teams (and teams in top-level European competition) run an inverted offense, where they post an excellent passing guard to generate good-quality shots off post passes. Just ask Villanova, and Jalen Brunson, who led it to two National Championships running this play!

TEAM	POSSESSIONS	PPP
ATL	4.4	0.85
BOS	7.8	0.79
BKN	9.3	0.96
CHA	6.9	0.91
CHI	8	0.82
CLE	8.8	0.8
DAL	6.8	0.85
DEN	5.3	0.8
DET	6.5	0.87
GSW	7.1	0.98
HOU	22.4	1.06
IND	5.7	0.79
LAC	7.2	0.93
LAL	9.4	0.91
MEM	6.2	0.86
MIA	5.1	0.78
MIL	9.6	0.98
MIN	7.7	0.84
NO	8	0.86
NYK	6.3	0.76
OKC	10.5	0.83
ORL	4.6	0.9
PHI	5	0.88
PHX	6.5	0.87
POR	7.7	0.9
SAC	8	0.83
SAS	7.4	0.95
TOR	7.7	0.94
UTA	3.6	0.83
WAS	7.8	0.84

FIGURE 36.3   Isolations Play Type Data.

ISOLATIONS

The decline in post ups does not prove that one-on-one basketball is not in the NBA anymore. It has just changed optics, and it is called isolation. While post ups decline, isolations remain about the same in terms of frequency in a team's playbook. Isolations in the 2018–2019 season (see Figure 36-3) were worth .87 PPP (even worse than post up plays). However, specific teams thrive off isolations. Houston used this play in 20% of its possessions during that season and had a whooping 1.06 PPP (if you are able to maintain this efficiency with this frequency rate, go for it!). Of course, responsible for these numbers is James Harden, who generated 1.11 PPP in isolation plays!

Before we close this chapter, it is very important to reiterate the discussion on the caveats of manual charting, and possible pitfalls. An isolation play is straightforward to spot and credit the points off of it. However, a "cut" play (which is the most efficient play type in the NBA) requires many things to go the offense's way to end up with a shot (e.g., off-ball movement, screens, off-balance defense, etc.). Then, and only then, a pass to a cutting player will be delivered, and this delivery will lead to a great quality chance. One thing that it will be interesting to have data on is how many times teams have tried to run a cut play and had the defense reject it (and hence, have never made it to the Synergy data). This is information only teams have, but one should certainly consider it when creating a game plan!

CHAPTER 37

SPORTVU, SECOND SPECTRUM, AND THE SPATIAL BASKETBALL DATA REVOLUTION

The advancements in computing technology during the past decade have enabled sports organizations to collect vast troves of detailed data that move beyond the traditional (or advanced) box score, or even the play-by-play data. We are literally able to replay a whole game just from data logs. At any time, we know the exact location of each player on the field or court. We saw in Chapter 27 that the NFL uses RFID sensors to collect this spatiotemporal information. The NBA has been collecting this type of information for much longer, since 2013, when the SportVU optical tracking system owned by STATS Inc. was installed in all 29 NBA arenas. Starting in the 2017–2018 season the NBA made an agreement with Second Spectrum to be the provider of this information. Contrary to the NFL player tracking data, NBA tracking data comes from an optical tracking system that makes use of six cameras residing at the rafters of the stadiums, which collect 25 "frames" per second. The raw data feed

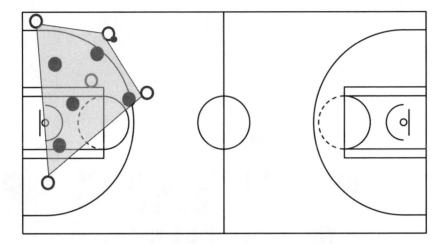

FIGURE 37.1   A Player Tracking Data Snapshot Visualization.

provided includes X, Y positioning of players as well as X, Y, and Z positioning of the ball, while Second Spectrum has added several layers of metadata, including play-type details (e.g., post ups and pick-n-roll metadata, etc.), shot quality, etc.

NBA front offices and researchers have analyzed these data for several years now and have made several advances. For one, this type of data has allowed us to start exploring things like floor spacing and defense, for which statistics do not exist in the traditional box score. For example, Figure 37-1 depicts a snapshot from player tracking data where the black circles correspond to the defending team, the white circles correspond to the offense, and the ball is depicted with the smaller dark circle. Watching the game film, one can (possibly in a subjective manner) decide whether the offense had a good spacing in this specific snapshot. However, having player tracking data allows us to now quantify this spacing. While several people might add "objectively" to "quantify this spacing," we prefer the word "consistently." After all, whatever definition one uses, it includes subjective elements. A typical way to do this is to calculate the area covered by the convex hull (see Chapter 27) defined by the locations of the five players on the court. In Figure 37-1, the convex hull defined by the offensive lineup is depicted with the shaded area, and, most importantly, this area can be calculated. This allows us to calculate how much more Steph Curry stretches the defense as compared to, say, Giannis. Anthony Sicilia, a Ph.D. student at the University of Pittsburgh, showed how one can replicate the adjusted plus/minus framework using as our dependent variable the defense/offense spread.[1]

EVALUATING DEFENSE

Teams have gone beyond this, trying to tackle previously largely ignored—mainly due to data availability—issues, such as defense evaluation. The Toronto Raptors were one of the early adopters of this technology and designed a system for *optimal* defensive positioning.[2] They analyzed the possessions from an expected points per possession perspective and developed "defensive ghost" players. These ghosts represent where the defender should be optimally positioned to minimize the expected points for the possession. Recently, researchers from Caltech, Disney Research, and STATS Inc. developed a similar automated data-driven ghosting system for soccer that is based on imitation learning.[3] These data-driven approaches currently identify the expected defense from the perspective of a league-average team, i.e., the objective function does not directly incorporate an expected points value.

Alexander Franks, Andrew Miller, Luke Bornn (ex-VP of Strategy and Analytics for the Sacramento Kings), and Kirk Goldsberry[4] (ex-VP for strategic research of the San Antonio Spurs, and

1. Anthony Sicilia, "On the application of convex hull based spatial metrics in the NBA," in Cascadia Symposium on Statistics in Sports, 2018 (poster).

2. http://grantland.com/features/the-toronto-raptors-sportvu-cameras-nba-analytical-revolution/

3. H. Le, P. Carr, Y. Yue, and P. Lucey, "Data-driven ghosting using deep imitation learning," 11th Annual MIT Sloan Sports Analytics Conference, 2017.

4. Alexander Franks, et al., "Counterpoints: Advanced defensive metrics for NBA basketball." 9th Annual MIT Sloan Sports Analytics Conference, 2015.

of *SprawlBall* fame) (FMBG) further used optical tracking data to develop defensive metrics for basketball. Tracking data allow us to identify who is guarding whom, and hence define metrics such as the total volume of attempts a defender faces, the degree to which a defender can reduce the effectiveness of the player he is guarding, and the (weighted) average of points scored against him. Central to the definition of these defensive metrics is identifying a defensive matchup at each point of a possession. To complete this task, FMBG[5] use regression to model the *canonical* (expected) position μ_{tk} of a defender of player k at time t. There are three independent variables that are related with the location of the players and the ball on the court:

$$\mu_{tk} = \gamma_O O_{tk} + \gamma_B B_t + \gamma_H H \tag{1}$$

where $\gamma_O + \gamma_B + \gamma_H = 1$, O is offensive player k's position at time t, B is the ball position at time t, and H is the location of the hoop. The weights finally learned are $\gamma_O = 0.62$, $\gamma_B = 0.11$, and $\gamma_H = 0.27$. Since the weights are expressed on a relative scale, the model *reveals* that defenders guard players more closely when they are near the basket, and defenders guard the ball carrier more closely compared to off-ball matchups (as one might have expected). Using these canonical positions, the evolution of the defensive matchups can be expressed through a hidden Markov model that can track defensive switches. Having inferred the defensive matchups through a possession, advanced defensive metrics can be defined. For instance, FMBG define the *disruption score* of defender p as:

$$D_p = \frac{\text{Observed points against}}{\text{Expected points against}} \tag{2}$$

---

5. A. Franks, A. Miller, L. Bornn, and K. Goldsberry, "Characterizing the spatial structure of defensive skill in professional basketball," *The Annals of Applied Statistics*, 9(1), 2015, 94–121.

The expected points against are calculated using the efficiency of the shooters that p defended, as $\Sigma_i(\text{eff}_{S(i)} \cdot \text{pts}(i))$, where $S(i)$ is the shooter in possession i, pts(i) is the point value of the shot in possession i, and $\text{eff}_{S(i)}$ is the historical efficiency of shooter $S(i)$. A value for $D_p < 1$ corresponds to a better-than-average defender, while $D_p > 1$ corresponds to a worse-than-average defender (with regard to the defined disruption score). Several other defensive metrics can be devised by utilizing the player tracking data, and this is clearly an area in basketball analytics that, while it has seen several developments during the last few years, is still under development.

DEEPHOOPS: REAL-TIME EXPECTED POINTS PER POSSESSION

Traditional box score statistics evaluate actions that end a possession or lead to a dead ball—e.g., shots taken or made, turnovers, blocks, fouls, etc. However, to reach this point there are several other actions that take place that it would be very useful to find a way to evaluate. For example, in a hammer set, how (much) does the hammer screen help the play? Box scores cannot provide such information. If there were a way to keep track of the expected points to be scored during a possession, then we could possibly start evaluating various actions on the court, such as screens (non-assist), passes, etc. Sicilia, Pelechrinis, and Goldsberry[6] (SPG for short) developed *DeepHoops*, a system that uses a deep learning model to predict the probability of a terminal action occurring within a window size r (typically very small). These terminal actions include taking a shot, committing a turnover, drawing a (shooting) foul, or nothing happening and the possession goes on. The input to this model is the whole trajectory of the players and the ball during the past T seconds. Once these probabilities are estimated every moment, the expected possession

---

6. A. Sicilia, K. Pelechrinis, and K. Goldsberry, "DeepHoops: Evaluating micro-actions in basketball using deep feature representations of spatio-temporal data," In the 25th ACM SIGKDD, 2019.

value can be estimated by taking the expectation of this distribution. For example, if the probabilities for the various actions at time τ are estimated as $\Pi = [\pi_{shot}, \pi_{shF}, \pi_{nshF}, \pi_{to}, \pi_{null}] = [0.21, 0.06, 0.03, 0.04, 0.66]$, and the corresponding values for these actions are $V = [v_{shot}, v_{shF}, v_{nshF}, v_{to}, v_{null}] = [0.91, 1.5, 1.5, 0, 1.09]$, then the expected points value for this possession at time τ is $EPV(\tau) = \Pi \cdot V = 1.045$. Vector V assigns a point value to each action. For instance, in this example $v_{shot} = 0.91$, which means that the expected points from a shot taken at time τ (with the specific player, player locations, defender(s), etc.) from the ball handler would have an expected value of 0.91. If there is a non-shooting foul and the team is in penalty, the value of this state is equal to the expected points from two free throw shots from the ball handler (in this case we assume a player with 75% at FTs). For the state where none of these happens, the value is equal to the average possession value (we used an average offensive rating of 109). The latter can be further adjusted by considering the specific teams and lineups on offense and defense. Figure 37-2 (taken from the original work) showcases an example of the output of these calculations for a Kevin Durant drive and kick-out to Ian Clark. We can see the snapshots of the play along with the running value of expected points and the terminal-action probability distribution at each moment. Frame 1 shows average expected points (ball starts in backcourt). Frame 2a shows Durant surrounded by defenders in the key, and at that point the model outputs an increased likelihood of both field goal attempt and shooting foul. Frame 2b shows the likelihood of shooting foul dropping as the ball is in flight to the corner. Frame 3 shows Ian Clark open in the corner, receiving the ball. While the model maintains low likelihood of shooting foul (no defenders nearby) the likelihood of a shot attempt is drastically increased now!

Using similar models, one can now start evaluating various action through the EPV added. For example, a pass that left from the passer's hand at t_1 and reached the receiver at t_2 has an EPV added value of $EPV(t_2) - EPV(t_1)$. Figure 37-3 presents the box plots for

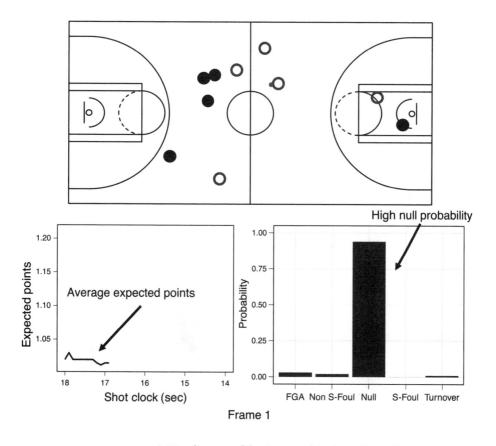

FIGURE 37.2   A Visualization of the Output of the DeepHoops System
(Sicilia et al., 2019).

three different types of passes in the SPG study, namely, backcourt
passes, assist opportunities (i.e., assists that lead to a shot—made or
missed), and all other passes. As we can see, backcourt passes add
pretty much 0 value, as expected, while assist opportunities add the
most value on average (approximately 0.2 EPV added). All other
passes have on average a 0.05 EPV added.

These models can further be combined with generative models
that model the anticipated (not the optimal) reaction or movement
of the defense to an offensive scheme and then inform the coaching staff about the expected efficiency of a designed play. Seidl et al.

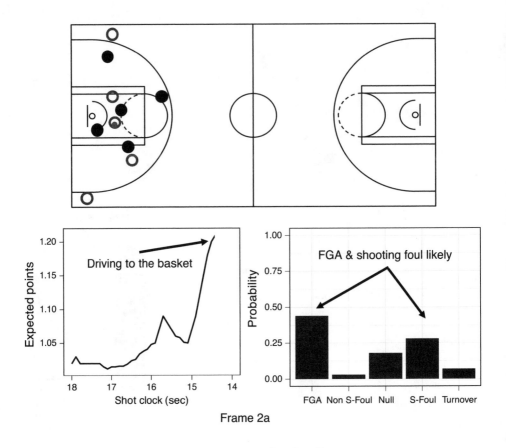

Frame 2a

FIGURE 37.2 (*continued*)

in their great work *Bhostgusters,*[7] developed such a model that uses imitation learning to predict the movement of the defense based on how players have reacted in the past. With enough data, one can also imagine these models taking into consideration the teams playing, or even the specific lineups matching up.

7. T. Seidl, A. Cherukumudi, A. Hartnett, P. Carr, and P. Lucey, "Bhostguster: Realtime interactive play sketching with synthesized NBA defenses," in MIT Sloan Sports Analytics Conference, 2018.

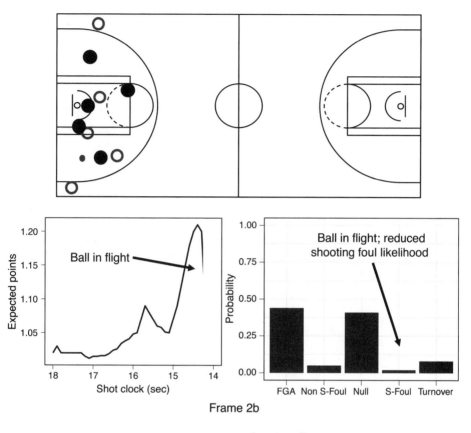

Frame 2b

FIGURE 37.2   (*continued*)

WHY ARE CORNER THREES SO EFFICIENT?

The left part of Figure 37-4 presents the league-average field goal percentage for the season 2016–2017 from the different areas on the floor. As we can see, there is an obvious relationship with distance, which becomes even more evident from the detailed data obtained from Basketball-Reference.com in Figure 37-5. Shots closer to the basket have a higher chance of going through the net. However, shots at the rim average 2 ∗ .63 = 1.26 points per shot and three-point shots average 3 ∗ .36 = 1.08 points per shot, while **all shots between a shot at the rim and a three-point shot average a pitiful 0.8 points**

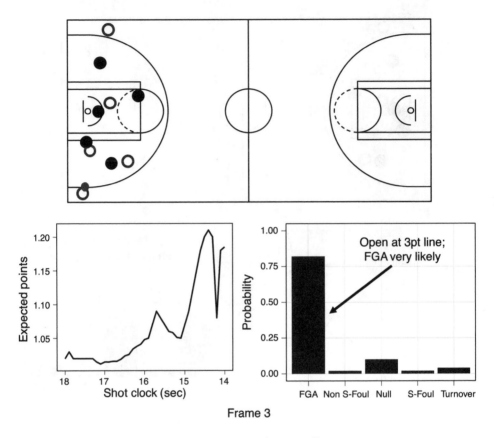

Frame 3

FIGURE 37.2 (*continued*)

per shot. This is further visualized in the right part of Figure 37-4.[8] Literally, *raw* and *efficiency*-based metrics paint a different picture! It should be clear then that long two-point shots are, on average, not a very good choice. An examination of the data shows that teams are moving away from these shots. In 2002–2003 teams were taking around 30 midrange shots per game (according to stats.nba.com).

8. Also, as pointed out by the great Pete Palmer in his review of our first edition, three-point shots are missed more often than long two-point shots, and this gives the offense a greater chance of getting another possession through an offensive rebound (about 27% of missed shots result in offensive rebounds).

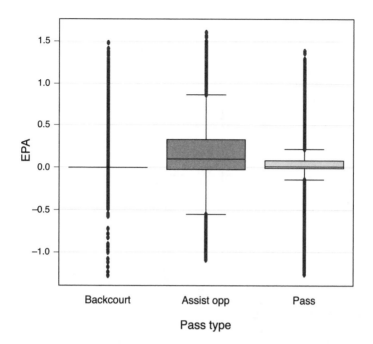

FIGURE 37.3 EPV Added for Different Types of Passes.

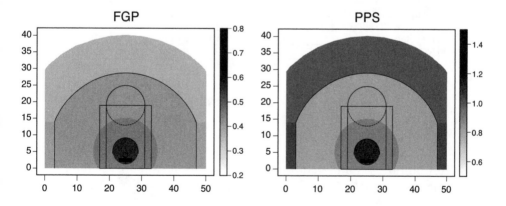

FIGURE 37.4 Three-point shots have the lowest field goal percentage (left figure), but provide the second-best points per shot (right figure).

	A	L	M	N	O	P
1		0–3 ft	3–10 ft	10–16 ft	16ft – 3P	3P
2	Atlanta Hawks	0.616	0.349	0.394	0.429	0.341
3	Boston Celtics	0.633	0.395	0.417	0.405	0.359
4	Brooklyn Nets	0.613	0.418	0.391	0.374	0.338
5	Chicago Bulls	0.6	0.449	0.393	0.377	0.34
6	Charlotte Hornets	0.61	0.4	0.419	0.38	0.351
7	Cleveland Cavaliers	0.666	0.362	0.428	0.425	0.384
8	Dallas Mavericks	0.637	0.444	0.441	0.391	0.355
9	Denver Nuggets	0.627	0.441	0.386	0.397	0.368
10	Detroit Pistons	0.646	0.384	0.445	0.416	0.33
11	Golden State Warriors	0.69	0.408	0.43	0.469	0.383
12	Houston Rockets	0.667	0.391	0.375	0.374	0.357
13	Indiana Pacers	0.625	0.464	0.417	0.418	0.376
14	Los Angeles Clippers	0.661	0.419	0.423	0.439	0.375
15	Los Angeles Lakers	0.633	0.397	0.379	0.39	0.346
16	Memphis Grizzlies	0.58	0.416	0.385	0.38	0.354
17	Miami Heat	0.63	0.412	0.387	0.389	0.365
18	Milwaukee Bucks	0.644	0.402	0.377	0.387	0.37
19	Minnesota Timberwolves	0.645	0.437	0.392	0.402	0.349
20	New Orleans Pelicans	0.627	0.417	0.427	0.38	0.35
21	New York Knicks	0.591	0.381	0.44	0.419	0.348
22	Oklahoma City Thunder	0.614	0.418	0.391	0.379	0.327
23	Orlando Magic	0.616	0.427	0.413	0.39	0.328
24	Philadelphia 76ers	0.638	0.376	0.384	0.344	0.34
25	Phoenix Suns	0.629	0.412	0.402	0.374	0.332
26	Portland Trail Blazers	0.596	0.418	0.441	0.417	0.375
27	Sacramento Kings	0.642	0.421	0.412	0.402	0.376
28	San Antonio Spurs	0.644	0.429	0.391	0.431	0.391
29	Toronto Raptors	0.631	0.45	0.449	0.407	0.363
30	Utah Jazz	0.651	0.413	0.427	0.363	0.372
31	Washington Wizards	0.629	0.429	0.426	0.442	0.372
32	**League Average**	**0.631**	**0.414**	**0.412**	**0.403**	**0.358**

FIGURE 37.5   2016–2017 NBA Shooting.

This number plummeted to a little less than 10 shots per game in 2019–2020. This is not to say that the midrange game is dead (as many who oppose quantitative analyses in sports believe it is). It is just that the midrange game is left to the *elite* players (the Durants and Leonards of the league), while less involved players spend their time on court on more *valuable* areas (not only providing higher efficiency shots but also improving spacing overall).

One thing to notice is that even among the three-point shots there is a difference in the FG% depending on whether they are above the break or from the corner, with the corner threes being made at a 38.8% rate, while the ones above the break have an FG% of 34.7%. This means that a corner three is worth 1.16 points per shot, while shots above the break are worth 1.04 points per shot. This is a net difference of 12.3 points per 100 shots! Why is this though? People have been pointing out that the shorter distance in the corners is responsible for this difference.[9] While this certainly should play a role in the difference observed, we would not expect this to be the sole explanation. Luckily, there are ways to explore whether this is the case. First, we can build a logistic regression model (similar to the one in Chapter 21 for the probability of a successful FG in the NFL) for the probability of a shot being made based solely on the distance of the shot from the basket. According to the model, the corner threes are expected to have a higher FG% when compared to above-the-break threes. However, their expected difference is approximately 1.5%, which is smaller than the 4% observed in practice.[10] Digging a bit deeper in the data, one of the differences we observed is the distance between the shooter and the closest defender across different shot types. In particular, corner three pointers are on average more open as compared to above-the-break three pointers, making them higher quality shots. What drives this difference though? Assists! Assisted shots are, on average, more open compared to unassisted shots. More than 90% of corner three-point shots are assisted (including also potential assists for shots that were not made, but if they were would be logged as assisted), while shots above the break are

9. https://bleacherreport.com/articles/2146753-whos-responsible-for-the-nbas-corner-three-revolution

10. Also the actual expected values in the FG% are different between the model and the actual data. The reason for this is essentially that the observed data are biases from the fact that better shooters will take three-point shots regularly, while the model is built using shots from all players everywhere on the floor, impacting the absolute results.

TABLE 37.1

Shorter distance explains only part of the difference in the field
goal % between different types of three. Corner threes are also
more open on average compared to above-the-break threes

	Actual FG% (from raw data)	Expected FG% (from logistic model)	Distance of Closest Defender (feet)
Corner 3	38.7%	35.4%	6.4
Above-the-Break 3	34.7%	33.8%	5.9

assisted at a rate just above 70%. Midrange shots are assisted at an
even lower rate, not helping their efficiency case.

In order to be certain that there exists a causal relationship between the assists and the FG% difference between corner and above-
the-break things, we would need a randomized control trial (similar
to how medical treatments are evaluated), which is the gold standard
for causal inference. For instance, we could have half of the games
played on a court where the distance from the three-point line is the
same everywhere, and half of them played on a *regular* NBA court.
Also, crucially, these games should be assigned a court randomly.
Now, as you can possibly imagine, even in the era of the NBA bubble,
we cannot do that. However, here we have a setting that resembles a
natural experiment, that is, an observational study where the study
subjects are exposed to the experimental and control conditions by
factors outside the control of the investigator (e.g., nature). In our
case, FIBA competitions do not have a discrepancy in the three-point
line between the corner threes and above the break. We analyzed
data from FIBA's Champions League (an international European
competition), and the same pattern emerged; corner three shots are
more efficient compared to above the break, and are assisted at a
much higher rate! Even though this is not by any means a formal
natural experiment study, it provides additional evidence that the
distance is not the driving force of the difference in the efficiency.

NATURAL EXPERIMENT

A natural experiment is an observational study where the *treatment* of interest is assigned seemingly randomly based on factors outside the control of the analyst (e.g., a public policy), and this is exploited to answer a particular question. Natural experiments are used often in fields like epidemiology and social and political sciences, where many times randomized control trials are costly or even unethical to run. One thing to remember is that in natural experiments the treatment is still not assigned randomly to the groups examined, and hence there can be unmeasured confounding. Exactly because of that reason, natural experiments will never be able to answer a causal question without a shred of a doubt. Hence, natural experiments are *quasi experiments*.

If this is the case, how are these shots generated? We should give all the credit to Kirk Goldsberry for bringing up this question to us during a chat in an MIT Sloan Conference and further discussing the issue later. We were interested in what Kirk termed as the *shooter-defender choreography* for corner three shots. We focused on the shooter and the defender and, in particular, their location for the last four seconds before the shot. Using data from approximately 600 games in the 2016–2017 season, we clustered all trajectories of the players and obtained the 10 clusters shown in Figure 37-6. In this figure, the different points represent the shooter and his defender for the four seconds prior to the shot. Two clusters (Stationed-RC3 and Stationed-LC3) include half of the corner three shots taken! In these instances, a player is (literally) anchored in the corner and waits for a drive-n-kick, while his defender is lingering between the basket (e.g., to double team the penetration) and the shooter. This indecisiveness of the defender seems suboptimal. Is it though?

To examine the answer to this we turn to game theory (see Chapter 22). Consider the simplified game shown in Figure 37-7. For the

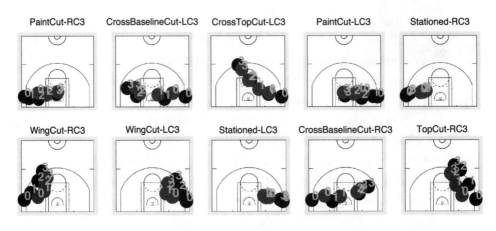

FIGURE 37.6 The "choreography" between a corner three shooter and his defender. Half of the corner threes involve a shooter anchored in the corner waiting for the pass.

offense, a player is anchored at the corner waiting for a pass from the driving player. The defender is d feet away from the corner, trying to decide whether to double team the penetration or commit to the corner three shooter. We have a zero-sum game again, where the offense has two strategies, namely, kick-out-pass or drive to the basket, and the defense has to choose the value of d (between 1 and 22). In order to solve this game, we need to define the payoff matrix. We essentially need to identify the points per shot from the corner as a function of the distance of the closest defender. We can use Second Spectrum's qSQ[11] to identify the league-average points per shot as a function of the distance of the closest defender. We can also simulate the impact of a second defender in a penetration close out through a parametrized model. For example, we consider that a second defender at distance d from the corner (and hence $22 - d$ from the player penetrating) will reduce the expected points per shot of the drive by a factor of $\left(1 - \dfrac{1}{\alpha^{(22-d)}}\right)$. Different values of

11. Y. H. Chang, R. Maheswaran, J. Su, S. Kwok, T. Levy, A. Wexler, and K. Squire, "Quantifying shot quality in the NBA," in *Proceedings of the 8th Annual MIT Sloan Sports Analytics Conference*, 2014.

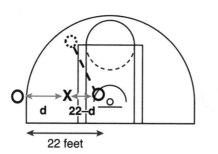

Defense strategy:
Distance to the corner
$d \in \{1, 2, ..., 21\}$

Offense strategy:
{drive, pass-corner3}

FIGURE 37.7   A Simplified Game for the Interactions between the Corner Three Shooter and the Defender.

α correspond to different levels of impact on a double team. For example, for $\alpha = 2$ the payoff matrix (for the offense) is presented in Table 37-2.

We solved the game for several values of α between 1 and 2, and Figure 37-8 shows the Nash equilibrium mixed strategy for the corner defender for two specific values ($\alpha = 1.3$ and $\alpha = 1.9$). As we can see, the Nash equilibrium strategy essentially says that the defender should commit to one of the two options (either guarding the corner or double teaming the penetration). Depending on the value of α (which controls the effectiveness of the double team), the fraction of time for each option changes. However, this simple game theory model says that one should not choose the "in-between" defenses. Now the expected value of d for each of the Nash equilibrium solutions is around 13 feet from the corner, which is close to the average (12.3 feet) of the observed distances of the defender from the corner three shooter. However, as we can see in the third part of Figure 37-8, the distribution is very different than the Nash equilibrium mixed strategy.

Obviously this is a very simplified model, which should be adjusted to specific situations (ability of corner three shooter, ability

TABLE 37.2

The Payoff Matrix for the Simplified Corner
Three Game

		Offensive Strategies	
		Corner 3	Drive
Defensive	d = 21	1.22	0.82
Strategies	d = 20	1.22	0.834
	d = 19	1.22	0.87
	d = 18	1.22	0.938
	d = 17	1.22	1.028
	d = 16	1.22	1.092
	d = 15	1.22	1.132
	d = 14	1.22	1.176
	d = 13	1.22	1.202
	d = 12	1.22	1.216
	d = 11	1.22	1.234
	d = 10	1.219	1.252
	d = 9	1.219	1.27
	d = 8	1.218	1.27
	d = 7	1.215	1.27
	d = 6	1.21	1.27
	d = 5	1.201	1.27
	d = 4	1.182	1.27
	d = 3	1.144	1.27
	d = 2	1.068	1.27
	d = 1	0.915	1.27

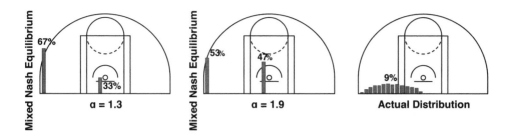

FIGURE 37.8   Nash Equilibrium and Actual Distribution of the
Defender's Distance to the Corner.

of a player to finish the drive, defensive ability of the defender, etc.)
and does not account for things such as ability to make a pass
to the corner when double teamed in a penetration. However, it
provides us with useful insights on how teams should rethink their
defensive strategy. In fact, the 76ers—famed for their interest in
incorporating analytics in their approach—have been reported[12]
to be cognizant of this: "Numbers can only tell you so much, but
they can really tell you how to guard certain teams," Simmons
said. "We know what other teams are good at and what they are
not good at. We usually play to that. People sometimes think we're
stupid, but they don't know what we know. To be fair, some of
the information and actions are counterintuitive. For instance:
when you play a team with deadly corner three-point shooters,
it's smarter to stay on your man in the corner and not help when
an opponent drives down the lane, and that can look foolish for
the unwashed."

The interesting question here might be whether players have any
cognitive biases on how they assess spatial risk. For instance, guarding closely a player in the corner might seem as if the defender is not
covering much of the court area where the game is played, ignoring
the fact though that not every area is equally important!

12. https://www.inquirer.com/philly/sports/sixers/sixers-76ers-philadelphia
-analytics-process-numbers-nba-playoffs-miami-heat-brett-brown-bryan-colangelo
-20180418.html

CLUSTERING

Clustering is an instance of an unsupervised learning task, where the goal is to group objects or data points together such that *similar objects* will be *assigned to the same group*. While the problem definition is easy to comprehend and conceptualize, it is extremely challenging. In fact, it has been mathematically proven to be an *ill-posed* problem. The artifact of this is that different algorithms and/or different definitions of similarity will provide different answers.

Tracking data are here to stay in the NBA. Teams have embraced the insights they are bringing, and pretty much every team in the league has dedicated analysts working on these data, to glean as much insight as possible. Tracking data (combined with other biosensor measurements) can also potentially open the way for tackling even more important problems such as injury prevention. While attempts in forecasting injuries, such as a Grade 2 MCL sprain, look more like the attempts to forecast earthquakes, it might become possible to obtain an accurate risk assessment for other types of non-contact injuries with the right information at hand. For example, one of the 2019 NBA Hackathon finalists attempted to develop a metric—using player tracking data—to estimate the fatigue of a player during a game (including dead ball times, timeouts, etc.). Tracking data are changing the game and keep driving innovation!

CHAPTER 38

IN-GAME BASKETBALL DECISION MAKING

In this chapter, we will see how we can use basic concepts of decision making under uncertainty to help basketball coaches improve their in-game decision making.

- In game 1 of the first round of the Eastern Conference 2001 playoffs, the 76ers led the Pacers by 2. The Pacers had the ball with five seconds to go. Should the Pacers have attempted a two pointer to tie or a three pointer to win?
- During game 6 of the 2005 Western Conference semifinals, the Dallas Mavericks led the Phoenix Suns by three points with five seconds to go. Steve Nash is bringing the ball up the court. Should the Mavericks foul Steve Nash or allow him to attempt a game-tying three pointer?
- At the end of a quarter, when should a team try and go 2-for-1?
- The Warriors were playing the Cavs and they had the ball at the end of the first quarter with 15 seconds left and were playing for the last shot. Should the Cavs have fouled them?

- Kentucky was playing UCLA and the score was tied with 15 seconds left. Kentucky had the ball and was in the 1 and 1. Should UCLA have fouled them?
- Is Hack a Dwight or Hack a Jordan a good idea?

WW was fortunate enough to be at the previously mentioned Pacers and Mavericks games. Reggie Miller hit a game-winning three as the buzzer went off in the Pacers–76ers game. Steve Nash hit a game-tying three pointer and the Suns went on to eliminate the Mavericks in a double overtime thriller.

Let's now use mathematics to analyze the optimal strategy choice in all these exciting situations.

TRAILING BY TWO: SHOULD WE GO FOR TWO OR THREE?

To begin, let's assume we have the ball and trail by two points with little time remaining in the game. Should our primary goal be to attempt a game-tying two pointer or to go for a buzzer-beating three pointer to win the game? This situation has often been used as a question in Microsoft job interviews (thanks to our friend Norm Tonina for sharing this fact with us). We assume our goal is to maximize the probability that we win the game. To simplify matters we assume that no foul will occur on our shot and the game will end with our shot. To make the proper decision we will need to estimate the values of the following parameters:

- PTWO = Probability that a two pointer is good. For the entire 2019–2020 season, PTWO was around .52.
- PTHREE = Probability a three pointer is good. For the entire 2019–2020 season, PTHREE was around .36.
- POT = Probability that we will win the game if the game goes to overtime. It seems reasonable to assume POT to be near .5.

If we go for two we will win the game only if we hit the two pointer and win in overtime. Events are said to be independent if knowing that one event occurs tells us nothing about the probability of the other event occurring. **To find the probability that independent events E_1, E_2, . . . , E_n will all occur we simply multiply their probabilities.** Since hitting a two pointer and then winning in overtime are independent events, the probability of winning the game if we go for two is $PTWO * POT = (.52)(.5) = .26$.

If we go for three, we win if and only if we make the three pointer. This will happen with probability $PTHREE = .36$. Thus it seems like we have a much better chance of winning the game if we go for three! Most coaches think this is the risky strategy and go for two.

Of course, our parameter estimates may be in error,[1] so we should perform a sensitivity analysis to determine how much our parameters have to change for our optimal decision to change. The standard error associated with a fraction (such as the two-point shot field goal percentage) is given by $\sqrt{\dfrac{p \cdot (1-p)}{N}}$, where N is the number of observations and p is the value of the fraction. Given that the value of N is approximately 60,000 for the three-point shots and even larger for the two-point shots, we can see that the standard error of the field goal percentages is small (approximately 1–2%). However, if our analysis is still sensitive to these parameters, a small change in them can change the final decision. A standard way of performing a sensitivity analysis is to hold all but one parameter constant at its most likely value, and determine the range of the remaining parameters for which our optimal decision (going for three) remains optimal.

We find that as long as PTWO <.72 we should go for three if PTHREE = .36 and POT = .5. We find that as long as PTHREE >.26 we should go for three if PTWO = .52 and POT = .5. We find that if POT <.36/.5 = .72 we should go for three if PTWO = .52 and PTHREE = .36. This last point is particularly important, since

---

1. Or one can claim that *crunch time* shooting is going to be tougher to make.

assuming the chance of winning in OT is 50% might not be appropriate. For instance, a team that was the heavy favorite prior to the game may have a much higher than 50% chance of winning in OT, and hence it might be better off going for the two-point shot. In contrast, a heavy underdog that has managed to be in a one possession game at the end might be better off taking the three. It should come without saying that each situation is unique, and while there might be a general footprint that one might have for in-game decisions, these need to be adjusted according to the specifics of each case. Another context to factor in the decision is the fact that these scenarios are situations under pressure, and therefore the league-average numbers might be too optimistic.

The sensitivity analysis is important during actual in-game situations. During the 2020–2021 regular season, the Wizards were visiting the Nuggets, who got a defensive rebound with six seconds left and down two points. The Nuggets did not get a timeout and pushed the ball, having a 4-on-1 fast break. A two-point shot had a high probability (PTWO >0.72) of going in. In actuality, all four players lingered outside the three-point line. As Mike Malone commented, this was not what he wanted. This play fueled opponents of analytics who believe "numbers are destroying the game." The numbers would have recommended finishing the fast break with a shot at the rim and taking the game to overtime. Also, the Nuggets were a seven-point pregame favorite, which means that the Nuggets had more than a 50–50 chance at winning in overtime. It is rare that you get a uniformly applicable answer to a question, but numbers help you sort through the data and determine the correct decision for the actual situation.

LEADING BY THREE: SHOULD WE FOUL OR PLAY DEFENSE?

We now turn our attention to the question of whether a team with a three-point lead should foul the other team when little time remains on the clock. This is a much tougher question, and two analysts,

Adrian Lawhorn (2006)[2] and David Annis (2006),[3] have analyzed this situation and concluded that the defensive team should foul the team with the ball. Let's briefly discuss their logic.

Lawhorn begins by looking at actual data. He begins by assuming that the current possession will be the last possession of regulation. He finds that with <11 seconds left, teams that trailed by three points hit 41 out of 205 three pointers. This means that by letting your opponent shoot it has a 20% chance to tie the game and therefore a 10% chance of winning the game (assuming that each team has the same chance to win in overtime). If we foul the other team, Lawhorn assumes that the only way the trailing team can tie is to make the first free throw and then miss the second free throw intentionally. Then Lawhorn assumes that the trailing team must either tip the ball in or kick the ball out and hit a two pointer to tie the game. Lawhorn calculates that the trailing team has roughly a 5% chance of tying the game or around a 2.5% chance of winning the game. Lawhorn does not consider the coach's ultimate nightmare: the trailing team hits the first free throw and after rebounding an intentionally missed second shot the trailing team hits a three pointer to win! NBA teams hit around 77% of their free throws and rebound around 14% of missed free throws. Suppose the trailing team can hit 30% of its three pointers. If the trailing team elects to take a three pointer after rebounding a missed three pointer, then the probability that it will win is $.75(.14)(.3) = .03$ or 3%. This is slightly better (if at all significant) than the 2.5% chance of winning if we attempt a two pointer. Therefore, fouling the trailing team reduces their chances of winning from 10% to 3%, and seems like a good idea. Annis comes to a similar conclusion. **The key assumption made by Lawhorn and Annis is, of course, that the game ends after the current possession.** In reality, if we foul the other team after their free throws, it will foul us and probably get the ball back. If the trailing team makes

---

2. http://www.82games.com/lawhorn.htm

3. David H. Annis, "Optimal end-game strategy in basketball," *Journal of Quantitative Analysis in Sports*, 2(2), 2006.

both free throws and we miss one out of two, then the trailing team can win the game with a three or tie with a two! The problem is that we cannot know if the current possession is the last possession. Lawhorn found 32 games where a team trailed by three points with <11 seconds left and the leading team intentionally fouled the trailing team. In seven of these games (21.9%), the trailing team tied the game. Although this is a small sample, the data indicates that when multiple possessions are a possibility, it is not clear that fouling is the correct strategy.

For a project in WW's sports and math class, Kevin Klocke looked at all NBA games from 2005 through 2008 in which a team had the ball between 1 and 10 seconds left and trailed by three points. The leading team did not foul 260 times and won 91.9% of the games. The leading team did foul 27 times and won 88.9% of the games. This evidence seems to indicate that fouling does not provide any significant benefit in terms of the chance of winning when you are three points ahead.

A great empirical study of college basketball by John Ezekowitz[4] found that fouling and not fouling basically yielded the same results. Of the 52 teams that committed a foul, six lost the game for a winning percentage of 88.46%. Of the 391 teams that did not foul, 33 lost the game for a winning percentage of 91.56%. **This study indicates that there is no significant difference between fouling and not fouling in this situation.**

We have performed our own simulations to further examine this situation. In our simulations, we assume that the game does not end with the current possession. However, we assume that if the leading team does not choose to foul, the trailing team will take its last shot closer to the end of the game in order to get a better look at the three. Hence, we assume that the three-point shot field goal percentage is the league average .36 (rather than the .20 that we calculated above). So if the leading team does not foul, it will

4. https://harvardsportsanalysis.wordpress.com/2010/08/24/intentionally-fouling-up-3-points-the-first-comprehensive-cbb-analysis/

lose the game 18% if the times. Fouling provides a large number of paths that can lead to either team winning. However, only a few paths lead the trailing team to winning the game. While we do not make the assumption that this is the absolutely last possession of the game, in order to keep the analysis tractable we have to make some assumptions:

- If a team trails by two at any time and has the last shot, it will attempt a three-point shot (as per the analysis above).
- If the trailing team is led to the free throw line, there is still enough time for fouling *once* the opponent in order to stop the clock.
- In case the trailing team gets the offensive rebound from its own free throw shot, and is still trailing by three, it can move the ball quickly outside the three-point line and take a three-point shot for tying the game (without giving a chance to the opponent to foul it). The same is true if the trailing team inbounds the ball after the leading team scores FTs.
- If at any point during the simulation a team takes the lead by four points or more, then we consider the chances of the team losing negligible.

Someone might (rightly so) oppose these assumptions. Nevertheless, these assumptions are made as the best-case scenario for the trailing team (or the worst-case scenario for the leading team) when the leading team chooses to foul. With these assumptions in mind, we can simulate this situation through Monte Carlo (see Chapter 4). We simulated this situation 10,000 times and on average the trailing team won 7.3% of the times. Figure 38-1 presents the distribution for the win probability of the trailing team when fouled down three. For these simulations we have used $PFT = 0.77$, $PTWO = 0.52$, $PTHREE = 0.36$, $POFR = 0.23$, and $POT = 0.5$. For the offensive rebound probability we have used the overall offensive rebound rate (rather than only the FT offensive rebound rate we mentioned before), since these give the best-case scenario for the trailing team. Of

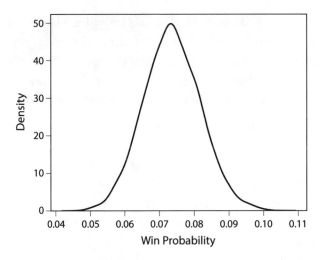

FIGURE 38.1   On average the trailing team wins the game 7.3% of the time
when the opponent fouls it while leading by 3.

course, changes in the values for these parameters will alter this win
probability, but one can repeat the simulations with the appropriate
input parameters for the setting examined.

AT THE END OF A QUARTER, WHEN SHOULD A TEAM TRY AND GO 2-FOR-1?

Jeff Ma's fascinating book *The House Advantage* addresses this issue.
You probably know Jeff Ma from his experience on the M.I.T. black-
jack team that was described in *Bringing Down the House* and the hit
movie *21*. The key to answering the 2-for-1 question is collecting the
proper data. Based on data he collected, Ma estimated the average
points a team that just gave up possession is outscored by for the rest
of the quarter as a function of the time remaining in the quarter. For
example, he found that if you give up the ball with 40 seconds left,
you get outscored by an average of 0.6 points for the rest of the quarter, but if you give up the ball with 32 seconds left, you get outscored

by around only 0.3 points per quarter. Therefore, teams should aim to shoot with 32 seconds left in the quarter. This gives you a good chance of getting the ball back with enough time to create a good shot. One interesting caveat here is that this assumes that the game and shot clock are both stopped when a basket is scored. While true for the NBA, for international competitions (e.g., FIBA) the game clock might still be running before the inbound pass (while the shot clock is stopped).

If we were to pick a single analytics innovation that made it into a game strategy it would be the 2-for-1. Most probably, this is due to the simple message: take a shot at around 32 seconds left in the quarter. However, teams seem to have gone far with it. The premise of the 2-for-1 is that taking a shot at around 32 seconds allows for the possibility of a second good look before the end of the quarter. However, the first look must be good as well. Currently, many teams, even if they get the ball back with 36 seconds left, will scramble and take a (fairly bad quality) shot. Key for the conclusion of an analysis to apply is to understand the mechanisms studied. Forcing shots to just force them is not going to reap any benefits.

SHOULD YOU FOUL THE TEAM WITH THE BALL AT THE END OF THE QUARTER?

The Warriors are playing the Cavs and the Warriors have the ball at the end of the first quarter with 15 seconds left and are playing for the last shot. Should the Cavs foul them? During 2016, the Warriors averaged 1.13 points per possession and the Cavs averaged 1.09 points per possessions. It's harder to make a basket at the end of a quarter, so let's assume that if the Warriors take the last shot, they will average one point per possession (and the Cavs 0.9 points per possession). The Warriors made 79% of their foul shots in 2016. If you foul the Warriors, you give up an average of $2 * 0.79 = 1.58$ points and get back 0.9 points. This is a net disadvantage of $0.9 - 1.58$ points. $= -0.68$ points. If we let the Warriors take the last shot, the

Cavs are down an average of one point. This analysis implies the Cavs should foul. If every team followed this fouling strategy, the fans would be bored to death, so we hope this strategy does not catch on!

KENTUCKY IS PLAYING UCLA AND THE SCORE IS TIED WITH 15 SECONDS LEFT. KENTUCKY HAS THE BALL AND IS IN THE 1 AND 1. SHOULD UCLA FOUL IT?

Ken Pomeroy's well-known college basketball site Kenpom.org did an excellent study on this matter.[5] He found that if Kentucky is not fouled, UCLA has a 35% chance of winning. If UCLA fouls a player with FT% less than 71%, then UCLA's win probability is higher than 35%! Therefore, the fouling strategy *may* make sense. Of course, a risk-averse coach will probably not try this controversial strategy.

IS HACK-A-JORDAN[6] A GOOD IDEA?

We have all been bored to death watching teams foul poor foul shooters such as Dwight Howard or DeAndre Jordan. The idea is that if the foul shooter makes, say, 40% of his free throws, then fouling the player yields an average of 0.8 points per possession, while not fouling gives up a little more than one point per possession. Again, we are indebted to John Ezekowitz[7] for an excellent analysis of this issue. Ezekowitz found that this superficial analysis ignores the chance that the fouled team will get an offensive rebound. When the chance of an offensive rebound is factored into the analysis, we find that the fouling strategy yields little or no benefit. Of course, the hacking strategy may disrupt a team's offensive rhythm, but we have not found any analysis that examines this angle.

---

5. http://kenpom.com/blog/studying-whether-to-foul-when-tied-part-2/

6. Well, this can be a quick no now since DeAndre has shot a whooping 68% since he left LA!

7. https://fivethirtyeight.com/features/intentionally-fouling-deandre-jordan-is-futile/

In general, fouling at the end of a game[8] to extend the time clock is a strategy that, while not used extensively in the NBA, is used in European (and international in general) basketball. Mike Beuoy has created a *hackulator*[9] that provides recommendations on when to start fouling based on the score differential and FT%. Hopefully, the Elam ending[10]—used at the 2020 and 2021 NBA All-Star Games—will gain more traction and *save* us from free throw hacking.

8. While this is in general a bit different from specifically targeting a bad FT shooter, the strategy and results are consistent.

9. http://stats.inpredictable.com/nba/hackulator.php

10. https://thetournament.com/elam-ending

PART IV

OTHER SPORTS

CHAPTER 39

SOCCER ANALYTICS

Soccer—or football as it is known outside the US—is the *king of sports*. This means that several people are working professionally or in an amateur fashion to analyze the game. Despite the worldwide interest in soccer, the status of soccer analytics was not at the level one might have expected a few years back. However, it has advanced rapidly over the last few years mainly due to the availability of player tracking data. In this chapter, we will review some of the most well-studied problems in soccer as well as the current state of the art.

SHOTS ON GOAL DO NOT MATTER

A typical soccer box score includes information about the total shots (on goal), fouls, corner kicks, yellow and red cards, and possession time. Unfortunately, these metrics barely describe important information about the game and the performance of the teams. For example, a team that significantly out-shoots its opponent might be expected to have dominated the game and won easily. However, there is no correlation between the shots differential and the actual goal differential,[1] i.e., winning. Similarly, possession time is at best

1. http://www.americansocceranalysis.com/home/2014/12/18/shots-confusion-in-correlations

weakly correlated with winning.[2] Therefore, there is a need for developing metrics that can explain (and predict) better what happened in the game.

One of the problems with using the number of shots is that not all shots are created equally. Shots from inside the box are typically of higher quality than shots 40 yards away from the goal line. To capture this, the notion of expected goals (xG) has been developed. The basis for xG is a probabilistic model that captures the probability that a shot will end up into a goal. Various features can be used for modeling the goal probability, such as distance to the goal, shot angle, etc. We have developed our own goal probability model[3] using a logistic regression model (see Chapter 21) and shot data from the 2016 MLS regular season. The independent variables for the model and the corresponding coefficients are presented in Table 39-1. As we can see, the distance to the goal line as well as the angle of the shot, that is, the angle created by the two straight lines connecting the shot location with each of the vertical goal posts, are significantly correlated with scoring, while a kick (either with the left or right foot) has better chances ending with a goal as compared to a header (the reference class for the variable "type").[4]

If we want to evaluate similar models, that is, models where we are interested in the quality of the probabilistic predictions rather than the binary answer (yes or no), classification accuracy is not the right metric. We would be better off if we calculated the model's probability **calibration curve**. Let us assume that we have two differ-

2. http://www.telegraph.co.uk/sport/football/competitions/champions -league/10793482/Do-football-possession-statistics-indicate-which-team-will-win -Not-necessarily.html

3. A. Fairchild, K. Pelechrinis, and M. Kokkodis, "Spatial analysis of shots in MLS: A model for expected goals and fractal dimensionality," *Journal of Sports Analytics*, 4(3), 2018, 165–174.

4. We would like to note here that our model is very simplistic and it is only meant for illustrative purposes. To be used from a team or a league to facilitate decision making, several other features, including defender locations and other information from player tracking data (e.g., player orientation), should be incorporated.

TABLE 39.1

Goal Probability Logistic Regression Model
(Significance Levels: **5%, ***1%)

Variable	xG		
	coef	*SE*	*sig*
Constant	−1.1	0.58	
Assist (Other)	−0.2	0.28	
Assist (Pass)	−0.1	0.25	
Assist (Self)	−.02	0.29	
Type (Left)	1.1	0.3	***
Type (Right)	1.1	0.28	***
Play (Set piece)	0	0.2	
x	0.11	0.05	
y	−0.01	0.01	
Angle	0.02	.007	***
Distance	−0.19	.062	***
Observations	1,114		
Log Likelihood	−553.7		
Deviance	1,107.5		

ent models, M_1 and M_2, that assign a goal probability of 10% and 45%, respectively, to shot s. Assuming the typical threshold of 50% for *predicting* whether shot s will result into a goal or not, both models will have the same accuracy (they will both answer, "No," there will not be a goal scored). However, the expected goals will be vastly different (.10 with M_1 and .45 with M_2). What is the right value for the probability? 45% or 10% (or something else)? Ideally, we would like to have the same exact shot being taken several times, and measure the fraction of times that it results in a goal. However, this is not possible, as should be evident to everyone, and hence, we take a different approach. With S_{oos} being the set of shots over which we will evaluate the model (i.e., the out-of-sample test set), we start by grouping or binning the shots in S_{oos} based on their predicted probability $\hat{p}$. Then for each of these groups we calculate the fraction of shots that indeed resulted in a goal, say, $\tilde{p}$. This can be thought of as the observed goal probability for the shots in the bin. Ideally, for a model that captures the goal probability accurately, we would like to have $\hat{p} = \tilde{p}$ for all the groups. The calibration curve then plots on the x-axis the predicted probability in each bin and on the y-axis the observed probability in each bin. Hence, if the probability model is well calibrated, the validation curve will be close to the line $y = x$. Figure 39-1 presents the validation curve for the model developed, where we can see that statistically it is indistinguishable from the 45° line, translating to a well-calibrated probability model.

To translate this goal probability model to an xGs metric, we describe the sequence of shots taken by a team through a Poisson binomial distribution. A Poisson binomial distribution is the sum of independent (but not necessarily identically distributed) Bernoulli trials (see Chapter 16 for the connection between Poisson and binomial distributions). The mean μ and standard deviation σ for a Poisson binomial distribution are given by:

$$\mu = \sum_{i=1}^{N} \pi_i \tag{1}$$

$$\sigma^2 = \sum_{i=1}^{N} \pi_i (1 - \pi_i) \tag{2}$$

where N is the number of trials and π_i is the probability of success of the ith trial. Therefore, with π_i being the probability of goal for shot i the expected goals of a team are simply the mean μ of the corresponding Poisson binomial distribution.

The expected goals model can then be used in a box score setting,[5] to evaluate teams over a season, etc. For example, we can classify the teams in our dataset based on how better or worse they perform

---

5. FiveThirtyEight provides information on xG for a select set of teams or games, while other data providers such as StatsBomb have released some of their xG data to the public.

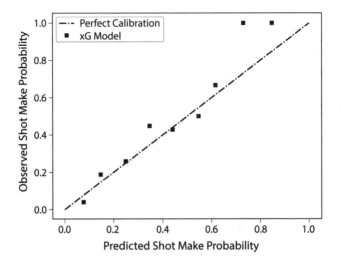

FIGURE 39.1   The Expected Goals Model Validation Curve.

over the expectation. More specifically, we can have the following definitions:

$$\text{Offensive efficiency} = \frac{G_{T,+} - xG_{T,+}}{G_{T,+}} \tag{3}$$

$$\text{Defensive efficiency} = \frac{G_{T,i} - xG_{T,-}}{G_{T,-}} \tag{4}$$

where $G_{T,+(-)}$ (>0) captures the number of goals scored (allowed) by team T, and $xG_{T,+(-)}$ is the expected goals scored (allowed) by T. Figure 39-2 depicts the efficiencies for each MLS team as calculated by our dataset.

Here we have presented a very simplified version of an expected goals model. There are several additional features that can be integrated to improve the quality of the model. For instance, one feature that has been shown to provide important contextual information is[6] the defensive formation, e.g., how many defenders are within a

---

6. P. Lucey, A. Bialkowski, M. Monfort, P. Carr, and I. Matthews, "'Quality vs quantity': Improved shot prediction in soccer using strategic features from spatio-temporal data," in 9th Annual MIT Sloan Sports Analytics Conference, 2015.

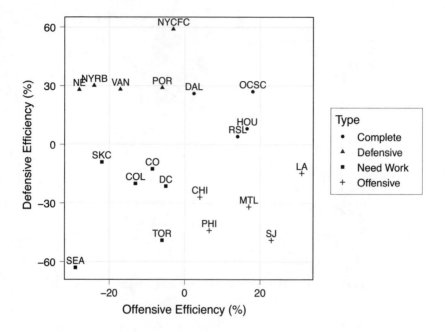

FIGURE 39.2 Offensive and Defensive Efficiency of MLS Teams as Captured through the Expected Goals Model.

predefined range from the shooter, what is the distance of the closest defender, etc. This type of information is available through tracking data.

XG ADDED: POSSESSION-BASED MODEL

One of the problems with the current statistics in soccer—even with an *advanced* statistic such as xG—is that they mainly capture the performance of players that take shots (e.g., strikers). However, it is rare that the person who took the shot really generated the opportunity himself or herself. How can we assign credit (or blame) to other players involved in the possession? Mackay[7] developed a framework that attempts to assign a value to each interaction (mainly passes)

7. N. Mackay, "Predicting goal probability of possessions in football," Vrije Universiteit Amsterdam, Technical Report, 2017.

within a possession. To achieve this, a possession-based model is necessary, where essentially a probability of a possession ending up in a goal needs to be calculated based on the current status of the possession. Intuitively, a passing play is a good one if the new state for the offense has an increased chance of scoring a goal. One could use a shot-based model, i.e., if the location of the pass is a location for a better shot or not; however, this does not consider situations— which appear often during a game—where setting back the play is the better option than forcing a pass to a player in a good location but with low completion percentage.

Mackay essentially developed an xG model but for a possession rather than a shot. While shots, as we saw earlier, are better when taken closer to the goal and from a straight angle, a possession can still be good if you have a sharp 180° angle to the goal line (i.e., you are practically almost over the corner line). Figure 39-3 shows in lighter shade the locations on the field where on average possession is better than a shot, and vice versa for the darker shade locations. It is evident that locations with a sharp angle are better locations to

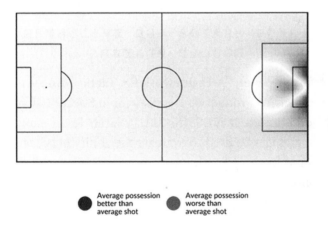

FIGURE 39.3 Locations that are not good for taking a shot are good for generating future chances, and hence, passing in these locations is valuable (figure obtained from Mackay, 2017).

keep the possession going, while locations in front of the goal are better for taking the shot.

With such a possession-based model, one can create an xG-added per pass, based on the improvement in the xG for the current possession. This is a metric that can assign credit to different participants in a possession and not only to the person who took the shot or assisted.

Decroos, Bransen, Van Haaren, and Davis[8] (DBVD) generalized this approach by providing a model that evaluates every action on the pitch (e.g., dribbles, passes, crosses, etc.). This model essentially captures how each action A impacts the probability of scoring a goal in the *near future* (the latter is defined as a score being observed within a window of k actions after action A). The score probability added can then be used to provide player ratings by simply aggregating over all the actions taken from a player (and then normalizing per a time unit to make sure that the rating is not biased by the play time). DBVD also weight on the endless debate of all soccer fans by reporting that Messi (and not Ronaldo) is the best player in the world!

MARKOV CHAINS AND EVALUATING SOCCER PLAYERS

In her brilliant paper, "A Framework for Tactical Analysis and Individual Offensive Production Assessment in Soccer Using Markov Chains" (presented at NESSIS, 2011), Sarah Rudd showed how Markov chains can be used to evaluate the ability of soccer players. The ideas put forth by Sarah Rudd are very similar with those of a possession-based model; however, they are based on a discretized representation of the game using Markov chains. Before discussing

8. T. Decroos, L. Bransen, J. Van Haaren, and J. Davis, "Actions speak louder than goals: Valuing player actions in soccer," in the 25th ACM SIGKDD, 2019, pp. 1851–1861.

Rudd's approach, we need to give the reader some knowledge of Markov chains.

MARKOV CHAINS

First we define a finite state stochastic process as a process that describes the random changes over time of a system that is always in one of N states (labeled 1, 2, . . . N). A finite state Markov chain is a special case of a finite state stochastic process. At times $t = 0$, 1, . . . , a finite state Markov chain is in one of N states (1, 2, . . . , N). Suppose at time t the chain is in state i. Then, **irrespective of the full past history of the chain,** the state at time $t + 1$ will be j with probability p_{ij}, i.e., it depends only on the state at time t. The fact that the future of the process only depends on its past history through the current state (and not, say, how it reached this current state) is known as the *Markov property.* The matrix of transition probabilities is called a transitional probability matrix.

As an example of a Markov chain, let's consider the classic gambler's ruin problem. Suppose we have $2 and we flip a biased coin that has a 60% chance of coming up heads. On each coin flip we win $1 if the coin comes up heads and lose $1 if the coin comes up tails. We play until we reach our goal of $4 or are bankrupt (have $0). Once we enter $0 or $4, we stay there forever, so $0 and $4 are examples of what we call *absorbing states.* For reasons that will soon become apparent, we will write the transition matrix for a Markov chain with the absorbing states (if any) listed last. Any state that is not an absorbing state is called a transient state, since sooner or later you will leave a transient state and (as in the MTA) never return.

Figure 39-4 shows the transition probability matrix, for our gambler's ruin example and for each transient state, the chance of ending up in each absorbing state.

	E	F	G	H	I	J	K	L	M	N	O	P
3												
4		Transition Matric										
5			$1	$2	$3	$0	$4					
6		$1	0	0.6	0	0.4	0					
7		$2	0.4	0	0.6	0	0					
8		$3	0	0.4	0	0	0.6					
9		$0	0	0	0	1	0					
10		$4	0	0	0	0	1					
11												
12		0	0.6	0								
13	Q	0.4	0	0.6								
14		0	0.4	0								
15												
16	R	0.4	0									
17		0	0			$(I-Q)^{-1}$	1.461538	1.153846	0.692308	{=MINVERSE(F24:H26)}		
18		0	0.6				0.769231	1.923077	1.153846			
19							0.307692	0.769231	1.461538			
20	I	1	0	0				$0	$4			
21		0	1	0		$(I-Q)^{-1}R$	$1	0.584615	0.415385	{=MMULT(K17:M19,F16:G18)}		
22		0	0	1			$2	0.307692	0.692308			
23							$3	0.123077	0.876923			
24	I-Q	1	-0.6	0								
25		-0.4	1	-0.6								
26		0	-0.4	1								

FIGURE 39.4   Analysis of Gambler's Ruin Example.

For any Markov chain with absorbing states, we define the matrix Q to be the portion of the transition matrix involving transitions between transient states, and the matrix R to be the portion of the transition matrix involving transitions from transient to absorbing states. Figure 39-4 shows both Q and R. *The matrix $(I - Q)^{-1}R$ now gives for each transient state the chance of ending up in each absorbing state.* Cell range F24:H26 gives the matrix $I - Q$. To determine the inverse of the matrix $I - Q$ we need to **array enter** a formula involving the MINVERSE function. To array enter a formula we must proceed as follows:

1. Select the range of cells (here K17:M19) that should be populated with the results of the formula.
2. Type the array formula in the upper-left hand corner of the selected range. Here we type = MINVERSE(H24:J26) in cell K17.
3. **Instead of hitting the Enter key, we must hit the Control + Shift + Enter** key combination. If we have Office 365,

then we need not hit Control + Shift + Enter and we don't need to select multiple cells.

Finally, we need to compute the matrix product $(I - Q)^{-1}R$ by array entering the appropriate MMULT array function. This matrix product will be a 3×2 matrix, so we select the range L21:M23 and with the cursor in L21 type in the formula = MMULT(K17:M19, F16:G18) and then hit the keystroke combination Control + Shift + Enter. We find the following probabilities of reaching $0 and $4.

- If we begin with $1 we have a 58% chance of ending with $0 and a 42% chance of making it to $4.
- If we begin with $2 we have a 31% chance of ending bankrupt and a 69% chance of ending up with $4.
- If we begin with $3 we have a 12% chance of ending with $0 and an 88% chance of ending with $4.

	Open on Side Defense my ball	Guarded on Side Defense my ball	Open Middle Defense my ball	Guarded Middle Defense my ball	Open on Side Defense opp ball	Guarded on Side Defense opp ball	Open Middle Defense opp ball	Guarded Middle Defense opp ball	Middle of Field with Ball	Middle of Field opp ball	Open on Side Offense my ball	Guarded on Side Offense my ball	Open Middle offense my ball	Guarded Middle offense my ball	Open on Side Offense opp ball	Guarded on Side Offense opp ball	Open Middle offense opp ball	Guarded Middle offense opp ball	Opp Corner Kick	Opp enalty Kick	Us Corner Kick	Us Penalty Kick	We score	Opponent Score
Open on Side Defense my ball	0	0.1	0.1	0.1	0.1	0	0.03	0.03	0.4	0.11	0	0	0	0	0	0	0	0	0	0	0	0	0	0
Guarded on Side Defense my ball	0	0.08	0.08	0.08	0.08	0.1	0.06	0.06	0.35	0.15	0	0	0	0	0	0	0	0	0	0	0	0	0	0
Open Middle Defense my ball	0	0.1	0.1	0.1	0.1	0	0.03	0.03	0.4	0.11	0	0	0	0	0	0	0	0	0	0	0	0	0	0
Guarded Middle Defense my ball	0	0.08	0.08	0.08	0.08	0.1	0.06	0.06	0.35	0.15	0	0	0	0	0	0	0	0	0	0	0	0	0	0
Open on Side Defense opp ball	0.03	0.03	0.03	0.03	0	0.2	0.15	0.15	0.12	0.03	0	0	0	0	0	0	0	0	0.2	0.04	0	0	0	0.04
Guarded on Side Defense opp ball	0.04	0.04	0.12	0.12	0.13	0	0.09	0.16	0.16	0.02	0	0	0	0	0	0	0	0	0.07	0.03	0	0	0	0.02
Open Middle Defense opp ball	0.02	0.03	0.09	0.08	0.1	0.1	0.09	0.09	0.13	0.04	0	0	0	0	0	0	0	0	0.1	0.05	0	0	0	0.09
Guarded Middle Defense opp ball	0.04	0.04	0.08	0.08	0.1	0.1	0.09	0.09	0.17	0.03	0	0	0	0	0	0	0	0	0.08	0.04	0	0	0	0.07
Middle of Field with Ball	0	0	0	0	0	0	0	0	0	0.2	0.1	0.1	0.1	0.1	0.1	0.1	0.1	0.1	0	0	0	0	0	0
Middle of Field opp ball	0.1	0.1	0.1	0.1	0.1	0.1	0.1	0.1	0.2	0	0	0	0	0	0	0	0	0	0	0	0	0	0	0
Open on Side Offense my ball	0	0	0	0	0	0	0	0	0	0	0.18	0.18	0.2	0.18	0.05	0	0.03	0.04	0	0	0.07	0.03	0	0
Guarded on Side Offense my ball	0	0	0	0	0	0	0	0	0	0	0.15	0.14	0.2	0.15	0.07	0.1	0.07	0.07	0	0	0.06	0.04	0	0
Open Middle Offense my ball	0	0	0	0	0	0	0	0	0	0	0.14	0.14	0.1	0.12	0.05	0.1	0.05	0.04	0	0	0.12	0.08	0.1	0
Guarded Middle offense my ball	0	0	0	0	0	0	0	0	0	0	0.16	0.16	0.2	0.1	0.07	0.1	0.07	0.07	0	0	0.08	0.04	0	0
Open on Side Offense opp ball	0	0	0	0	0	0	0.08	0.24	0.04	0.04	0	0.04	0.14	0.1	0.14	0.1	0	0	0	0	0	0	0	0
Guarded on Side Offense opp ball	0	0	0	0	0	0	0.1	0.2	0.05	0.05	0.1	0.05	0.12	0.1	0.12	0.12	0	0	0	0	0	0	0	0
Open Middle offense opp ball	0	0	0	0	0	0	0.09	0.23	0.04	0.04	0	0.04	0.14	0.1	0.14	0.1	0	0	0	0	0	0	0	0
Guarded Middle offense opp ball	0	0	0	0	0	0	0.1	0.18	0.07	0.05	0.1	0.05	0.12	0.1	0.12	0.12	0	0	0	0	0	0	0	0
Opp Corner Kick	0	0	0	0	0.2	0.2	0.15	0.15	0.03	0.03	0.03	0.03	0	0	0	0	0	0	0	0.06	0	0	0	0.12
Opp Penalty Kick	0	0	0.3	0	0	0	0	0	0	0	0	0	0	0	0	0	0	0	0	0	0	0	0	0.7
Us Corner Kick	0	0	0	0	0	0	0	0	0	0	0.15	0.15	0.2	0.15	0.06	0.1	0.04	0.02	0	0	0	0	0.1	0.1
Us Penalty Kick	0	0	0	0	0	0	0	0	0	0	0	0	0	0	0.3	0	0	0	0	0	0	0	0.7	0
We Score	0	0	0	0	0	0	0	0	0	0	0	0	0	0	0	0	0	0	0	0	0	0	1	0
Opponent Score	0	0	0	0	0	0	0	0	0	0	0	0	0	0	0	0	0	0	0	0	0	0	0	1

FIGURE 39.5   Soccer Transition Probability Matrix.

So what do absorbing Markov chains have to do with evaluating soccer players? At any point in time we can denote the state of the soccer game to be based on the location of the ball on the pitch (divide the pitch into zones) and the team in possession of the ball. Of course, there are states for corner kicks and penalty kicks. The two absorbing states correspond to each of the teams scoring a goal. The file Soccermarkov.xlsx contains an illustrative transition matrix, shown in Figure 39-5.

	F	G	H
		Chance We score next	Chance Opponent Scores Next
95			
96	Open on Side Defense my ball	0.627	0.373
97	Guarded on Side Defense my ball	0.616	0.384
98	Open Middle Defense my ball	0.627	0.373
99	Guarded Middle Defense my ball	0.616	0.384
100	Open on Side Defense opp ball	0.512	0.488
101	Guarded on Side Defense opp ball	0.555	0.445
102	Open Middle Defense opp ball	0.499	0.501
103	Guarded Middle Defense opp ball	0.522	0.478
104	Middle of Field with Ball	0.693	0.307
105	Middle of Field opp ball	0.596	0.404
106	Open on Side Offense my ball	0.756	0.244
107	Guarded on Side Offense my ball	0.75	0.25
108	Open Middle offense my ball	0.779	0.221
109	Guarded Middle offense my ball	0.749	0.251
110	Open on Side Offense opp ball	0.671	0.329
111	Guarded on Side Offense opp ball	0.678	0.322
112	Open Middle offense opp ball	0.672	0.328
113	Guarded Middle offense opp ball	0.681	0.319
114	Opp Corner Kick	0.462	0.538
115	Opp Penalty Kick	0.188	0.812
116	Us Corner Kick	0.787	0.213
117	Us Penalty Kick	0.902	0.098

FIGURE 39.6 Soccer Absorption Probabilities.

For each transient state, we computed the chance of being absorbed into the "we score state" or the "opponent scores state." These chances are calculated in the file Soccermarkov.xlsx and are shown in Figure 39-6.

Based on this matrix, we can estimate the value added by various movements of the ball in a soccer match. For example, if a player on our team has the ball in the middle of the field, we see from row 104 the value of the situation is .69 − .31 = .38 goals. This follows because in this situation, on average, on the next goal in the match we will outscore the opponent by .38 goals. If our player passes the ball successfully to an open player in the middle of the offense side of the pitch, the value to our team has increased to .72 − .22 = .50 goals, so this ball movement was worth .50 − .38 = .12 goals. If we can appropriately partition this increment of .12 goals between the receiver of the pass and the player who made the pass, we have the beginnings of a player evaluation method that is more closely tied to all events during the match than simply looking at goals, assists, steals, etc. Of course, the absorbing Markov chain technique can be applied to other sports such as hockey or lacrosse, where ball or puck movement is of critical importance. Obviously, the critical part here is to define the states and identify the transition probabilities, the latter of which could be done by processing play-by-play or tracking data.

PENALTY KICKS AND GAME THEORY

Penalty kicks have provided some of the most unforgettable and iconic moments in the sport. For those old enough to remember the 1994 World Cup, the view of Roberto Baggio after missing the penalty in the final, which sealed the win and the trophy for Brazil, is still vivid and has even been used in many successful advertising campaigns. Penalty misses are more devastating since 75–80% of the penalties end up in a goal. So how does a goalie decide which area to jump (if at all), and the kicker where to shot?

Top-league professional teams scout their opponents' tendencies in penalty kicking. Some basic rules of thumb suggest that right-footed kickers will find it easier to kick to their left. Chiappori, Levitt, and

Groseclose (2002)[9] (CLG) have modeled the decisions of the goalie and the kicker using basic game theory. The objective of the kicker is to maximize the probability of scoring a goal, while the goalie wants to minimize the same probability. Both players have the same set of strategies to choose from: kick/jump right, kick/jump left, kick/stay at the center. The "penalty kick" game is a zero-sum game, since the gain of one player is exactly balanced from the losses of the other player; e.g., if the kicker scores, the goal gets a utility of +1 goal, while the goalie gets a utility of −1 goal. The *payoff matrix* is shown in Table 39-2. P_L, P_R, π_L, π_R, and μ capture the probability of a goal being scored when the kicker K_i and goalie G_i pick their strategies accordingly.

TABLE 39.2
The Penalty Game Payoff Matrix (Figure obtained from Chiappori et al., 2002)

K_i	G_i L	C	R
L	P_L	π_L	π_L
C	μ	0	μ
R	π_R	π_R	P_R

When analyzing a game, the notion of a **Nash equilibrium** is important for ultimate decision making (see Chapter 22). CLG found that in the Nash equilibrium for this game, kickers should not kick to the center (and hence, goalies do not stay in the center) unless the scoring probability μ is large enough.[10] In this case they should perform a restricted randomization between strategies left and right. If μ is large enough then both players should perform a general randomization over all three strategies.

9. P.-A. Chiappori, Steven Levitt, and Timothy Groseclose, "Testing mixed-strategy equilibria when players are heterogeneous: The case of penalty kicks in soccer," *American Economic Review*, 2002, 1138–1151.

10. They show that the threshold is $\dfrac{\pi_L \pi_R - P_L P_R}{\pi_L + \pi_R - P_L - P_R}$.

Using data from the first-division leagues in France and Italy, CLG estimated empirically the payoffs for the penalty game, and shown in Table 39-3. In theory, the success percentage when the kicker kicks to the left should be the same regardless of whether the goalie stayed in the center or jumped to the right. However, in practice these success rates exhibit differences. Furthermore, kickers who choose to hit down the middle enjoy the highest average payoff (over 80%). This can be due to possible "selection bias" in the data, since kickers pick the middle only when they feel the probability of scoring is relatively high. CLG further performed a series of tests on the empirical data from the penalty kicks and found that the hypothesis that kickers and goalies choose their strategies optimally cannot be rejected. It seems that there are more things involved in a penalty kick than simple athletic talent.

TABLE 39.3

The Observed Payoff Matrix for the Penalty Game
(Figure obtained from Chiappori et al., 2002)

| Goalie | Kicker | | | Total |
	Left	Middle	Right	
Left	63.2	81.2	89.5	76.2
Middle	100	0	100	72.7
Right	94.1	89.3	44	73.4
Total	76.7	81	70.1	74.9

In a later study on penalty kicks, Bar-Eli, Azar, Ritov, Keidar-Levin, and Schein (BARKS)[11] found that goalkeepers exhibit a bias toward jumping. BARKS found from their dataset that staying in the middle provides the highest penalty save rate, that is, 33.3%, while jumping (left or right) provides an aggregate 13% save rate. Based

11. Michael Bar-Eli, Ofer H. Azar, Ilana Ritov, Yael Keidar-Levin, and Galit Schein, "Action bias among elite soccer goalkeepers: The case of penalty kicks," *Journal of Economic Psychology*, 28(5), 2007, 606–621.

on the probability matching principle,[12] and given that the kickers select the center 29% of the time per their data, the goalies should stay in the center approximately 29% of the time. However, they only stay in the middle for approximately 6.3% of the time! The authors offered an interesting theoretical explanation for this irrational behavior from the goalies based on Kahneman and Miller's norm theory.[13] Norm theory predicts that a negative outcome will be amplified following inaction. Therefore, if a goalie decides to choose to remain in the middle but the kicker kicks to the left or right (and scores), he—and certainly the fans and the media—will *feel* the loss more, compared to a case where the kicker kicks the penalty dead center (and scores) and the goalie jumps (left or right). Goalies (and athletes in general) are humans, too, and like every other human exhibit the same cognitive biases!

PROBABILITY MATCHING

Probability matching is a heuristic strategy in decision making in which predictions and choices are made according to *base rates*. In particular, if we have to choose between two options A and B, where

   (i) A is optimal with probability p > 0.5, and B is optimal with probability (1−p), and
   (ii) the difference in utility between A and B is constant,

then once p is learned (e.g., through repetitions of the game) the *optimal* behavior is to always choose option A. **However, in reality, players choose option A with probability p and B with probability (1−p), i.e., they match their actions to the probability of the strategy chosen as being optimal.**

---

12. M. F. Norman and J. I. Yellott, "Probability matching," *Psychometrika*, 31(1), 1966, 43–60.

13. D. Kahneman and D. T. Miller, D "Norm theory: Comparing reality to its alternatives," *Psychological Review*, 93, 1986, 136–153.

PLAYER TRACKING DATA

If we were to rank sports in order of importance of player tracking data, soccer would most probably be ranked at the top of the list. Soccer is a game of space, where on-ball events are a rarity for a player. In the words of Johan Cruyff, the Dutch star of the past for Ajax and Barcelona, "Players have the ball for three minutes, on average. So, the most important thing is: what do you do during those 87 minutes when you do not have the ball? That is what determines whether you're a good player or not." Current stats—even advanced ones like xG—evaluate on-ball events. However, these are only a small part of the beautiful game. Player tracking data can help better understand what players are doing when they do not have the ball in their feet. During the past few years, a lot of researchers have developed ways to evaluate these off-ball actions. For example, Fernandez and Bornn[14] used player tracking data to define new metrics related to pitch control and space occupancy and generation. Essentially, they are interested in identifying players who can generate space for their teammates (e.g., by attracting opponents during an attack, while space opens for teammates). In a similar direction, Spearman[15] (currently the lead data scientist for Liverpool) used spatiotemporal player tracking data to evaluate off-ball scoring opportunities (OBSO). This will allow us, for example, to identify players who were able to get themselves in a good position for potentially scoring but never got the ball in their feet (e.g., a striker who is near the penalty spot, but to whom the winger fails to deliver the pass). The proposed model is also able to assign credit to players for creating space in a scoring location even if the player never touches the ball. OBSO also improves the prediction of future player (team) scoring, as it can be used on top of expected goals from shots, or even possession

14. J. Fernandez and L. Bornn, "Wide open spaces: A statistical technique for measuring space creation in professional soccer," in MIT Sloan Sports Analytics Conference, 2018

15. W. Spearman, "Beyond expected goals," in MIT Sloan Sports Analytics Conference, 2018.

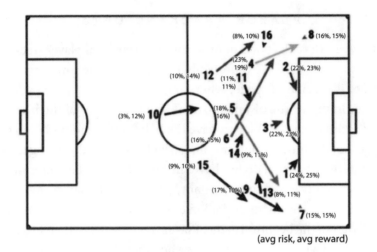

FIGURE 39.7 Prototype Passes and the Corresponding Average Risk-Reward (figure obtained from Power et al., 2017).

in general, to obtain a total of expected goals for a team or player. In a tangential study, a group from STATS Inc., led by Paul Power, developed models[16] using player tracking data for quantifying the risk and the reward from a pass. The risk is essentially the probability of the pass being completed to the intended target, while the reward of the pass is the probability that the pass will result in a shot within the next 10 seconds. It also identified specific pass clusters (Figure 39-7), along with their average risk-reward profile.

Research in this space is flourishing, and the set of studies mentioned above is just a sample of what people are doing today in soccer analytics. The sport is going to significantly benefit from these studies, and all big soccer clubs are trying to get with the times.

16. P. Power, H. Ruiz, X. Wei, and P. Lucey, "Not all passes are created equal: Objectively measuring the risk and reward of passes in soccer from tracking data," in ACM SIGKDD, 2017.

CHAPTER 40

HOCKEY ANALYTICS

While the *war* between the analytics and traditional folks in baseball was over long before 2014, in hockey things were still going the traditional way. However, everything changed during the summer of 2014. During that off season, teams took a big step—a leap of faith for some—and hired "amateur" hockey statisticians to take them through the new analytics era. This analytics revolution in hockey has not slowed down but, rather, accelerated, with the back-to-back Stanley Cup winners in 2016 and 2017, the Pittsburgh Penguins, leading the way with Sam Ventura, who was one of the creators of war-on-ice.com (the other creator, Andrew Thomas, was also hired by the Minnesota Wilds). In this chapter, we will show how analytics are used in the National Hockey League.

CORSI AND FENWICK STATISTICS

The (raw) plus-minus metric that we discussed in the context of basketball was actually first introduced in hockey in the 1950s. Being on the ice while your team scored a goal gives you a +1, while being on the ice while your team allowed a goal gives you a −1 (times when a team was shorthanded or in the power play are omitted). As we have seen, there are several problems with the raw version of plus-minus.

As Rob Vollman says in his book *Stat Shot* (2006), "plus-minus is like communism: extremely elegant in theory but not very good in practice." Some of the problems discussed for plus-minus in basketball are even more accentuated in hockey. For example, confounding effects from teammates are more pronounced in hockey, where specific lines tend to play together most of the time. Furthermore, the goalkeeper is (almost) all the time present in the lineup, and hence his quality will have a significant impact on the metric. Not to mention that additional bias in hockey plus-minus can be introduced by the zone start of a player's shift. Players who start their (typically 50 seconds) shift with a face-off in their defensive zone will be disadvantaged compared to starting with a face-off in the neutral zone or their offensive zone.

Adjusted versions of plus-minus have been used to control for some of these problems, and indeed they perform well when analyzing players over a longer period (e.g., 5–10 years). However, teams are interested in obtaining a good estimate of a player's performance over shorter periods of time (i.e., during a season over a span of several games).

One of the issues with using the plus-minus metric for short-term evaluation is its small sample size, since goals are not a commodity in hockey, leading to a metric with high variance. On average, only 8% of the shots at even strength in NHL result in a goal. This conversion rate means that the sample of shots at even strength is 12 times larger than the sample of goals, and could lead to a more robust metric! This is exactly what Corsi (also known as Shot Attempts or SAT) does. The Corsi value of a player is the difference between the shots taken by his team while he is on ice and the number of shots against his team during the same period. Corsi counts all shots at **even strength** play, i.e., shots on goal, missed shots, and blocked shots. Corsi sometimes is also expressed as a percentage, Corsi% = Corsi for/(Corsi for + Corsi against), where a value greater than 50% corresponds to a player better than average. For example, if the Pittsburgh Penguins took 1,250 shots on even strength with Sidney Crosby on the ice, while they had to defend 1,071 shots the (raw) Corsi is +179 while

TABLE 40.1

Top 10 Leaders in Corsi % over the 10-Season Period
2006–2007/2016–2017 in the NHL

Player	Corsi %	Corsi For	CoRsi Against
Patrice Bergeron (C)	58.27	8,632	6,183
Pavel Datsyuk (C)	57.74	6,254	4,577
Justin Williams (R)	57.49	8,232	6,086
Anze Kopitar (C)	57.24	9,383	7,008
Brand Marchand (L)	57.00	7,741	5,839
Drew Doughty (D)	56.18	11,469	8,946
Jonathan Toews (C)	55.89	8,541	6,742
Patrick Sharp (L)	55.64	7,674	6,118
Alec Martinez (D)	55.36	6,110	4,926
Joe Thorton (C)	55.25	9,100	7,370

the Corsi% = 1,250/(1,250 + 1,071) = 53.9%. Table 40-1 provides the top 10 leaders with at least 400 games played in Corsi% for the period between the 2006–2007 and 2016–2017 seasons.

Fenwick is a similar statistic but excludes blocked shots from the calculations. The reason for excluding blocked shots is that blocking a shot can be considered as a defensive skill, and hence no credit should be given to the offense for this shot (e.g., to a defensive player who closely covers the skater with the puck and does not allow him to take a good-quality shot).

When using metrics such as Corsi% and Fenwick%, if a team posts numbers at the range of 53–57%, this means that it exhibits a nonnegligible difference in the shot attempts for and against. Given the rarity of goals (as compared to shots), a similar percentage for the goals is going to translate to a much smaller absolute difference, which over a whole season can be the result of lucky bounces. For example, in the 2016–2017 season the average number of goals in a game was 5.54, with 6,810 goals scored in total. A team outscoring its opponents in the ratio

55–45% will have outscored its opponents by approximately +45 goals. Over a whole season, 45 goals in 82 games is not a particularly large number given the lucky bounces a puck takes through a single game. In contrast, the average NHL team took on average 2,860 unblocked shot attempts over the whole season. A team that outperforms its opponents 55–45% translates to a +572 Fenwick, which is significant even considering the lucky bounces of the puck.

During the 2016–2017 season, the Capitals had the best PTS% with .72 and a goal difference of +84. During the same season the Penguins—who eliminated the Capitals in the playoffs and went on to win the Stanley Cup—had a similar PTS%, .68, but almost half of the goal differential, i.e., +49. By looking just at the goal differential, one would have expected the Penguins to have had a much lower win percentage than the Capitals, which was not the case. However, if we look at the Fenwick for the Capitals and the Penguins, they are very similar, +153 and +138, respectively.

It should be clear that both goal- and shot-based metrics are important and can reveal different information. However, in the era of advanced analytics one should be careful about their usage. As mentioned in previous chapters, players are humans and humans respond to incentives. As such, if players know that the teams are considering shot totals (rather than just goals), they can simply start taking more shots (even if they are from a bad angle, longer distance, etc.). Of course, this will provide analysts with the problem of improving upon the metrics, and one possible solution to this problem is, instead of considering each shot equally, to consider its quality (e.g., through a goal probability model).

WHY ARE GOALIES PULLED EARLIER THAN EVER?

During round 2, game 3 of the 2016–2017 NHL playoffs, the Capitals led the Penguins in Pittsburgh 0–2, with 1:51 seconds left. At this point the Penguins had a 1% probability of winning and pulled Marc-Andre Fluery from the goal, leaving the net empty, and added an additional offensive player in the lineup. The Penguins ended up tying the game 2–2—they finally fell in OT 2–3 at a power play goal. Had the game been played 10 years earlier, Fluery would have most probably stayed an extra 25–30 seconds' under the net before being pulled out. That could have cost the Penguins critical time in their efforts to tie the game. In general, the data reveals that teams today pull the goalies earlier from the game compared to, say, 10 years back. During the 2014–2015 season, teams that were down two goals pulled their goalies with about 130 seconds remaining, as compared to 85 seconds in 2007–2008 season. With one goal difference, teams pulled their goalies on average 75 seconds before the end of the game during the 2014–2015 season, 25 seconds earlier compared to the 2007–2008 season. But how can one decide on the appropriate time to pull the goalie?

A simple approach would be to assume that the number of goals scored by the trailing team i with time t left in the game follows a Poisson distribution with mean $\lambda = \mu_t * \text{offRtg}_T * \text{defRtg}_L$, where μ_t is the league-average number of goals scored over a time period of t, offRtg_T is the offensive rating for the trailing team (T) with empty net, and defRtg_L is the defensive rating of the leading team (L) in the same situation (opponent with empty net). We will see in Chapter 46 how we can obtain team ratings in a similar way with soccer. For now the only thing we need to understand is what these ratings represent. For example, if a team has an offensive rating of $o = 1.07$ and a defensive rating of $d = 0.97$, it means that it scores 7% more goals than an average team and allows 3% less goals than an average team (an average team will have $o = d = 1$). However, these ratings are obtained using the final game score and do not differentiate between the strength state of the game (e.g., power play, even strength, etc.). The questions then become, how do we adjust the offensive and defensive ratings to account for the uneven situation on ice, as well as, how do we adjust the league average goals scored over a time period t? Starting with the latter, we can assume that the goals are scored uniformly at random within a game, and hence, if the league-

---

1. https://fivethirtyeight.com/features/nhl-coaches-are-pulling-goalies-earlier-than-ever/

average goals scored per 60 minutes is x, then $\mu_t = t * x/1800$, where 1,800 is used to express the Poisson rates per second. Adjusting the ratings is a little trickier. During this time period of t, the trailing team essentially plays in a power play with empty net. Therefore, its offensive rating can be adjusted based on its power play percentage. For example, if a team with offensive rating o has 23% power play percentage and an overall 10% shooting percentage, its offensive rating can be adjusted to $\text{offRtg}_T = o * 0.23/0.1$. Similarly, the defensive rating of the leading team is adjusted based on its power play killing percentage. If the team has a defensive rating d, with a power play killing percentage of 84% and an overall shoot save percentage of 92%, its defensive rating during the period of an opponent-pulled goalie is $\text{defRtg}_L = d * 0.84/0.92$.

However, the leading team can also score a goal, in which case we assume that its win probability will become almost equal to 1 (this assumption is not very far from reality). Again, we can assume that the goals scored by the leading team during the period that the opponent goalie is pulled follows a Poisson distribution with a mean $\rho = \mu_l * \text{offRtg}_L * \text{defRtg}_T$, where offRtg_L is the offensive rating of the leading team while playing against an empty net and defRtg_T is the defensive rating of the trailing team while playing without its goalie. It is hard to adjust the offensive and defensive ratings in this case since there do not exist similar situations within the game, and therefore we will adjust the mean of the distribution through parameter μ_l, which is the league-wide average number of goals scored by a team whose opponent has pulled the goalie. For this we will need to know the empty net goal (ENG) percentage for a team. The league average is approximately 5%, that is, 5% of the goals of a team are scored while their opponent has an empty net. Therefore, with μ being the overall league average number of goals scored (2.77 in the case of 2016–2017 season), and τ being the average time per game that a team plays with an empty net, μ_l can be expressed as $0.05 * \mu * (1800/\tau) * (t/1800)$. Note here, that τ is fairly small (around 20–30 seconds), since this is the average time playing against empty net over all the games.

Having these in place, how do we decide on the *best* time to pull the goalie when we are down one goal? One possible way to determine the *right* time to pull the goalie is to identify the time t that will provide an expected number of goals for the team trailing equal to 1. Anything less than 1 will not be enough to tie the game in expectation. Therefore, assuming an average team (i.e., ratings equal to 1), we have $2.77 * 0.91 * 2.3 * (t/1800) = 1$, which provides us with a value for t approximately equal to five minutes! It should be clear that this does not mean that the probability of tying the game is 1 in expectation! We should also estimate the probability of the trailing team scoring first. In order to estimate this we use the fact that the interarrival times of a Poisson process follow an exponential distribution with the same rate. The rate of the exponential distribution for the inter-arrival (scoring) times for the trailing team is .003, while that for the inter-arrival (scoring) times for the leading team playing with an empty net is .006 (with $\tau = 25$ sec). Consequently, the trailing team has approximately a .375 probability of scoring first. This probability does not depend on the time that the goalie is being pulled. Therefore, pulling the goalie five minutes before the end of regulation gives a .375 probability for the trailing team to tie the game. In contrast, pulling the goalie 130 seconds before the regulation (the 2014–2015 season average) gives an expected number of goals for the trailing team of 0.42, and a probability of tying the game of only $0.42 * 0.375 = 0.16$.

The above analysis is clearly not fine-grained, but it shows that coaches move in the right direction by pulling their goalies earlier than ever. Certainly, one can find in the literature more detailed approaches to estimating the time of pulling the goalie, such as in the work of Andrew Thomas.[2] The next step might be for coaches to think when is the best time to pull their goalie if the game is tied at the last minute (e.g., 30 seconds before the end of regulation, with a face-off at our offensive zone).

2. Andrew C. Thomas, "Inter-arrival times of goals in ice hockey," *Journal of Quantitative Analysis in Sports*, 3(3), 2007.

FACE-OFFS: SKILL OR LUCK?

One of the constant questions in sports is, how much of a team's or a player's success is due to skill and how much is due to luck? In hockey, this question comes up many times when fans and media talk about the face-offs. To answer this question, we follow a very elegant approach that originated in the sabermetrics community[3] and was *popularized* by Mauboussin's book *The Success Equation* (2012). This approach is based on a very simple mathematical equation, which states that for two independent random variables X and Y, the variance of their sum is equal to the sum of their variances, i.e., $var(X + Y) = var(X) + var(Y)$. In our case, the independent variables X and Y correspond to the skill and luck, respectively, associated with the observed face-off win percentages of the players, since the sum of these two variables includes everything observed.[4]

To further examine this, we downloaded data on the face-offs won and lost for each player for the 2016–2017 NHL season from hockey-reference.com.[5] Half of the players took less than three face-offs, while only 30% of the players took more than 40 face-offs. In order, to avoid skewing our results from the high variability from players taking very few face-offs, we considered only the players with at least 40 face-offs, which leaves us with a sample of 255 skaters. Table 40-2 shows the top and bottom five players with respect to face-off win %. We calculated the variance of the observed face-off win % (FOW %) for the players in our data to be equal to $var(observed) = 38.27$.

---

3. http://www.insidethebook.com/ee/index.php/site/comments/true_talent
_levels_for_sports_leagues/
4. Now this assumes that skill and luck are independent. While not very possible in the case of face-offs, as Sam Ventura said in one of our discussions, one could possibly imagine that players with greater skill put themselves in situations in which they can benefit from luck more often than less skillful players. This would break down the independence assumption.
5. Another great site with several hockey analytics resources (predictions, data, etc.) is http://moneypuck.com.

TABLE 40.2

Top and Bottom Five Players for Face-off Win %
(at least 40 Face-offs) for the 2016–17 NHL Season

	Player	Face-offs Won	Face-offs Lost	Face-off Win%
Top-5	M. Duchene	687	411	62.6
	A. Vermette	744	451	62.3
	P. Bergeron	1086	722	60.1
	F. Hamilton	75	51	59.5
	A. Andreoff	41	28	59.4
Bottom-5	R. Vrbata	17	30	36.2
	T. Brouwer	20	44	31.3
	N. Schmaltz	84	188	30.9
	J. Hayes	14	41	25.5
	P. Kane	7	44	13.7

To calculate the variance expected in a completely random (with respect to face-offs) NHL, we model each face-off a player faced as a Bernoulli trial with probability of success equal to 0.5 (i.e., a coin flip). For each player, we flip the coin as many times as the face-offs they had in the season and calculate the simulated (purely random) face-off win %. For example, Matt Duchene took 1,098 face-offs, from which he won 687 of them. To obtain an estimate for Duchene's FOW% in a purely random world, we flip a coin 1,098 times and keep track of how many times the coin lands on the "face-off win" side for Duchene. Our coin flip series provides a 49.7% FOW% for Duchene in a world of pure luck. Repeating this process for all the players allows us to calculate the FOW% for each player in a completely random league. We can then calculate the corresponding variance over all the players in this random world, which gives us var(luck) = 12.65. Simply put, the contribution of luck in face-off success (or lack thereof) is only about 33% (12.65/38.27) of the observed variance of the players' face-off win %. In other, words 67% of face-off success can be attributed to the player's skill!

ARE ALL POWER PLAYS CREATED EQUALLY?

Power play ability and play killing are important aspects of special team plays in hockey. Are there contextual factors that might *affect* the power play or kill ability of a team? For example, do carry over penalties, that is, penalties that span more than one period, have different statistical properties? We obtained play-by-play data from the 2014–2015 regular season using the great R package nhlscrapR and we extracted the penalties. Figure 40-1 visualizes the empirical distribution with respect to the timing of a penalty. We then analyzed the power play percentage for these penalties and compared it with the power play percentage for the penalties that start and end in the same period. We further break down the results based on the time S remaining between the penalty call and the intermission. Table 40-3 presents the results.

These results indicate that the power-play% of carry over penalties is lower as compared to the power-play% of non–carry over penalties. Power play success is based on continuous passing and puck possession. Anything that breaks this down has the potential

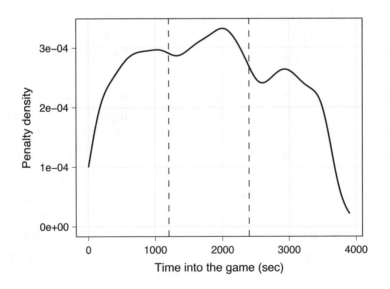

FIGURE 40.1   There appears to be a surge of penalties in the second half of the second period.

TABLE 40.3

Carry over penalties have lower success rate compared to
penalties that span through only one period

S (seconds)	Power-play% for Carry over Penalties	Power-play% for Rest of the Penalties	Statistical Significance (p-value of proportions test)
30	14.3	19.1	0.02
60	15.8	19.1	0.07
90	16.2	19.1	0.08
110	16.0	19.1	0.05

to reduce the power-play%. We have also included in the last column of Table 40-3 the p-value for the statistical test between the
two proportions. As we can see we can reject the hypothesis that
the two proportions are equal at the significance level $\alpha = 0.05$ for
almost all the cases.

However, what if carry over penalties tend to be penalties with
a larger player advantage (e.g., 5 vs. 3)? We calculated the average absolute difference in the number of skaters for carry over and
non–carry over penalties and there was not any statistical difference at the man advantage. Therefore, the difference observed in the
power-play% cannot be attributed to the number of skaters on ice.
Of course, we are leaving it to the coaching staff to decide whether
the 3.1% difference is important in terms of adjusting their power
play strategies (we think it is).

TWO PROPORTION Z-TEST

A two-proportion z-test examines the following hypotheses:

$$H_0 : p_1 = p_2$$
$$H_1 : p_1 \neq p_2$$

where p_1 and p_2 are two proportions (e.g., fraction of successful
Bernoulli trials). Assuming that the first proportion is obtained

through N_1 tries, while the second is obtained through N_2 tries, we calculate $Z = \dfrac{P_1 - P_2}{\sqrt{p \cdot (1-p) \cdot \left(\frac{1}{N_1} + \frac{1}{N_2}\right)}}$, where p is the overall proportion: $p = \dfrac{P_1 \cdot N_1 + P_2 \cdot N_2}{N_1 + N_2}$. This is the z-score of the test, so if $|Z| > 1.96$, we can reject H_0 at the 5% level. A $|Z| > 1.645$, rejects the null hypothesis at the 10% level.

Furthermore, we have also calculated the proportion of short-handed goals scored for the two different types of penalties. As we can see from the results in Table 40-4, there are no statistically significant differences observed.

TABLE 40.4

Carry over penalties do not alter the fraction of shorthanded goals as compared to the rest of the penalties

S (seconds)	Shorthanded Goal % for Carry Over Penalties	Shorthanded Goal % for Rest of the Penalties	Statistical Significance (p-value of proportions test)
30	3.4	3.1	0.81
60	3.2	3.1	0.94
90	3.4	3.1	0.75
110	3.1	3.1	0.96

Hockey analytics will probably go through a *second wave* of development with the collection and use of player tracking data starting from the 2021 season. Similarly to other sports where off-ball (puck in this case) events are crucial, hockey analysts are expected to benefit greatly from the analysis of these data.

CHAPTER 41

VOLLEYBALL ANALYTICS

Volleyball is one of those sports that appears to be amenable to analysis due to its structure—even more so than, say, basketball, and certainly soccer, American football, and hockey—but has not been a focal point for sports statisticians. This lack of volleyball analytics research might be due to the limited public datasets, or vice versa (this is a chicken and egg problem). In this chapter, we will explore some basic analytics tasks for volleyball.

PYTHAGOREAN THEOREM FOR VOLLEYBALL

We have seen in previous chapters that pretty much every sport has its own version of the (sports) Pythagorean Theorem that associates the *points* scored or allowed by a team and its number of wins. Depending on the sport, the notion of points is well defined (e.g., points in basketball, runs in baseball, goals in soccer and hockey, etc.). In volleyball, there are two quantities that one could consider when thinking about points. One is the total points scored and allowed, while the other is the sets won and lost. Using data from the Greek professional league for the past five seasons we identified the Pythagorean coefficient for points and sets as $\alpha_{points} = 10.8$ and $\alpha_{sets} = 1.4$, respectively. They both fit the data very well, but the

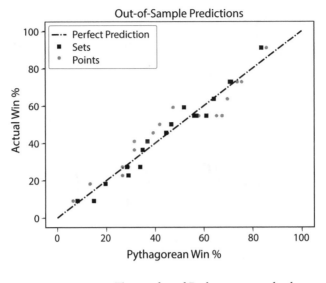

FIGURE 41.1 The sets-based Pythagorean tracks the actual win % closely.

sets' Pythagorean has (slightly) better performance out-of-sample. Figure 41-1 presents the predicted Pythagorean win % for a given points and set differential and the actual win % on a test set. As we can see, both versions follow the actual win % closely, but the sets' Pythagorean has slightly better performance (points are closer to the y = x line).

However, the Pythagorean predictions are typically end-of-season predictions, attempting to identify teams that might have been (un)lucky during the season and over- or underperformed in terms of win percentage. Can we use the Pythagorean predictions during a season for predicting the final win percentage of a team? Using data from the 2018–2019 season from the Greek national league and the sets' Pythagorean exponent learned above (the training set did not include the 2018–2019 season), we used the cumulative sets won and lost for each team after each week or round, and projected the end-of-season Pythagorean win percentage for each team. Figure 41-2 presents the mean absolute error after every league round. As we can see—and as it might have been

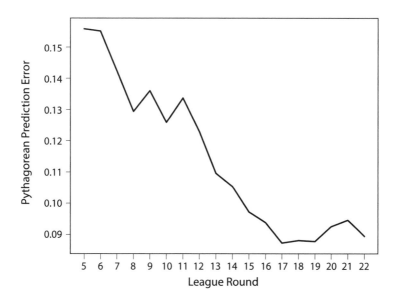

FIGURE 41.2   The sets-based Pythagorean in-season predictions improve as the season progresses.

expected—later in the season the projections improve. Some of the noise observed could be smoothed out by using more data, i.e., over more seasons.

BRADLEY–TERRY MODEL

The Pythagorean formula can help with the prediction of the number of total wins for a team but it does not typically provide predictions at a game level. To build team ratings we can assume that each set played between two teams, A and B, is a Bernoulli trial, where the probability of team A winning the set is given by $Pr(A > B) = p_{AB} = f(c, r_A, r_B)$, a function of the team's *abilities* (to be learned) and a constant that can capture the home advantage. The Bradley–Terry model parametrizes this probability (assuming A is the home team) as

$$\log\left(\frac{Pr(A \succ B)}{Pr(B \succ A)}\right) = c + r_A - r_B.$$

TABLE 41.1

Bradley–Terry Team Abilities for the 2018–2019
Greek Volleyball League (Up to Round 15)

Team	BT Abilities
Home Edge	0.3933
AEK	− 0.4983
Komotini	−0.9866
Kifisia	− 0.5554
Foinikas	0.3860
Pamvochaikos	0.0303
Iraklis	−0.2615
Ethnikos	− 0.5712
Olympiacos	1.0442
PAOK	0.9410
Panathinaikos	0.4715

So, the goal now is to identify the abilities r_i for every team based on the sets they have played and the outcome of each set. A data point will consist of 10 indicator variables (equal to the number of teams) and will correspond to a specific set played. All the indicator variables will be 0, except for the one corresponding to the home team of the data point, which will be +1, and the one for the visiting team, which will be −1 (similarly to the adjusted +/− setting for rating basketball players). The dependent variable will be 1 if the home team won the set and 0 otherwise. One key point here is that we only consider non–tie break sets. Tie breaks are different than the rest of the sets since they go up to 15 points instead of 25. One could create a separate rating for tie breaks, but this approach would suffer from data sparsity—there are only 24 tie breaks, which represent 7% of the sets in the season. Fitting the model using the data from the first 15 rounds of the 2018–2019 season in the Greek volleyball league, we get the team abilities presented in Table 41-1.

The BT abilities learned are simply strength coefficients that can be used to obtain win probabilities for future matchups. Using the abilities learned we obtained predictions for the rest of the games in the 2018–2019 Greek league. In particular, for every matchup we calculated the set of win probabilities for every team. For instance, if AEK were playing at home with Iraklis, its set win probability would be $\dfrac{e^{0.3933-0.4983-(-0.2615)}}{1+e^{0.3933-0.4983-(-0.2615)}} = 0.54$. Using this we can simulate the (potentially five) game sets by sampling a Bernoulli distribution with success probability 0.54. We can repeat this several times and obtain a win probability for each team. In case a single game simulation results in a tie break, we can assume a 50–50 win probability for each team or give the home team a baseline win probability. For example, 15 tie breaks out of the 24 (62.5%) in our dataset were won by the home team, while the Bradley–Terry model above implies a 59% win probability for a home team that matches up with another team of equal ability.

This approach exhibits a 72.5% accuracy in predicting the rest of the Greek league (rounds 16–22). We also calculate in Figure 41-3 the reliability curve for the predicted probabilities. As we discussed in Chapter 39, the reliability curve is a succinct way of evaluating the *quality* of the output of a probabilistic model. To reiterate, in order to plot or calculate the reliability curve, we can group together games for which we predict similar win probabilities for the home team. Given the continuous nature of the probabilities, we use bins to do the grouping (typically bins with a range of 0.05 or 0.1). The reliability curve then has a point (x_b, y_b) for every bin b calculated as follows: let G_b be the set of games in bin b and N_b be the number of such games. Then,

$$x_b = \frac{\sum_{i \in G_b} p_i}{N_b}$$

$$y_b = \frac{\sum_{i \in G_b} \mathbb{I}_i}{N_b}$$

where p_i is the win probability (for the home team) of game i and $\mathbb{I}_i$ is an indicator function that is 1 if the home team in game i won and 0 otherwise. When plotted, if the predicted probabilities are *well calibrated,* these points should be close to the y = x line, i.e., the predicted probabilities are close to the observed ones. As we can see from Figure 41-3, our predictions for the Greek volleyball league are well calibrated. However, the number of games predicted is fairly small and this is why we use only five bins. For the interested reader, Weisheimer and Palmer[1] have a nice overview of reliability curves in the context of weather forecast and their potential pitfalls.

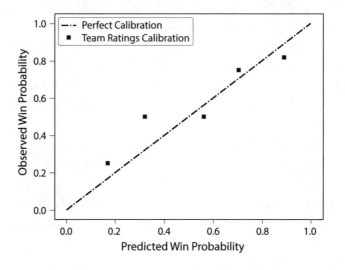

FIGURE 41.3 The output probabilities of the
Bradley–Terry model are well-calibrated.

Of course, the above is just one way to obtain team ratings. One could use the Bradley–Terry model directly on game outcomes (rather than sets), while more advanced models make use of Bayesian inference and truncated distributions to model the actual points won within a set.[2]

1. A. Weisheimer, and T. N. Palmer, "On the reliability of seasonal climate forecasts," *Journal of the Royal Society Interface*, 11(96), 2014, 20131162.
2. L. Egidi and I. Ntzoufras, "Modelling volleyball data using a Bayesian Approach," in Math Sport International Conference, 2019; V. Palaskas, I. Ntzoufras, and S. Drikos, "Bayesian modelling of volleyball sets," in Math Sport International Conference, 2019.

CHAPTER 42

GOLF ANALYTICS

In 2014 Mark Broadie of Columbia University revolutionized golf analytics with his book *Every Shot Counts* (2014). Via the PGA's ShotLink System, Mark had access to 15 million shots by PGA golfers. Using clever, yet easy-to-explain analysis, Mark developed many important insights that help us better understand what leads to a golfer's success (or failure!). As we will see, the key idea behind Mark's analysis is like John Dewan's approach to analyzing MLB fielders. In this chapter, we will try and summarize Mark's brilliant insights and analyses.

PUTTING

Suppose Jordan Spieth used 30 putts on the last round of his 2017 British Open win and Rory McElroy used 34 putts on the same round. Does this mean Jordan's putting performance was superior to Rory's? If Jordan's putts on each hole started three feet from the hole and all of Rory's puts started 30 feet from the hole, I am sure you would agree that Rory's putting was vastly superior to Jordan's.

Broadie's key idea is the concept of strokes gained. Figure 42-1 (see file Golf.xlsx) shows the average number of puts needed to hole out based on starting distance from the hole.

	I	J
25	Distance	Average Strokes
26	2	1.01
27	3	1.04
28	4	1.13
29	5	1.23
30	6	1.34
31	7	1.42
32	8	1.5
33	9	1.56
34	10	1.61
35	15	1.78
36	20	1.87
37	30	1.98
38	40	2.06
39	50	2.14
40	60	2.21
41	90	2.4

FIGURE 42.1   Average Puts
Needed to Hole Out.

For example, from 20 feet a PGA golfer needs on average 1.87 strokes to hole out. As an example of how to use this chart, suppose a golfer, Eric Taylor, hit four greens eight feet from the hole. Eric holed out in one putt on three of the holes and two-putted once. Since the average PGA golfer needed 1.5 putts to hole out from eight feet, Eric gained $3 * .5 = 1.5$ putts on the three greens when he needed one, and gained $1.5 - 2 = -.5$ on the green on which he two-putted. Therefore, Eric gained a total of one stroke on those four holes.

As another example, Tiger Woods in the round shown in Figure 42-2 lost one stroke compared to average PGA golfers.

You can measure difficulty of greens on a course by looking at strokes needed at each distance compared to the PGA average. Broadie found that Pebble Beach had the toughest greens: .77 strokes tougher per round than the average PGA course.

To identify the best putter over a given time period, simply find the golfer who gained the most strokes putting per round. During the years 2004–2012 Broadie found that Luke Donald (0.7 strokes

	I	J	K	L	M
3	A Tiger Woods Round				Total
4					−1.002
5	Hole	Starting Distance	Putts Used	Average Putts	Strokes Gained
6	1	4	1	1.13	0.13
7	2	3	1	1.04	0.04
8	3	3	2	1.04	−0.96
9	4	3	1	1.04	0.04
10	5	1	1	1	0
11	6	6	1	1.34	0.34
12	7	22	2	1.892	−0.108
13	8	45	2	2.1	0.1
14	9	6	2	1.34	−0.66
15	10	12	1	1.678	0.678
16	11	4	2	1.13	−0.87
17	12	42	2	2.076	0.076
18	13	15	2	1.78	−0.22
19	14	6	1	1.34	0.34
20	15	5	1	1.23	0.23
21	16	4	1	1.13	0.13
22	17	13	2	1.712	−0.288
23	18	13	2	1.712	−0.288

FIGURE 42.2   Tiger Gains One Stroke Putting.

gained putting per round) was the best putter with Tiger Woods (0.6 strokes gained putting per round), a close second.

We close our discussion of putting by noting that after adjusting for the distance of putts, amateurs take 4 more putts per round. Of course, their courses are easier.

EVALUATING TEE TO GREEN PLAY

The concept of strokes gained also can be applied to evaluate a golfer's ability on drives, approach shots, and the short game.

Broadie looked at how many shots are needed from each situation. A shot is a good shot if and only if after the shot you have reduced strokes needed by more than one stroke. The file Golf.xlsx and Figure 42-3 show the average strokes needed to hole out in various situations.

	C	D	E	F	G
1					
2	Average Shots Needed				
3					
4	Distance	Tee	Fairway	Sand	Recovery
5	20		2.4	2.53	
6	40		2.6	2.82	
7	60		2.7	3.15	
8	80		2.75	3.24	
9	100	2.92	2.8	3.23	3.8
10	120	2.99	2.85	3.21	3.78
11	140	2.97	2.91	3.22	3.8
12	160	2.99	2.98	3.28	3.81
13	180	3.05	3.08	3.4	3.82
14	200	3.12	3.19	3.55	3.87
15	220	3.17	3.32	3.7	3.92
16	240	3.25	3.45	3.84	3.97
17	260	3.45	3.58	3.93	4.03
18	280	3.65	3.69	4	4.1
19	300	3.71	3.78	4.04	4.2
20	320	3.79	3.84	4.12	4.31
21	340	3.86	3.88	4.26	4.44
22	360	3.92	3.95	4.41	4.56
23	380	3.96	4.03	4.55	4.66
24	400	3.99	4.11	4.69	4.75
25	420	4.02	4.15	4.73	4.79
26	440	4.08	4.2	4.78	4.84
27	460	4.17	4.29	4.87	4.93
28	480	4.28	4.4	4.98	5.04
29	500	4.41	4.53	5.11	5.37
30	520	4.54	4.66	5.24	5.3

FIGURE 42.3   Strokes Needed to Hole Out.

To illustrate the use of this chart, suppose Eric Taylor starts 394 yards from a hole. Eric hits a terrible tee shot of 114 yards. This shot gained $3.99 - 3.65 = 0.34$ strokes, but Eric used one stroke. Therefore, this tee shot added -0.66 strokes. Eric's second shot finished 62 yards away in the sand. Before his second shot, an average golfer needed 3.65 strokes to hole out, and after his second shot an average golfer needed 3.15 strokes. The second shot therefore gained $0.5 - 1 = -0.5$ strokes. Eric's third shot finished on the green 17 feet from hole. Before this shot an average golfer needed 3.15 strokes to hole out. After his third shot an average golfer would need 1.8 shots to hole out. Therefore, this shot gained $(3.15 - 1.8) - 1 = 0.35$ strokes. Finally, Eric made his 17-foot putt, and on it gained $(1.8 - 1) = 0.8$ strokes.

During the 2004–2012 period Tiger Woods was the best golfer and gained 2.8 strokes per round. Brodie broke this down as 0.6 strokes gained on driving, 1.3 shots gained on approaching, 0.3 shots gained on the short game, and 0.6 shots per round gained on putting. Broadie found that on average the top 10 golfers gained 20% of their strokes on driving, **45% on approach shots,** 20% on the short game, and only 15% on putting. As Brodie said at the 2014 Sloan Sports Analytics Conference, "You don't drive for show and putt for dough. It's really the long game that matters."

Using the idea of strokes gained, we can easily determine what it is worth to improve our driving ability. Averaging over all holes, Broadie found that for PGA golfers adding 20 yards of driving distance is worth about 0.75 strokes per round. Surprisingly, Broadie found that for amateurs **20 extra yards on drives is worth around three strokes per round.** Broadie also found that the difference between a 90 golfer and an 80 golfer broke down on average as follows: 2.5 strokes on driving, four strokes on approaching, 2.1 strokes on the short game, and 1.4 on putting.

ANALYTICS AND PUTTING STRATEGY

Basic physics can make the weekend golfer and PGA golfer a much better putter. To apply physics and analytics to putting the golfer needs to know the slope and speed of the green. A one-degree slope means that the green goes up 1.5 feet per 60 feet. A green's stimp reading describes the speed of the green. Stimp is how far a golf ball will go on level green when hit with 4 mph speed. Stimp readings usually vary between 7 and 13. A stimp reading of 9 is average on a public course, while a stimp reading of 11 is average on PGA courses.

Given the speed and slope of the green, Table 42-1 shows how hard to hit a 20-foot putt. For example, on a green with an uphill

TABLE 42.1
How Hard to Hit a 20-Foot Putt

Slope	Stimp 7	Stimp 9	Stimp 11	Stimp 13
Uphill 2	26	28	30	32
Uphill 1	23	24	25	26
Level 0	20	20	20	20
Downhill 1	17	16	15	14
Downhill 2	14	12	10	8

two-degree slope and a stimp 13 reading, we should hit a 20-foot putt like a 32-foot putt on slope 0.

When putting, should a golfer aim for the hole, aim short of the hole, or aim beyond the hole? If we make the reasonable assumption that our errors in putting distance (relative to target) are symmetrically distributed about the target, then aiming for the hole will result in 1/2 of our putts falling short, so the golfer should aim beyond hole. Broadie found that on 15-foot putts the best PGA putters left 14% of all puts short, while amateurs who averaged shooting 90 left 28% of all 15-foot putts short. This information reveals that duffers need to aim further beyond the hole!

TEE TO GREEN STRATEGY

Consider a 400-yard hole with out of bounds on the right side of the fairway and a rough on left. Where should a golfer aim her tee shot? If you are an 80s golfer aiming down center, you will average 4.7 strokes per hole, but aiming for the left edge of the fairway you will average 4.6. This change in strategy is worth 1.8 strokes per round. A 100 golfer should aim for the left rough. This saves .2 strokes per hole over aiming for the center of the fairway.

Now consider a par 4 that is so long that we cannot reach the green in two shots. Many golfers would strive for increased accuracy by shooting two 5 irons followed by a 3 iron. This strategy is

TABLE 42.2

Average Strokes Needed to Hole Out Based on Distance to
Green and Rough vs. Fairway

Type of Golfer	30 Yards Fairway	30 Yards Rough	80 Yards Fairway	80 Yards Rough
PGA	2.5	2.7	2.7	3
80s	2.7	2.8	3.1	3.2
90s	2.9	3.1	3.4	3.5

wrong because it gives up distance. In general, should a golfer try to lay up (be more accurate but end up farther from hole) or try and get closer? Table 42-2 shows that giving up a little accuracy to get closer to the hole is a worthwhile trade-off. This figure shows that being 30 yards away from the green in the rough is better than being 80 yards away from the green in the fairway.

CHAPTER 43

ANALYTICS AND CYBER ATHLETES

The Era of e-Sports

If someone asked you to name your favorite sport (or any sport for that matter) we are confident that your answer would include one of football (any type), basketball, baseball, hockey, cricket, tennis, or (beach) volleyball. Well, others too but certainly we would not expect you to mention electronic sports (e-sports). e-Sports is the name for **competitive gaming**. Yes, e-sports players play video games, while being watched by an audience (and are also getting paid for it). e-Sports is the fastest growing sport globally. Revenues have been increasing globally by 9.7% every year, while the League of Legends (LoL) World Championship in 2018 had approximately 99.8 million viewers (Super Bowl LII had 103.4 million viewers). There are several different types of electronic sports ranging from real-time strategy and multiplayer online battle arena games (e.g., League of Legends, DOTA, StarCraft, etc.) to actual sports video games (e.g., NBA 2K). Different e-sports require (and help develop) a variety of skills, many of which are different from those of traditional sports. At the same time e-sports can help the development of a variety of soft skills in education.[1] e-Sports are also friendly to game logging and hence are rich in data. Player training is certainly different than in traditional sports. Strength and conditioning coaches are now replaced by "mental coaches," while if there were an "e-Sports combine" equivalent, this would certainly include eye tracking and reaction times instead of a 40-yard dash. In this chapter we will give an introduction to some of the most popular e-sports and how data and analytics find applications in competitive gaming.

DOTA 2

DotA 2 is a Multiplayer Online Battle Arena (MOBA) game that includes two teams—the Radiant and the Dire. Each team consists of five players and occupies opposite parts of a map that is divided by a river, with three paths across it. The goal is to destroy the opponent's fortress (called Ancient) that protects its area, while also defending its own. Each of the players chooses one hero from a pool of 113 heroes (this is the "drafting" phase) who are associated with a specific set of features that essentially dictate their role.

Various websites, such as https://www.dotabuff.com/, provide some very basic statistics—think of them as the box score stats—for every hero:

- **Win rate:** the fraction of games won by teams that had the specific hero
- **Pick rate:** the fraction of games that the hero was picked by either the Radiant or the Dire
- **KDA ratio:** the ratio between the kills and assists a hero has over its deaths. This can be calculated either over a game or over all games a hero has played.

---

1. https://edtechnology.co.uk/Blog/esports-encourges-skills-development-in -education/

These statistics by themselves do no reveal too much. Heroes have their own abilities but their *efficiency* also depends on how the player controlling them makes use of them.

If you are not well versed on DotA 2—just as we are not—you might not expect to see a "home field" advantage in the game. However, it appears there is one! In particular, using a match dataset from approximately 40,000 games[2] that includes various pieces of information about each team (player levels, heroes picked, etc.), Radiant teams win 52.5% of the games. This number is not adjusted for player level or heroes chosen. So it is possible that the benefit is higher for starters or novice players and disappears for pro-level players (similar to the advantage of playing white in chess). The main reason for this *Radiant edge* is the map advantage the Radiant team gets for being positioned at the bottom left half of the map. In terms of the home edge in traditional sports, this percentage would rank lower than all the four major sports in the US. There are several reasons for this. For instance, Dire also gets the advantage of last pick of hero, which lets the team choose a hero that matches the strengths and weaknesses of the heroes chosen from the Radiant team. Also, throughout the years, various patches for the game have been released that had an impact on this edge (sometimes reducing it, sometimes increasing it).

Focusing on specific heroes we can now start looking at the statistics over a variety of splits. For instance, we can see how the win rate of a hero relates with the team the hero is in and the level of the player controlling the team. We consider Pudge, the hero with the highest pick rate at the time of writing, according to dotaduff.com. The overall win rate for Pudge in our dataset is 49.7%. However, if we split it according to which team Pudge belongs to, Radiant teams with Pudge have a win rate of 53.6%, while Dire teams with Pudge have a win rate of 46.3%. In fact, a proportions test further supports that this difference is not purely due to chance. Furthermore, the level of the players controlling Pudge is statistically not different in

2. https://www.kaggle.com/c/mlcourse-dota2-win-prediction/data

TABLE 43.1

Win Probability for a Team That Has Picked Pudge

	Pudge Team Win		
Variable	*coefficient*	*SE*	*p-value*
Intercept	−0.17	0.03	<0.001
Level	0.004	0.003	0.24
Team (Radiant)	0.32	0.06	<0.001
Level*Team (Radiant)	−0.003	0.005	0.48
N	18,395		

the two groups (level diff 0.01, p-val = 0.91). We further build a logistic regression for modeling the win probability of a team that has Pudge, using as covariates the team and the level of the player controlling Pudge. Table 43-1 shows our results, where, as we see, even when adjusting for the level of the player, the win rate of Pudge is higher when he is on the Radiant team. Now in our dataset we only have players in levels 1–25, which can be considered as a starter level. For more experienced players, this gap in the win rate for heroes depending on the team they are in might be—and most probably is—much smaller.

Given all these detailed data, one can start looking at things such as how different heroes interact together in a team. This analysis can also further differentiate between the levels of players competing, and could possibly help with draft selection (all athletes—cyber or not—have to go through it). The setting seems to be *ideal* (in italics because the setting in reality is never ideal) to apply a +/− approach. The difference here—as compared to the basketball setting for example—is that every hero can be in any of the (only) two teams. The number of possible five-hero teams is much larger than the number of games and hence our data are extremely sparse. In practice, many heroes have low pick rates and can be considered as "replacement heroes." Given the fact that Radiant exhibits a minor strategic advantage, we can set up our data assuming that Radiant heroes play at home, while Dire is the visiting team. Every game provides a "stint" where the outcome is +1 if Radiant wins and 0 if Dire wins. With this setup, the (adjusted) +/− that we obtain can be interpreted as the +/− win probability added for each hero. We use ridge regression with cross-validation to identify the best regularization parameters, and Figure 43-1 presents the top and bottom five heroes in our data. Also note that the intercept we obtained from the regression is .523, which is essentially the Radiant edge we identified above.

Hero	+/−
Riki	0.072
Abaddon	0.06
Spectre	0.058
Zuus	0.057
Skeleton King	0.044
Templar Assassin	−0.079
Tiny	−0.084
Puck	−0.087
Alchemist	−0.094
Wisp	−0.097

FIGURE 43.1   Top and Bottom
Five Heroes Based on Adjusted +/−.

Let us say that we want to make win probability predictions based solely on the heroes on each team. If Dire consists of five average players, and Riki is on the Radiant along with other four average players, then Radiant is projected to win with a .523 + .072 = .595 *probability*.[3] Evaluating our model on a test set (out-of-sample),

---

3. To be technically correct, this is not exactly a probability score. Instead, we should use this *total rating* for the game and pass it through a probability calibration method such as Platt scaling. This is used to obtain probability estimates when a classification's model output does not have an inherent probabilistic interpretation. We will talk more about this in Chapter 47, but for now let us assume that the total rating is the actual probability. As we will see, even this total rating score is fairly accurate in terms of win probability predictions. Problems will appear when these scores are too high or too low, and can essentially be larger than 1 or smaller than 0, which, however, never occurred in our dataset.

its accuracy is 58.5%, that is, it can correctly predict the winner in 58.5% of the time. Figure 43-2 further presents the calibration curve for our probabilistic predictions. As we can see, the predictions follow the observations pretty closely. One interesting thing to observe is that the distribution of the predicted probabilities is less spread compared to predictions for other sports, which can translate to higher uncertainty of the outcome. Also, we would like to reiterate that adjusted +/− is more of a descriptive statistic, rather than a predictive one, not to mention that the player controlling the hero is also crucial for predictions. *Coaching* is certainly more important in MOBA compared to other sports. A predictive model should also consider the actual players and their *quality*. Therefore, there is certainly a lot of room for improving these predictions.

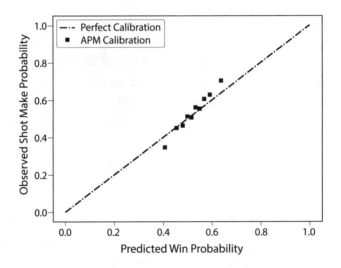

FIGURE 43.2 Predicted probabilities from the simple adjusted +/− model for DotA 2 heroes follow closely the observed ones.

While we have discussed here a few examples on what an analyst can start looking for in the game of DotA 2, there are several additional things that will be of interest, especially for professional players. For example, specific strategic decisions of players can be analyzed and potentially exploited. Similarly to teams in basketball or football that have their own playbook, DotA 2 players (might) tend to use the same ward spots[4] frequently. Knowledge of these spots can allow an opponent to reliably destroy these wards.

NBA2K LEAGUE

In 2018, the NBA became the first US professional sports league that operates an e-sports league. The NBA2K league is a joint venture between the NBA and Take-Two Interactive. The structure of the league is very similar to that of the NBA. There is an NBA2K combine that leads up to the draft before the season starts. There are currently 21 teams with a six-player roster. Every player competes as a unique character who follows a specific archetype. For example, a point guard can choose between being a playmaking shot creator (shot creation and passing or ball handling skills) or a playmaking sharpshooter (3PT shooting and passing or ball handling skills). Unfortunately, there is very limited amount of data in the public sphere to analyze these archetypes and how they mesh together. For example, is it sustainable to have a backcourt that consists of all three pure sharpshooters, but lacks in shot creation for teammates?

Brendan Donohue—the league's managing director—revealed some details on how the data from the combine were used.[5] Data on points scored, rebounds, and assists, but also on dribble move counts, jump shot timing, etc., were collected and analyzed by an external data analytics firm to evaluate the hopeful draftees. During the 2019 combine windows, analyzing the data revealed that male players were not passing the ball to their female teammates.[6] This was very crucial, since if you do not get the ball, you cannot show many things in the game (at least in offense). The result of this was

---

4. Wards are items that allow players to have vision of the enemy and spy on them.

5. https://2kleague.nba.com/news/post-combine-thank-you-and-congratula tions-from-brendan-donohue/

6. https://www.espn.com/esports/story/_/id/26146486/warriors-draft-first -woman-nba-2k-league

for analysts to start focusing more on what a player did once he or she got the ball. This eventually led to the Warriors Gaming (the 2K team operated by the Golden State Warriors) drafting Chiquita Evans, the first woman ever drafted into the NBA2K league!

With rising concerns about long-term health issues stemming from repeated exposure to contact sports, this might also seem as a "safer" alternative (until it's not). e-Sports are here and are growing. The amount of data they generate is unprecedented, and so are the possibilities for analysis. It only remains to be seen what the future holds for e-sports (and how hackers will create cheating scandals in the leagues).

PART V

SPORTS GAMBLING

CHAPTER 44

SPORTS GAMBLING 101

In this chapter, we will review (through a Q&A format) the basic definitions and concepts involved in football, basketball, and baseball gambling.

In the 2017 Super Bowl the Patriots were favored by three points over the Falcons and the predicted total points for the game was 59 points. How could I bet on these odds?

Theoretically the Patriots being favored by three points means the bookies think that there is an equal chance that the Patriots would win by more than three or fewer than three points (in the next chapter we will see this may not be the case!). We often express this line as Patriots -3 or Falcons $+3$, because if (Patriots' points -3) > 0, a Patriots bettor wins, while if (Falcons' points $+3$) > 0, a Falcons bettor wins.

Most bookmakers give 11–10 odds. This means that if we bet "a unit" on the Patriots to cover the point spread (i.e., win the game with more than three points) then we win \$10 if the Patriots indeed win by more than three points. If the Patriots win by fewer than three points we pay the bookmaker \$11. If the Patriots win by exactly three points, the game is considered a "push" and no money changes hands. Total points bets work in a similar fashion. If we bet the "over" on a totals bet we win \$10 if more than 59 points are scored, while we lose \$11 if the total points scored is fewer than 59 points. If exactly 59 points is scored, the totals bet is a "push" and no money changes hands. Similarly, if a bettor takes the under, the bettor wins if total points scored is fewer than 59 and loses if total points scored is greater than 59. Most gamblers believe the totals line (in this case 59 points) is the most likely value of the total points scored in the game. Basketball point spread betting and totals betting work in an identical fashion to football betting.

HOW CAN BETTORS MAKE MONEY GAMBLING?

Let p = probability that a gambler wins a point spread bet. If $10p - 11(1-p) = 0$, our expected profit on a bet equals 0. We find that $p = 11/21 = .524$ makes our expected profit per bet equal to 0. Therefore, if we can beat the spread or totals more than 52.4% of the time, we can make money! Suppose we are excellent at picking games and can win 57% of our bets. What would be our expected profit per dollar invested? Our expected return per dollar invested is $(.57(10) + .43(-11))/11 = 8.8\%$. This implies that a bettor who can pick 57% winners against the spread can make a pretty good living betting. Picking 57% winners against the spread in the long run is virtually impossible, however. If we believe we have a probability $p > .524$ of winning a bet what percentage of our bankroll should we bet on each gamble? In Chapter 49 we will use the famous Kelley growth criterion to answer this question.

HOW DO BOOKMAKERS MAKE MONEY?

Until Steven Levitt's (2004) brilliant article on NFL betting (to be discussed in Chapter 45), the prevailing wisdom was that bookmakers tried to set the line so half the money was bet on each side. If this is the case, the bookmaker cannot lose! For example, suppose one bettor bets $10 on the Patriots −3 (Patriots to win by three points) and another bettor bets $10 on the Falcons +3. Then unless the game is a push, the bookie pays one bettor $10 and collects $11 from the other bettor and is guaranteed a profit of $1. The bookmaker's mean profit per dollar bet is called vigorish or "the vig." In our example, $10 + $11 = $21 are bet and the bookmaker wins $1, so the vig is $1/21 = 4.8\%$. In our example, the bookmaker makes a riskless profit of 4.8%. We will see in the next chapter that a smart bookmaker can take advantage of gambler biases and make an expected profit (with some risk, however) exceeding 4.8%.

HOW DOES THE MONEY LINE WORK?

The money line enables a bettor to bet on who wins a game or an event straight up, without involving the margin of victory. For example, the money line on the 2017 NBA Finals was: Warriors −240/ Cavaliers +200. For any money line bet, the team with the negative number is the favorite and the team with the positive number is the underdog. The meaning of this money line is that to win $100 on the Warriors you must bet or risk $240. If I place this bet and the Warriors win the series, I win $100, but if the Warriors lose the series I lose $240. If I bet $100 on the Cavaliers to win the series and the Cavaliers win the series, I win $200. If the Cavaliers lose the series, I lose $100. Let p be the probability that the Warriors will win the series. A risk-neutral gambler (this means a gambler who makes decisions based on expected profit) who believes that $100p − 240(1−p) > 0$ would bet on the Warriors, while a gambler who believes that $200(1−p) − 100p > 0$ would bet on the Cavaliers. Solving for the value of p that satisfies each inequality, we find that gamblers who feel the Warriors have a chance of winning greater than $240/340 = 71\%$ would bet on the Warriors, while gamblers who feel that the Warriors have a chance $p < 200/300 = 67\%$ chance of winning would bet on the Cavs. If we assume that the true probability of the Warriors winning was the average of 67% and 71% (69%) and also assume that bettor estimates of the Warriors' chances of winning are symmetrically distributed about 69%, then we would expect an equal number of bettors to bet on the Cavaliers and the Warriors. Suppose one gambler bets on the Warriors and one on the Cavs. If the Warriors win, the bookmaker breaks even by paying the Warriors bettor $100 and collects $100 from the Cavs bettor. If the Cavs win the bookmaker wins $240 by collecting $240 from the Warriors bettor and loses $200 to the Cavs bettor. If the Warriors true chance of winning the series is 69% then the bookmaker's expected profit per dollar bet is given by $(.69(0) + .31(40))/(240 + 100) = 3.6\%$.

In Chapter 47 we will learn how to use point spreads to estimate probabilities of a team winning a game, an NBA playoff series, or the NCAA tournament.

As another example, in the 2017 Super Bowl the money line was Patriots −120 and Falcons +100. That means if we bet the Patriots to win and they lose, we lose $120, and if the Patriots win, we win $100. If we bet the Falcons to win and they win, we win $100, and if the Falcons lose, we lose $100.

HOW DOES BASEBALL BETTING WORK?

The starting pitchers play a critical role in determining the winner of a game. Therefore, the baseball gambling line is only valid if the listed starting pitchers start the game. For example, on July 28, 2017, the World Series Champions Cubs were playing the Milwaukee Brewers in a crucial game. The pregame odds are shown in Table 44-1.

This information means the Cubs were playing at the Brewers with Quintana starting for the Cubs and Suter for the Brewers. As

TABLE 44.1
Odds on Cubs at Brewers July 28, 2017

Date and Time	Teams (Away Team first)	Starting Pitchers	Money Line	Total Runs Line	Favorite −1.5 Runs Line
July 28, 2017	Cubs	Jose Quintana	−149	Over 9 −125	+100
8:10 PM	Brewers	Brent Suter	+133	Under 9 +105	−120

before, the team with the negative entry in the money line is the favorite and the team with the positive entry in the money line is the underdog. If we bet $149 on the Cubs and they win, we win $100, and if the Cubs lose, we lose $149. If we bet $100 on the Brewers to win and they win, we win $133, and if the Brewers lose, we lose $100. Following the logic in our Warriors–Cavs example, the interested reader can show that a gambler would bet on the Cubs if she believes the Cubs' chance of winning exceeds $149/249 = 60\%$ and a gambler would bet on the Brewers if she believes the Cubs' chance of winning is less than $133/233 = 57\%$.

The totals runs part of the table is analogous to the total points betting line in football or basketball. The Over 9 runs bet plays the role of favorite. If we bet $125 on the Over, we win $100 if more than nine total runs are scored in the game. If fewer than nine runs are scored, then we lose $125. On the underside of the bet, we lose $100 if more than nine runs are scored. If fewer than nine runs are scored, we win $105. If exactly nine runs are scored, then no money changes hands.

If we bet $100 on the Cubs −1.5 runs bet, and the Cubs win by two or more runs, we win $100; if the Cubs lose or win by one run, we lose $100. Similarly, if we bet $120 on the Brewers +1.5 runs and the Brewers win or lose by one run we win $100. Otherwise the Brewers bettor loses $120.

WHAT IS AN ARBITRAGE BETTING OPPORTUNITY?

Often, different bookmakers or Internet betting sites have lines on games that differ slightly. In rare cases, a combination of bets exists (called an arbitrage opportunity) that guarantees you a riskless profit. For example, suppose two different bookies had the following lines on the 2020 Super Bowl:

Bookie 1 Chiefs − 122 49ers + 112
Bookie 2 Chiefs − 135 49ers + 125

Since Bookie 1 offers better odds on the Chiefs and Bookie 2 offers better odds on the 49ers, we will bet on the Chiefs with Bookie 1 and the 49ers with Bookie 2. Suppose we bet x with Bookie 1 on the Chiefs and suppose you bet $100 with Bookie 2 on the 49ers. If the Chiefs win, your profit is $100 * (x/122) - 100$. This will be greater than 0 if $x > \$122$. If the 49ers win your profit is $125 - x$, which is greater than 0 if $x < 125$. This implies that by betting $100 on the 49ers and between $122 and $125 on the Chiefs we can lock in a sure profit. For example, betting $123.50 on the Chiefs and $100 on the 49ers locks in a sure profit of $1.23. The problem with an arbitrage opportunity is that the line can move before you finish placing all the needed bets. For example, if after betting $100 on the 49ers with Bookie 2 the Chiefs line with Bookie 1 moves to −130 before we can place our bet, then an arbitrage opportunity no longer exists.

In real life, arbitrage is surprisingly common in soccer betting. In soccer, you can bet on the home team to win, the away team to win, or the match to result in a draw. In his Yale undergraduate thesis, Avery Schwartz[1] found many examples of soccer arbitrage. To illustrate the idea, consider the betting odds for a soccer game shown in Figure 44-1. For example, if we bet $10 with Bookie 1 on the home team to win and the home team wins, then we receive $2 * (\$10)$ and lose $10 for a profit of $10. The file Soccerarb.xlsx (see Figure 44-1) shows how a bettor with $100 to bet can use the Excel Solver to find a maximum guaranteed profit.

For each possible match outcome, the bettor will bet with the bookie offering the best odds, so we enter trial bet values in cells G13, H13, and I14. Then in cells F17:F19 we compute the profit earned for each possible match outcome. Cell F20 computes the total amount bet, and with the formula = MIN(F17:F19) cell F21 computes the bettor's guaranteed profit. The use of the MIN function makes our Solver model nonlinear. The MIN function often gives the ordinary GRG solver engine trouble. For this reason, we

1. https://economics.yale.edu/sites/default/files/files/Undergraduate/Nominated%20Senior%20Essays/2015-16/Schwartz_Avery_SeniorEssay%202016.pdf

	D	E	F	G	H	I
6			Odds			
7				Home Win	Draw	Away Win
8			Bookie 1	2	4	3.5
9			Bookie 2	1.5	3	5
10						
11			Bet			
12				Home Win	Draw	Away Win
13			Bookie 1	$52.63	$26.32	$0.00
14			Bookie 2	$0.00	$0.00	$21.05
15						
16	Profit Formula	Outcome	Profit			
17	=G13*G8-100	Home Win	$5.26	>=	0	
18	=H8*H13-100	Draw	$5.26	>=	0	
19	=I14*I9-100	Away Win	$5.26	>=	0	
20	=SUM(G13,H13,I14)	Total Bet	$100.00			
21	=MIN(F17:F19)	Guaranteed	$5.26			

FIGURE 44.1   Finding a Soccer Arbitrage.

will use the GRG multistart engine. The GRG multistart engine tries many combinations of starting values for the changing cells and finds the best solution based on each set of starting values. Then Solver returns the best of the best answers found over all combinations of starting, changing cell values. **The GRG multistart engine requires both upper and lower bounds on all changing cells.** Here all changing cells are between $0 and $100. Our Solver settings are as follows:

Target Cell: Maximize F21.
Changing Cells: Bet amounts in G13, H13, and I14.
Constraints:
- All changing cells between 0 and 100.
- Profit for each match outcome (cells F17:F19) >= 0. **These constraints ensure that no matter how the match turns out, our bets cannot lose money.**
- Total amount bet (cell F20) = $100.

After running Solver, we find that a profit of $5.26 can be guaranteed by betting

- $52.63 with Bookie 1 on a home team win.
- $26.32 with Bookie 1 on a draw.
- $21.05 with Bookie 2 on an away team win.

If the optimal target cell value is $\Leftarrow$ $0, then no arbitrage opportunity exists.

WHAT IS A PARLAY?

A parlay is a selection of two or more bets all of which must win for the parlay to pay off. If any of the bets result in a push, no money changes hands. An example of a two-bet parlay would be taking the Colts −4 to beat the Patriots and the Bears −6 to beat the Saints. You can combine totals bets with point spreads and even combine bets involving different sports. For example, in a two-team parlay we have a ½ chance of winning each bet so our chance of winning the parlay (ignoring a push on either bet) is $(1/2)^2 = .25$. A 3–1 odds would be fair because then our expected profit on a $100 bet would be $.25(300) - .75(100) = \$0$. With an actual payout of 2.6–1 our expected profit on a $100 bet is $.25(260) - .75(100) = -\$10$, which is an average house edge of −10%. The more the number of teams in the parlay, the larger the house edge. The true odds and the typical payout on parlays are shown in Table 44-2.

TABLE 44.2
Parlay Betting Payoffs

Number of Bets	Actual Odds	Standard Payout Odds	House Percentage Edge
2	3–1	2.6–1	10%
3	7–1	6–1	12.50%
4	15–1	12–1	18.75%
5	31–1	25–1	18.75%
6	63–1	35–1	43.75%

Our calculation of the house edge assumes that the bets are independent, that is, the outcome of one bet does not affect the outcome of the other bet. For example, the results of bets on the point spreads of two different games would be independent. If we were to choose a two-bet parlay involving the Chiefs –7 points over the Broncos and the total points over on a line of 44 points, these bets might not be independent. Our logic might be that if the Chiefs cover the point spread, then Patrick Mahomes must have had a good day and the total points is more likely to go over 44. Looking at it another way, if the Chiefs fail to cover, it was probably a bad day for Mahomes, and our over bet has little chance of winning. This is an example of a correlated parlay, because the outcomes of the bets composing the parlay are correlated. Suppose that if the Chiefs cover there is a 70% chance the total will go over 44 points while if the Chiefs do not cover there is only a 30% chance that the total will go over 44 points. Then our chance of winning the parlay is .5(.7) = .35, which is far better than our chance of winning a two-bet parlay composed of independent bets. For this reason, most bookmakers will not take correlated parlays.

WHAT ARE TEASERS?

Teasers are similar to parlay bets, but the bettor can adjust the line by a predetermined amount of time. So let's assume that we have the following odds:

Game 1: Chargers –8
Game 2: Titans +3

A seven-point teaser for the above games makes the new lines of the bets:

Game 1: Chargers –1 (–8 + 7)
Game 2: Titans +10 (+3 + 7)

For an x-point teaser, we essentially add +x to the original line. In order to win the teaser **we need to win both bets to collect.** If we place the above seven-pointer teaser bet and take the Chargers and Titans, we win if and only if the Chargers win by more than one point and the Titans lose by nine or fewer points. If either game ends with a tie against the revised point spread, the teaser is called off and no money changes hands. Otherwise we lose the teaser bet. Here are some examples of how this teaser might play out.

- Chargers win by 2 and Titans lose by 3: We win the teaser.
- Chargers win by three and Titans lose by 12: We lose the teaser.
- Chargers win by 1: The teaser is a push and no money changes hands.
- Chargers lose by 1 and Titans win by 5: We lose the teaser.

Teasers usually involve 6, 6.5, or 7 points. The betting site https://www.sportsbookreview.com/best-sportsbooks/football-teasers/ gives odds from many bookies on teasers. An example is shown in Table 44-3. For example, if we bet a two-team seven-point teaser and we win the teaser, then we win $100. If we lose the teaser, we lose $135. If we bet a four-team six-point teaser and all four teams cover their revised points, we win $265. If no games push and we do not cover all four revised spreads, we lose $100. During the years 2000–2005 a team covered a seven-point teaser 70.6% of the time, pushed 1.5% of the time, and lost 27.9% of the time.

TABLE 44.3

Teaser Payoffs

	C	D	E	F	G	H
8	Points	2 team	3 team	4 team	5 team	6 team
9	5.0	103	190	315	490	745
10	5.5	100	180	300	465	710
11	6.0	−110	165	265	410	610
12	6.5	−120	150	240	365	550
13	7.0	−135	135	215	320	460
14	7.5	−150	120	185	270	380
15	8.0	−165	110	168	245	345

Let's determine our expected profit on a two-team, seven-point teaser with a $100 bet. We begin by figuring out the probability that we win the teaser, push, or lose the teaser. **We assume that the outcomes of the individual teaser bets are independent events. That is, if we cover one game involved in the teaser, this does not affect our probability of covering any other game involved in the teaser.** Now we can compute our probability of winning a two-team teaser bet.

- We win the teaser with probability $.706^2 = .498436$.
- We push if exactly one game is a push or both games push. This occurs with probability $(.015) * (1 - .015) + (1 - .015) * (.015) + (.015)^2 = .029775$.
- We lose the teaser with probability $1 - .498436 - .029775 = 0.471789$.

Our expected profit on the teaser is $(\$100) * (.498436) + 0 * (.029775) - \$135(.471789) = -\$13.85$. Therefore, in a two-team teaser the bookie has an edge of $-\$13.85 / \$135 = 10.1\%$. The interested reader can show that the bookie's edge increases as more teams are involved in the teaser. Overall, teaser bets might **appear** easier to win—since the bettor is allowed to change the original line—but at the end of the day they are still parlay bets (with smaller payouts) that give the bookmaker an edge anywhere between 10% and 20%.

CHAPTER 45

FREAKONOMICS MEETS THE BOOKMAKER

Recall from Chapter 44 that if a bookmaker gives 11–10 odds on NFL point spread bets and sets a line so that half the money is bet on each side, then the bookmaker is guaranteed to make a riskless 4.8% profit.

Steven Levitt of *Freakonomics* fame (2020) showed in a 2004 article that bookmakers can exploit bettor biases to make an expected profit exceeding 4.8% per dollar bet. Levitt obtained bettor records for 20,000 bettors during the 2001 NFL season. He found that much more than 50% of all money is bet on favorites and less than 50% on underdogs! When the home team was favored, 56.1% of the bets were on the favorite and 43.9% of the bets were on the underdog. When the visiting team was favored, 68.2% of the bets were on the favorite and 31.8% of the bets were on the underdog. These results are inconsistent with the widely held belief that spreads are set to balance the amount of money bet on the underdog and favorite. We will see that if more money is bet on favorites, and favorites cover the spread less than half the time, then bookmakers can earn an expected profit exceeding 4.8%. For Levitt's sample, this does turn out to be the case. In Levitt's sample, bets on home favorites win 49.1% of time. Bets on home underdogs win 57.7% of time. Levitt found that bets on visiting favorites win 47.8% of time while bets on road underdogs win 50.4% of time. For this set of game, favorites are not a good bet. This data indicates that the line on favorites is inflated to take advantage of the bias of bettors toward favorites. For example, when setting the line for Super Bowl LIII the bookmaker may have really thought the Patriots were only 1.5 points better than the Rams. Since bettors are biased toward the favorite, the bookmaker might set the line at Patriots −2.5. Since the true situation is that the Patriots are 1.5 points better than the Rams, a bet on Patriots −2.5 has less than 50% chance of winning. Due to the bettor's bias toward favorites, lots of money would still come in on the Patriots. Since the true point spread should be 1.5, the Patriots have less than a 50% chance of covering the 2.5-point spread. This means that on average the bookies would do better than the sure profit rate of 4.8% that they would be guaranteed if an equal amount of money were bet on each side of the line.

So how does the bookmaker do in Levitt's sample? We find that the bettors win 49.45% of their bets and lose 50.55% of their bets. On average the bookmaker earns $.4945(-10) + .5055(11) = 61.56$ cents per $10 bet. If bettors win half their bets, then of course bookmakers make on average $.5(-10) + .5(11) = 50$ cents per $10 bet. For this dataset the bookies' apparent slight edge (49.45% vs. 50% wins) in winning bets translates to a $61.56/50 = 23\%$ increase in mean profits.

Levitt checked to see if the surprising failure of favorites to cover the spread had occurred during previous seasons. During 1980–2001 48.8% of home favorites and 46.7% of visiting favorites covered the spread. This means that a bettor could have made money by simply betting on home underdogs!

We analyzed 2002–2016 point spreads and found the following results (see file NFLspreads.xlsx):

- Visiting teams that were underdogs by at least 10 points covered 53.08% of the time in 373 games. In these games if we shift the line to make the visiting team an underdog by one point less than the actual spread (for example, change an 11-point visiting underdog to a 10-point visiting underdog), then the visiting team covers exactly 50% of the time!! This indicates that the bookies shift the line to take advantage of the fact that people over-bet large home favorites.

- Visiting teams that were favored to win by at least eight points covered only 44.52% of the time in 155 games. In these games if we shift the line to reduce the point spread by even two points (for example, make a visiting 10-point favorite an eight-point favorite), the favored visiting team covers the spread only 48.3% of the time. This shows that large home underdogs still appear to be good bets. Shifting the line by 2.5 points is needed to cause a big visiting favorite to have an exactly 50–50 chance of covering the spread.

CHAPTER 46

RATING SPORTS TEAMS

Despite Levitt's work, most gamblers believe that when bookies set point spreads their goal is to have half the money bet on each team. If I bet $10, for example, on a 7.5-point favorite to cover the spread, then I win $10 if the team covers, but I lose $11 if the favorite does not cover. If the favorite covers the points spread 1/2 the time then on average each $10 bet results in an expected profit of $(1/2)($10$)+(1/2)(-$11)=-0.50. This implies that a bettor loses on average $0.50/$10.50 or $1/21 per dollar bet. Repeating the exercise from Chapter 44, and assuming we bet the same amount on each game, to break even we would have to win a fraction p of our bets where

$$p(10)+(1-p)(-11)=0.$$

The value $p=11/21=.524$ solves this equation. Thus, to win money on average we must beat the point spread at least 52.4% of the time.

Most bookmakers have power ratings on NFL and NBA teams.[1] These ratings can be used to set point spreads for which the favorite

---

1. In baseball, you bet on a team to win. The probability of a baseball team winning depends heavily on the starting pitchers. We will ignore baseball in our discussions because of this added complexity.

has approximately a 50% chance of covering the spread. For example, if the Colts have a power rating of +10 and the Browns have a power rating of −4, we would expect on a neutral field the Colts to win by $10 − (−4) = 14$ points. Of course, teams play better at home. Home edges are considered to be three points for the NFL, NBA, and college football and four points for NCAA men's basketball.[2] We will see later in the chapter how to estimate the home edge for a given set of games. Using the NFL home edge of three points the bookies would favor the Colts by $14 + 3 = 17$ points at home and favor the Colts by $14 − 3 = 11$ points at Cleveland. Predictions created from power ratings usually create (in the absence of injuries) a "fair line" in the sense that the favorite and underdog have an equal chance of covering the prediction.

We can now use the simple "point spread setting" system described in the last paragraph to fit power ratings to a set of game results. The file Nfl2016.xlsx contains most of the work for this chapter. We will now show various approaches that can be used to determine power ratings for each team and estimate the NFL home edge for the 2016 season. We will constrain our ratings so they average to 0. This implies that a team with a power rating of +5 is five points better than average while a team with a power rating of −7 is 7 points worse than average.

LEAST SQUARES RATINGS

We begin by finding the home edge and set of team ratings that best fit the 2016 regular season NFL scores. We will use the Excel Solver to determine the home edge and team ratings that minimize the sum of the squared errors implied by our prediction of each regular season game. Our work is in the worksheet Least squares. Figure 46-1 contains our final regular season power ratings while Figure 46-2 excerpts the data for the season's first 16 games. We determine NFL power ratings and league home edge as follows:

---

2. Home edge, while still existing, is on the decline: https://theconversation.com/what-really-causes-home-field-advantage-and-why-its-on-the-decline-126086

	B	C	D
3	mean	−0.0003125	=
4	Team	Rating	Rank
5	Arizona Cardinals	1.59	13
6	Atlanta Falcons	8.48	2
7	Baltimore Ravens	1.54	14
8	Buffalo Bills	−0.33	21
9	Carolina Panthers	−1.00	22
10	Chicago Bears	−7.50	28
11	Cincinnati Bengals	1.04	16
12	Cleveland Browns	−10.09	30
13	Dallas Cowboys	6.97	3
14	Denver Broncos	4.05	6
15	Detroit Lions	−1.40	24
16	Green Bay Packers	2.83	9
17	Houston Texans	−2.63	26
18	Indianapolis Colts	0.37	18
19	Jacksonville Jaguars	−4.97	27
20	Kansas City Chiefs	5.60	4
21	Los Angeles Rams	−11.09	31
22	Miami Dolphins	−2.40	25
23	Minnesota Vikings	0.94	17
24	New England Patriots	9.29	1
25	New Orleans Saints	1.54	14
26	New York Giants	2.13	10
27	New York Jets	−8.52	29
28	Oakland Raiders	3.26	8
29	Philadelphia Eagles	3.80	7
30	Pittsburgh Steelers	4.74	5
31	San Diego Chargers	0.06	19
32	San Francisco 49ers	−11.21	32
33	Seattle Seahawks	2.13	10
34	Tampa Bay Buccaneers	−0.19	20
35	Tennessee Titans	−1.01	23
36	Washington Redskins	1.97	12

FIGURE 46.1   NFL 2016 Least Squares Ratings.

Step 1: Enter trial ratings (in range C5:C36) and a trial home edge (in cell E1).

Step 2: Column G gives the game number, Column H the home team, Column I the away team, Column J the points scored by the home team, and Column K the points scored by the away team. For example, in the first game, Denver was home against Carolina and won 21–20.

	G	H	I	J	K	L	M	N
5	GameNumber	Home Team	Away Team	Home Points	Away Points	Home Margin	Prediction	Squared Error
6	1	Denver Broncos	Carolina Panthers	21	20	1	7.616229145	43.77448809
7	2	Jacksonville Jaguars	Green Bay Packers	23	27	−4	−5.235159954	1.525620113
8	3	Baltimore Ravens	Buffalo Bills	13	7	6	4.43585114	2.446561655
9	4	Philadelphia Eagles	Cleveland Browns	29	10	19	16.4575564	6.464019479
10	5	Houston Texans	Chicago Bears	23	14	9	7.439498561	2.43516474
11	6	Kansas City Chiefs	San Diego Chargers	33	27	6	8.105981645	4.435158646
12	7	Atlanta Falcons	Tampa Bay Buccaneers	24	31	−7	11.24243976	332.7866083
13	8	Tennessee Titans	Minnesota Vikings	16	25	−9	0.616576471	92.47854302
14	9	New Orleans Saints	Oakland Raiders	34	35	−1	0.84678444	3.410612769
15	10	New York Jets	Cincinnati Bengals	22	23	−1	−6.995063337	35.94078441
16	11	Seattle Seahawks	Miami Dolphins	12	10	2	7.097827482	25.98784504
17	12	Dallas Cowboys	New York Giants	19	20	−1	7.407371309	70.68389232
18	13	Indianapolis Colts	Detroit Lions	35	39	−4	4.330119587	69.39089234
19	14	Arizona Cardinals	New England Patriots	21	23	−2	−5.129595689	9.794369176
20	15	Washington Redskins	Pittsburgh Steelers	16	38	−22	−0.205640642	474.9940998
21	16	San Francisco 49ers	Los Angeles Rams	28	0	28	2.440708483	653.2773829

FIGURE 46.2   First 16 Games of the NFL 2016 Season.

Step 3: In column L, copying from L6 to L7:L277 the formula =IFERROR(J6-K6,"") determines the number of points the home team won by. In game 1, Denver won by $21 - 20 = 1$ point. Note a negative number in column F means the home team lost. For example, in game 2, Jacksonville lost at home to Green Bay by four points. Note that Excel's powerful IFERROR function evaluates the formula before the comma, and if the formula does not yield an error the formula's value is returned, else what comes after the comma (a blank space) is entered into the cell. Thus in L22 our formula returns a blank space. The IFERROR function allows us to copy our formulas from row 6 to row 277 simply by clicking the left mouse after selecting the formula in L6 and obtaining the cross in the lower right-hand corner of cell L6. **It is important to note that Excel ignores cells containing text or blanks when doing calculations on a cell range.**

Step 4: Copying from M6 to M7:M277 the formula

$$=IFERROR(\text{Home\_edge} + VLOOKUP$$
$$(\text{H6,lookpoints,2,FALSE}) -$$
$$VLOOKUP(\text{I6,lookpoints,2,FALSE}),"")$$

generates a prediction (based on our power ratings and home edge) for the home margin of each game. The prediction is simply

(Home edge) + (Home team rating) − (Away team rating).

The term VLOOKUP(H6,lookpoints,2,FALSE) finds the name in cell H6 (Denver) in the first column of the range lookpoints (B5:C36) and then goes across to the second column of the range to pick off the trial value for Denver's rating. Similarly, the term VLOOKUP(H6,lookpoints,2,FALSE) finds the name in cell I6 (Carolina) in the first column of the range lookpoints (B5:C36) and then goes across to the second column of the range to pick off the trial value for Denver's rating. The 2 ensures that the VLOOKUP returns a value from the second column of the table range lookpoints. The argument FALSE is needed to ensure that the VLOOKUP returns an exact match to the team name.

For game 1 our prediction (based on ratings given in Figure 46-2) would be for Denver to win by $2.57 + (4.05) − (−1.00) = 7.62$ points.

Step 5: In Column N, copying from N6 to N7:N277 the formula

$$=\text{IFERROR}((\text{L6-M6})\wedge 2,\text{""})$$

computes our squared forecast error (actual home margin − predicted home margin)^2 for each game. For example, in game 1, the home team won by 1 and our prediction was that they would win by 7.62. Therefore, our squared forecast error was $(7.62 − 1)^2$.

Note that a positive forecast error means the home team did better than predicted while a negative error means that the home team did worse than predicted. Intuitively, it seems like the sum of the forecast errors should be 0. This would imply that on average we over-forecast by as much as we under-forecast. It can be shown that this is indeed true.

Step 6: Now we can use the Excel Solver to change the home edge and team ratings to minimize the sum of the squared forecast errors. We will constrain the average team rating (computed in cell C3) to equal 0. **Note that minimizing the sum of squared errors ensures that positive and negative errors do not cancel each other out.** All regression equations estimated in this book were also computed by minimizing squared errors. When the Solver completes its magic, we can be sure that the team ratings and home edge shown in Figure 46-1 do a better job of fitting the scores than any other set of ratings and home edge. Note that the Patriots and Falcons ranked 1 and 2, and as expected, they both cruised through to win their respective conference titles. The 49ers, Rams, and Browns were the three worst teams.

Note that to simplify our presentation we considered games such as the two games played in London as home games for the designated home team. We obtained a home edge of 2.57 points.

EVALUATING STRENGTH OF SCHEDULE

We can now use our ratings to calculate the schedule strength faced by each team. We simply use Excel's SUMIF function to average in the worksheet Schedule Strength the ability of all opponents. For example, consider the formula in cell E5

$$(SUMIF(Home\_Team,B5,Away\_Ability) + SUMIF(Away\_Team,B5,Home\_Ability))/16.$$

The first term finds each row in which the Home_Team column is Arizona and adds up the ability of the Away_Team Arizona is playing in that row's game. The second term finds each row in which the Away_Team is Arizona and adds up the ability of the Home_Team Arizona is playing in that row's game.

Figure 46-3 gives the schedule strength faced by each team. The Broncos faced the toughest schedule (1.8 points per game more difficult than average) while the Patriots faced the easiest schedule (2.69 points easier per game than average.)

RANKING TEAMS BASED ON MEAN ABSOLUTE ERRORS

Minimizing squared errors gives lots of weight to games with unexpected outcomes. As an alternative, we can simply take the absolute value of the error for each game and minimize the sum of absolute errors (often called MAD for Mean Absolute Deviation). For example, if in one game the home team wins by five more than expected and in another game the home team wins by five less than expected, these games would contribute $|5| + |-5| = 10$ points to our target cell. Note that with an absolute value criterion positive and negative errors do not cancel out. **Minimizing the sum of absolute errors gives less weight to unusual games and more weight to a team's typical performance.** Figure 46-4 and the worksheet show the ranking of the teams using the minimization of absolute errors criterion.

	B	C	D	E	F
3	mean	1.14795E−05	=	1.14795E−05	
4	Team	Rating	Rank	Schedule Strength	Rank
5	Arizona Cardinals	1.47	15	−1.95	31
6	Atlanta Falcons	8.59	2	0.08	18
7	Baltimore Ravens	1.63	13	0.16	16
8	Buffalo Bills	−0.55	21	−1.66	29
9	Carolina Panthers	−1.16	23	1.06	6
10	Chicago Bears	−7.37	28	0.02	19
11	Cincinnati Bengals	1.06	16	0.43	13
12	Cleveland Browns	−10.03	30	1.66	2
13	Dallas Cowboys	6.95	3	−0.19	21
14	Denver Broncos	4.08	6	1.80	1
15	Detroit Lions	−1.36	24	−0.61	25
16	Green Bay Packers	2.88	9	0.12	17
17	Houston Texans	−2.59	26	0.46	12
18	Indianapolis Colts	0.40	18	−0.79	26
19	Jacksonville Jaguars	−4.94	27	0.18	15
20	Kansas City Chiefs	5.61	4	0.73	9
21	Los Angeles Rams	−11.13	31	−0.51	23
22	Miami Dolphins	−2.43	25	−1.38	28
23	Minnesota Vikings	0.97	17	−0.29	22
24	New England Patriots	9.25	1	−2.69	32
25	New Orleans Saints	1.53	14	0.59	10
26	New York Giants	2.15	10	0.53	11
27	New York Jets	−8.55	29	−0.18	20
28	Oakland Raiders	3.27	8	1.33	4
29	Philadelphia Eagles	3.84	7	1.59	3
30	Pittsburgh Steelers	4.75	5	0.25	14
31	San Diego Chargers	0.07	19	0.88	7
32	San Francisco 49ers	−11.26	32	−0.57	24
33	Seattle Seahawks	2.09	11	−1.79	30
34	Tampa Bay Buccaneers	−0.21	20	0.73	8
35	Tennessee Titans	−0.97	22	−1.16	27
36	Washington Redskins	1.99	12	1.17	5

FIGURE 46.3   2016 Team Schedule Strengths.

	B	C	D	E	F
4	Team	MAD Rating	Rank MAD	Least Squares Rating	Least Squares Rank
5	Arizona Cardinals	−0.64	20	1.59	13
6	Atlanta Falcons	12.74	1	8.48	2
7	Baltimore Ravens	0.97	17	1.54	15
8	Buffalo Bills	−2.98	25	−0.33	21
9	Carolina Panthers	3.49	10	−1.00	22
10	Chicago Bears	−8.13	29	−7.50	28
11	Cincinnati Bengals	−2.51	24	1.04	16
12	Cleveland Browns	−14.67	32	−10.09	30
13	Dallas Cowboys	7.38	3	6.97	3
14	Denver Broncos	5.70	5	4.05	6
15	Detroit Lions	−1.38	21	−1.40	24
16	Green Bay Packers	3.78	8	2.83	9
17	Houston Texans	−2.37	23	−2.63	26
18	Indianapolis Colts	−3.83	26	0.37	18
19	Jacksonville Jaguars	−6.13	27	−4.97	27
20	Kansas City Chiefs	9.19	2	5.60	4
21	Los Angeles Rams	−7.66	28	−11.09	31
22	Miami Dolphins	−1.82	22	−2.40	25
23	Minnesota Vikings	3.52	9	0.94	17
24	New England Patriots	7.12	4	9.29	1
25	New Orleans Saints	3.00	13	1.54	14
26	New York Giants	1.18	16	2.13	11
27	New York Jets	−9.96	30	−8.52	29
28	Oakland Raiders	4.00	7	3.26	8
29	Philadelphia Eagles	2.49	14	3.80	7
30	Pittsburgh Steelers	3.15	12	4.74	5
31	San Diego Chargers	0.04	18	0.06	19
32	San Francisco 49ers	−13.65	31	−11.21	32
33	Seattle Seahawks	1.18	15	2.13	10
34	Tampa Bay Buccaneers	3.23	11	−0.19	20
35	Tennessee Titans	−0.57	19	−1.01	23
36	Washington Redskins	4.11	6	1.97	12

FIGURE 46.4   NFL 2016 Team Ratings Based on Absolute Errors.

Note that the MAD criterion drops the Patriots from #1 to #4, and, amazingly, the Panthers move up 12 positions. The fundamental difference in ranking teams by least squares vs. ranking teams based on MAD is that least squares emphasizes outliers or atypical team performances much more than the MAD criterion. This is because in least squares we square each error, so a big error's influence on the target cell is greatly increased (and further amplified by the short season in the NFL). Another way of saying this is to say rankings based on MAD are governed more by a team's typical performance than by its best or worst games. MAD has the disadvantage that more than one set of rankings may minimize the target cell. Extensive research by our colleague Jeff Sagarin shows that ratings based on least squares are significantly better at predicting future game outcomes than ratings based on MAD.

What is another way to avoid *overreacting* to outliers? We saw in Chapter 30 that regularization can help, by essentially shrinking the coefficients in a regression (in our case shrinking the team ratings). Again, the *right* amount of regularization can be identified through a cross-validation set. The worksheets provided obtain end-of-season ratings, and hence we cannot evaluate the predictive performance of these ratings. However, one can learn the ratings using weeks 1 through K and predict the results of week K + 1 (out-of-sample) in order to get a sense of their predictive performance.

EVALUATING TEAM OFFENSES
AND DEFENSES

Bookmakers also allow you to bet on the total points scored in a game. This is called the over/under number. For example, in Super Bowl LI the over/under was 59 points. This means that if you bet over you win if 60 or more points are scored and if you bet under you win if 58 or fewer points are scored. If 59 points are scored the game is a push. As you might remember the Patriots won by a score of 34–28, so the over bet was victorious.

We can obtain total points predictions by computing an offensive and a defensive rating for each team. Our work is in the worksheet Offense Defense. Our changing cells are the home edge, the average number of points scored by a team in a game, and each team's offensive and defensive rating. We will constrain the average of all teams' offensive and defensive ratings to equal 0. A positive offensive rating means a

	B	C	D	E	F	G
4	Team	Offense	Defense	Overall	Offense Rank	Defense Rank
5	Arizona Cardinals	2.37	0.78	1.59	11	16
6	Atlanta Falcons	10.53	2.05	8.48	1	23
7	Baltimore Ravens	−1.07	−2.61	1.54	18	9
8	Buffalo Bills	1.84	2.17	−0.33	12	24
9	Carolina Panthers	−0.20	0.80	−1.00	16	17
10	Chicago Bears	−5.20	2.30	−7.50	28	26
11	Cincinnati Bengals	−1.48	−2.53	1.04	20	11
12	Cleveland Browns	−5.21	4.88	−10.09	29	30
13	Dallas Cowboys	4.06	−2.91	6.97	5	7
14	Denver Broncos	−2.02	−6.06	4.05	22	1
15	Detroit Lions	−1.34	0.06	−1.40	19	14
16	Green Bay Packers	4.87	2.04	2.83	3	22
17	Houston Texans	−5.30	−2.68	−2.63	30	8
18	Indianapolis Colts	3.09	2.72	0.37	8	27
19	Jacksonville Jaguars	−2.69	2.28	−4.97	25	25
20	Kansas City Chiefs	1.15	−4.45	5.60	14	5
21	Los Angeles Rams	−9.46	1.63	−11.09	32	19
22	Miami Dolphins	−0.60	1.81	−2.40	17	21
23	Minnesota Vikings	−2.65	−3.59	0.94	24	6
24	New England Patriots	4.27	−5.02	9.29	4	3
25	New Orleans Saints	6.79	5.25	1.54	2	31
26	New York Giants	−3.23	−5.36	2.13	26	2
27	New York Jets	−5.55	2.97	−8.52	31	28
28	Oakland Raiders	3.52	0.25	3.26	6	15
29	Philadelphia Eagles	1.27	−2.53	3.80	13	10
30	Pittsburgh Steelers	2.77	−1.97	4.74	10	12
31	San Diego Chargers	3.04	2.98	0.06	9	29
32	San Francisco 49ers	−3.70	7.51	−11.21	27	32
33	Seattle Seahawks	−2.40	−4.53	2.13	23	4
34	Tampa Bay Buccaneers	−1.51	−1.31	−0.19	21	13
35	Tennessee Titans	0.73	1.74	−1.01	15	20
36	Washington Redskins	3.32	1.35	1.97	7	18

FIGURE 46.5   NFL 2016 Offensive and Defensive Team Ratings.

team scores more points than average while a negative offensive rating means a team scores fewer points than average. A positive defensive rating means a team gives up more points than average while a negative defensive rating means a team gives up fewer points than average.

We set up predictions for the number of points scored by the home and away teams in each game. Then we choose our changing cells to minimize the sum over all games of the squared errors made in predicting the home and away points scored. We predict the points scored by the home team in each game to be

(League average constant) + .5(Home edge) +
(Home team offensive rating) + (Away team defensive rating).

We predict the number of points scored by the away team in each game to be

(League average constant) − .5(Home edge) +
(Away team offensive rating) + (Home team defensive rating).

Note that we divide up the home edge equally among the home team's points and the away team's points. After minimizing the sum of the squared prediction errors for home and away points scored in each game, we find the results shown in Figure 46-5. For example, we find that the Falcons had the best offense, scoring 10.53 more points than average. The Broncos had the best defense, yielding 6.06 fewer points han average. The Rams had the worst offense, scoring 9.46 fewer points than average, while the 49ers had the worst defense, yielding 7.51 more points than average. We note that for each team the team's overall rating = team offense rating − team defense rating.

RANKING TEAMS BASED JUST ON WINS AND LOSSES

College football and college basketball rankings have long sparked controversy. Until 2015, the Bowl Championship Series (BCS) picked two teams to play for the college football championship. A key element in the BCS final ranking is an averaging of several computer rankings of college football teams. The BCS dictated that the scores of games cannot be used to rank teams. Only a team's win-loss record can be used to rank teams (see Chapter 58 for further discussion of the College Football Playoff, which has replaced the BCS).

The BCS believed that allowing game scores to influence rankings would encourage the top teams to run up the score. When the NCAA chooses and seeds its great basketball tournament, it utilizes several computer ranking systems as part of its selection process.[3] Again, since the NCAA does not want teams to run up the score, the NCAA basketball selection committee currently truncates margin of victory at 10 points when calculating its computer rankings. Using a technique like logistic regression (described in Chapter 21 when we tried to predict the chance of making a field goal based on the yard line), we can rate NFL teams based simply on their record of wins and losses. Our changing cells will be a rating for each team and a home edge cell. Following our Chapter 21 development, we assume the probability p of the home team winning the game can be determined from

$$\mathrm{Ln}\frac{p}{1-p} = \text{Home rating} - \text{Away rating} + \text{Home edge}. \qquad (1)$$

Rearranging equation (1) we find[4]:

$$p = \frac{e^{\text{Home rating} - \text{Away rating} + \text{Home edge}}}{e^{\text{Home rating} - \text{Away rating} + \text{Home edge}} + 1} \qquad (2)$$

To estimate the ratings for each team, we use the method of maximum likelihood estimation. We simply choose the team ratings and home edge that maximize the probability of the actual sequence of wins and losses we observed. To illustrate the idea of maximum likelihood, suppose we want to estimate Shaq's chance of making a free throw and we observe Shaq making 40 of 100 free throws. Let p = probability Shaq makes a free throw. Then $1-p$ is probability Shaq misses a free throw. The probability of observing 40 made and 60 missed free throws is $\frac{100!}{60! \times 40!}p^{40}(1-p)^{60}$. The constant term is

3. https://www.cbssports.com/college-basketball/news/what-experts-who -met-with-ncaa-say-about-changes-to-tourney-selection-process/

4. Most probably you are able to spot the fact that this is essentially a Bradley–Terry model (see Chapter 41), which is essentially a logistic regression model.

simply the number of ways to place 40 made free throws in a sequence of 100 free throws. To maximize this probability, simple calculus (or the Excel Solver) shows we should set p = .40. Thus we would estimate Shaq is a 40% foul shooter. This estimate certainly agrees with our intuition.

In a similar fashion, we choose the team ratings and home edge to maximize the probability of observing the results we have seen in the data. For individual games we have:

- For each game won by the home team the probability of observing this result is equal to probability that the home team wins (from equation 2).
- For each game won by the away team the probability of observing this result is equal to probability that the away team wins, i.e., 1 – probability home team wins.
- For each game that ends up in a tie we consider it to be half a win and half a loss and we use .5 * probability home team wins + .5 * probability away team wins as the likelihood of a tie.

Assuming that the games are independent, we can obtain the probability of observing these results by multiplying the probabilities for each game. We can then choose the team ratings and the home edge that maximize the product of this probability.[5] After using the Excel Solver to perform this optimization, we find the team ratings and home edge shown in Figure 46-6. Our work is in the worksheet Win Loss. The Solver performs better when under Options on the GRG Nonlinear tab we check the Multistart option. For Solver to use the Use Multistart option, we need to set lower and upper bounds for each changing cell. In most seasons constraining

5. Actually, we have the Solver maximize the sum of the natural logarithms of these probabilities. This is computationally convenient and is equivalent to maximizing the product of the probabilities. Maximizing the log-likelihood helps Excel quickly find the optimal solution, because we are maximizing a concave function.

	B	C	D	E
1			Home edge	0.46000
2				
3	mean	−4.5742E−09	=	0
4	Team	Rating	Rank	
5	Arizona Cardinals	−0.47	23	
6	Atlanta Falcons	0.96	7	
7	Baltimore Ravens	−0.01	20	
8	Buffalo Bills	−0.58	25	
9	Carolina Panthers	−0.50	24	
10	Chicago Bears	−1.64	30	
11	Cincinnati Bengals	−0.37	22	
12	Cleveland Browns	−2.78	32	
13	Dallas Cowboys	1.73	2	
14	Denver Broncos	0.69	9	
15	Detroit Lions	0.31	15	
16	Green Bay Packers	0.72	8	
17	Houston Texans	0.47	10	
18	Indianapolis Colts	0.09	18	
19	Jacksonville Jaguars	−1.39	28	
20	Kansas City Chiefs	1.55	3	
21	Los Angeles Rams	−1.54	29	
22	Miami Dolphins	0.32	13	
23	Minnesota Vikings	0.07	19	
24	New England Patriots	1.85	1	
25	New Orleans Saints	−0.17	21	
26	New York Giants	1.04	5	
27	New York Jets	−1.18	27	
28	Oakland Raiders	1.44	4	
29	Philadelphia Eagles	0.11	17	
30	Pittsburgh Steelers	1.04	6	
31	San Diego Chargers	−0.65	26	
32	San Francisco 49ers	−2.45	31	
33	Seattle Seahawks	0.39	11	
34	Tampa Bay Buccaneers	0.32	14	
35	Tennessee Titans	0.29	16	
36	Washington Redskins	0.36	12	

FIGURE 46.6   NFL 2016 Ratings Based on Wins and Losses.

the home edge to be between 0 and 1 and constraining each team's rating to be between −5 and +5 works fine. **If you use GRG Multistart and you ever find that your Solver optimal solution comes very close to an upper or lower bound, you should relax the bound until the Solver does not come close to the bound.**

By our win-loss rankings, the Patriots were the best team, the Cowboys were second, and the Chiefs third. The 49ers are the next to worst team, with the Browns being the worst team. Using these ratings, how would we predict the chances of the Patriots beating the Falcons in Super Bowl LI? Using equation (2) we would estimate the chances of the Patriots beating the Falcons (remember there is no home edge in the Super Bowl) as $e^{1.85 - 0.96}/(1 + e^{1.85 - 0.96}) = 0.79$. Compare this with the implied win probability from the betting markets that was equal to 0.58!

The astute reader has probably anticipated a possible problem with rating teams based solely on their wins and losses. Suppose Harvard is the only undefeated college team. Also suppose the usual powers like Alabama, Clemson, and Ohio State are 11–1. Almost nobody would claim that Harvard is the best team. They simply went undefeated by facing a relatively easy Ivy League schedule. If we run our win-loss ranking system, a sole undefeated team will have an infinite rating. This is unreasonable. David Mease (2003) came up with an easy solution to this problem: introduce a fictitious team (call it Faber College) and assume that **each real team had a 1–1 record against Faber.**[6] Now running our win-loss system usually gives results that would place a team like Harvard behind the traditional powerhouses.

We note that our points-based and win-loss-based rating systems can easily be modified to give more weight to more recent games. Simply give a weight of 1 to the most recent week's games, a weight of λ to last week's games, a weight λ^2 to games from two weeks ago, etc. Here λ must be between 0 and 1. For pro football, $\lambda = .95$ seems to work well, while for college football $\lambda = .9$ works well (in terms of future predictions). Essentially, $\lambda = .9$ means last week's game counts 10% less than this week's game. The value of λ can be optimized to give the most accurate forecasts of future games through cross-validation. The Sagarin ratings are generally

6. This approach is called Laplace smoothing, which we will discuss in more detail in Chapter 59.

	B	C	D	E	F
3	mean	0.00031	0		
4	Team	Offense	Defense	Overall	Rank
5	Arizona Cardinals	2.46	0.90	1.56	12
6	Atlanta Falcons	11.85	0.95	10.90	2
7	Baltimore Ravens	−1.46	−2.57	1.11	14
8	Buffalo Bills	1.96	4.20	−2.24	24
9	Carolina Panthers	−0.11	0.61	−0.72	22
10	Chicago Bears	−5.44	2.19	−7.63	28
11	Cincinnati Bengals	−1.83	−2.65	0.82	15
12	Cleveland Browns	−5.59	4.85	−10.44	30
13	Dallas Cowboys	3.86	−1.69	5.55	3
14	Denver Broncos	−2.07	−6.46	4.39	6
15	Detroit Lions	−2.51	0.08	−2.59	25
16	Green Bay Packers	6.20	2.03	4.17	7
17	Houston Texans	−3.58	−3.10	−0.48	21
18	Indianapolis Colts	3.12	2.56	0.56	17
19	Jacksonville Jaguars	−2.86	2.05	−4.91	27
20	Kansas City Chiefs	0.42	−4.72	5.14	4
21	Los Angeles Rams	−9.62	1.32	−10.94	31
22	Miami Dolphins	−2.05	1.68	−3.73	26
23	Minnesota Vikings	−2.92	−3.72	0.80	16
24	New England Patriots	8.54	−4.36	12.90	1
25	New Orleans Saints	6.83	5.15	1.68	11
26	New York Giants	−4.25	−3.86	−0.39	20
27	New York Jets	−5.87	2.59	−8.46	29
28	Oakland Raiders	2.61	0.70	1.91	10
29	Philadelphia Eagles	0.93	−2.45	3.38	8
30	Pittsburgh Steelers	2.17	−2.65	4.82	5
31	San Diego Chargers	3.07	2.96	0.11	18
32	San Francisco 49ers	−3.85	7.10	−10.95	32
33	Seattle Seahawks	−2.11	−5.31	3.20	9
34	Tampa Bay Buccaneers	−1.47	−1.52	0.05	19
35	Tennessee Titans	0.71	1.63	−0.92	23
36	Washington Redskins	2.87	1.51	1.36	13

FIGURE 46.7   Ratings for Forecasting Super Bowl LI.

considered the best set of team rankings. They incorporate a proprietary weighted least squares algorithm. Sagarin's ratings for the
current season of pro and college football and basketball (and other
sports) may be found at http://www.usatoday.com/sports/sagarin

.htm. Sagarin's ratings for past seasons can be found at http://www
.usatoday.com/sports/sagarin-archive.htm.

USING COMMON SENSE TO MODIFY
A RATING SYSTEM

Suppose that, going into the 2017 Super Bowl, we want to accurately set a point spread and totals line. Recall that Tom Brady
missed the first four games due to Deflategate. We also omitted
the playoff games in our previous work. In trying to predict the
Super Bowl, we can ignore the games Brady missed, include the
playoff games, and give twice as much weight to the playoff games
as regular season games. Our work is in the worksheet Super Bowl
and Figure 46-7. Note that the Patriots' rating has increased to 12.9
points better than average. Our prediction for the Super Bowl is
shown in Figure 46-8.

	J	K	L
1	Pats	32.43	=mean+C24+D6
2	Falcons	30.43	=C6+D24+mean

FIGURE 46.8   Super Bowl LI Forecast.

Our model would have set a totals line of 62.5 points and favored
the Patriots by two points.

SOCCER RATINGS

While the high-level approach for rating soccer teams is similar to
what we have discussed, there are some challenges associated with
the rarity of goal scoring (as compared to scoring in basketball and
football) that we need to adjust/account for. In particular, if we use
our additive model to predict Premier League Soccer games, we
might predict negative goals scored when a bad offensive team plays
a good defensive team. An alternative approach is to use a multipli-

	C	D	E	F	G	H	I
1		1.56	1.31			SSE	967.595
2	HomeTeam	AwayTeam	Home Goals	Away Goals	Homepre	Awaypre	Squared Error
3	Arsenal	Sunderland	0	0	2.206574	0.685281	5.33858
4	Fulham	Norwich	5	0	1.714075	0.94732	11.6947
5	Newcastle	Tottenham	2	1	1.227479	1.903396	1.41291
6	QPR	Swansea	0	5	0.795072	1.176245	15.2532
7	Reading	Stoke	1	1	1.029313	1.107344	0.01238
8	West Brom	Liverpool	3	0	1.314664	1.636298	5.51783
9	West Ham	Aston Villa	1	0	1.838359	1.076671	1.86207
10	Man City	Southampton	3	2	2.188904	0.689373	2.37562
11	Wigan	Chelsea	0	2	1.088859	2.38716	1.33551
12	Everton	Man United	1	0	1.345671	1.410237	2.10826
13	Chelsea	Reading	4	2	2.996572	0.722735	2.63827
14	Aston Villa	Everton	1	3	0.957365	1.737844	1.59486
15	Chelsea	Newcastle	2	0	2.957209	0.774972	1.51683
16	Man United	Fulham	3	2	2.703301	0.976856	1.13485
17	Norwich	QPR	1	1	1.146255	0.811705	0.05685
18	Southampton	Wigan	0	2	1.871322	1.309321	3.97888
19	Swansea	West Ham	3	0	1.437967	0.970361	3.38155

FIGURE 46.9   2012–2013 Premier League Results.

cative based rating system. The workbook Premierleague.xlsx contains our work. Figure 46-9 shows the scores from the 2012–2013 Premier League matches.

We use the following changing cells in our multiplicative ratings model:

- Mean goals scored in a match.
- A multiplicative home edge value. A home edge of 1.1, for example, means that we increase our predicted goals for the home team by 10% and divide our predicted goals for the away team by 1.1.
- For each team, we have an offensive goal rating and a defensive goal rating. For example, as shown in Figure 46-10, Arsenal has an offensive goal rating of 1.43 and a defensive goal rating of 0.71. This means that after adjusting for opponents played, our best estimate is Arsenal scores 43% more goals than an average Premier League team and gives up 29% less goals than an average Premier League team.

	K	L	M	N	O
1					
2	mean	1.38			
3	home	1.12	average	1	1
4			Team	Off	Def
5			Arsenal	1.43	0.71
6			Fulham	0.96	1.09
7			Newcastle	0.84	1.34
8			QPR	0.57	1.04
9	use multistart		Reading	0.79	1.36
10	under		West Brom	0.99	1.02
11	GRG		West Ham	0.88	1.01
12	options		Man City	1.25	0.63
13	need		Wigan	0.94	1.36
14	bounds		Everton	1.05	0.71
15	on changing		Chelsea	1.42	0.75
16	cells		Aston Villa	0.86	1.35
17			Man United	1.60	0.83
18			Norwich	0.71	1.15
19			Southampton	0.89	1.13
20			Swansea	0.92	0.90
21			Tottenham	1.16	0.94
22			Liverpool	1.30	0.86
23			Stoke	0.66	0.84
24			Sunderland	0.78	0.99

FIGURE 46.10   Premier League 2012–2013 Offense
Defense Ratings.

To obtain these ratings we proceeded as follows:

- Predict the goals scored by the home team in each match by copying from G3 to G4:G382 the formula

$$=\text{mean} * \text{home} * \text{VLOOKUP}(C3, \text{lookup}, 2, \text{FALSE}) * \text{VLOOKUP}(D3, \text{lookup}, 3, \text{FALSE}).$$

- Predict the goals scored by the away team by copying from H3 to H4:H382 the formula

$$=\text{mean} * \text{VLOOKUP}(C3, \text{lookup}, 3, \text{FALSE}) * \text{VLOOKUP}(D3, \text{lookup}, 2, \text{FALSE}) / \text{home}.$$

- Compute the total squared error in predicting home and away goals in each match by copying from H3 to H4:H382 the formula $= (E3-G3)^2 + (F3-H3)^2$.
- Compute the total sum of squared errors in cell I1 with the formula $=SUM(I3:I382)$.
- Copying from cell N3 to O3 the formula $=AVERAGE(H5:N24)$ computes the average of the team offensive and defensive ratings. We will constrain these averages to equal 1.
- We will use the GRG Multistart Solver to solve for the home edge and team ratings that minimize the sum of our squared prediction errors. The target cell is to minimize I1 and our changing cells are L2 (mean goals scored), L3 (home edge), and N5:O24 (team ratings). The constraints N3:O3 = 1 ensure that the average of the team ratings equals 1. Since we are using the GRG Multistart Solver, we include upper and lower bounds on each changing cell. Our lower bound is 0 for each changing cell and we use an upper bound of 3. Of course, if Solver's optimal solution causes a changing cell to come closer to the upper bound, then we relax the bound. Figure 46-10 displays our final ratings.

Ranking the teams is not a trivial matter. If we assume a team is playing against a opponent with average offensive and defensive ability, then the teams should be ranked by offensive rating divided by defensive rating, because this ordering would rank teams by the predicted scoring ratio a team would achieve against an average team.

Figure 46-11 shows a sample prediction for Chelsea at Reading. Rounding off, we would predict Chelsea to win by 2–1.

We can use our predicted score to predict the outcome of the game. Events that occur rarely, such as number of car accidents a driver has in a year, number of defects in a product, and number of goals scored in a soccer game, have often been accurately modeled by the Poisson random variable (see Chapter 16 for a discussion of the Poisson random variable). For a Poisson random variable with mean λ, the probability that the random variable equals x is given by $\lambda^x e^{-\lambda}/x!$.

	L	M	N	O
26	**Predict Chelsea at Reading**			
27				
28	Chelsea 1.42 .75			
29	Reading .79 1.36			
30	Home edge 1.12			
31	Mean goals 1.38			
32				
33	Chelsea			
34	2.38	=1.38*(1/1.12)*(1.42)*1.36		
35	Reading			
36	0.916	=1.38*1.12*0.79*0.75		
37				
38	Chelsea 2.38 to .92			

FIGURE 46.11   Prediction for Chelsea at Reading.

For example, if we predict a team to score 1.4 goals, there is a probability $1.4e^{-1.4}/2! = .242$ of scoring 0 goals. The Excel formula POISSON(x,mean,False) gives the probability that a Poisson random variable with mean λ takes on the value x. Suppose we are given the predicted number of goals scored by each team. In the worksheet Probability of the Premier League we use the following formulas to predict the probability of each team winning a game and a tie. We assume no team will score more than 10 goals in a game. Then

$$\text{Probability of tie} = \sum_{i=1}^{10} (\text{Prob team 1 scores i goals}) \cdot$$
$$(\text{Prob team 2 scores i goals})$$

$$\text{Probability team 1 wins} = \sum_{i=1}^{10} (\text{Prob team 1 scores i goals}) \cdot$$
$$(\text{Prob team 2 scores} < i \text{ goals})$$

$$\text{Probability team 2 wins} =$$
$$1 - \text{Probability of a tie} - \text{Probability team 1 wins}.$$

Figure 46-12 (see the worksheet Probability) calculates the probability of each match outcome for Chelsea at Reading.

	A	B	C	D	E	F	G	H
1								
2			Predicted Goals					
3		Chelsea	2.38					
4		Reading	0.92					
5								
6					Goals	Chelsea	Reading	Chance Chelsea Wins with this Many Goals
7	Chance Chelsea Wins	0.698			0	0.0925506	0.3985190	
8	Chance of Draw	0.174			1	0.2202704	0.3666375	0.087782
9	Chance Reading Wins	0.128			2	0.2621217	0.1686533	0.200564
10					3	0.2079499	0.0517203	0.194186
11					4	0.1237302	0.0118957	0.12194
12					5	0.0588956	0.0021888	0.058744
13					6	0.0233619	0.0003356	0.023353
14					7	0.0079430	0.0000441	0.007943
15					8	0.0023631	0.0000051	0.002363
16					9	0.0006249	0.0000005	0.000625
17					10	0.0001487	0.0000000	0.000149

FIGURE 46.12   Computing Probabilities for Chelsea at Reading Match.

We find that Chelsea has a 70% chance of winning, that Reading has a 13% chance of winning, and that there is a 17% chance of a draw.

The above calculations assume that the two Poisson distributions modeling the goals scored by each team are independent. In reality, the two distributions are weakly correlated. Despite the small correlation (typically less than 0.1), the underlying mis-specified model can have several significant impacts. For example, even under small correlation between the goals scored from each team, when using the two independent Poisson distributions the probability of a tie is typically underestimated.[7] Karlis and Ntzoufras,[8] suggested that one way to overcome this problem when predicting the probability of a tie, a home, or an away win is to directly model the difference of

7. D. Karlis and I. Ntzoufras, "Analysis of sports data by using bivariate Poisson models," *Journal of the Royal Statistical Society*: Series D (The Statistician), 52(3), 2003, 381–393.

8. Dimitris Karlis and Ioannis Ntzoufras, "Bayesian modelling of football outcomes: Using the Skellam's distribution for the goal difference," *IMA Journal of Management Mathematics*, 20(2), 2009, 133–145.

the goals scored by the two teams. Given that the goals scored by a team are modeled through a Poisson distribution, their difference can be modeled through a Skellam distribution. The benefit of using the Skellam is that it has the same form regardless of the correlation between the two Poisson distributions. With X and Y being two independent Poisson distributions, the distribution for $Z = X - Y$ is

$$\Pr[z] = e^{\lambda_1 + \lambda_2} \cdot \left(\frac{\lambda_1}{\lambda_2}\right)^{z/2} \cdot I_z\left(2\sqrt{\lambda_1 \lambda_2}\right)$$

where I_z is the modified Bessel function. Even if X and Y are correlated Poisson distributions (and hence, they follow a bivariate Poisson distribution), their difference still follows the above distribution. Simply put, if we model directly the goal margin difference, then the correlation between the goals scored from the two teams will not impact our predictions.

We have implemented this model in the workbook Premier League::2012–2013 Skellam. We start by expressing the parameters of the Skellam distribution as a function of the ratings to be learned (columns):

$$\lambda_1 = \text{mean}_g \cdot \text{home}_e \cdot o_h \cdot d_a$$

$$\lambda_2 = \frac{\text{mean}_g \cdot o_a \cdot d_h}{\text{home}_e}$$

	N	O	P	Q	R	S
30	Goal Margin		Goal Margin		Goal Margin	
31	−10	2.93979E−06			10	7E−13
32	−9	1.94632E−05			9	2.13E−11
33	−8	0.000116089			8	5.83E−10
34	−7	0.000616322			7	1.42E−08
35	−6	0.002868662			6	3.04E−07
36	−5	0.011478038			5	5.6E−06
37	−4	0.038448472			4	8.62E−05
38	−3	0.103858808			3	0.00107
39	−2	0.213717501			2	0.010119
40	−1	0.304309642	0	0.24707	1	0.066216
41	Chelsea wins	0.675435935	Draw	0.24707	Chelsea loses	0.077498

FIGURE 46.13   Computing Probabilities for Chelsea at Reading Match Using the Skellam Model.

We can then find the ratings (o_h, o_a, d_h, d_a)—and the mean$_g$ and home$_e$—that maximize the likelihood of observing the specific goal margins using the Skellam PDF. Using this model, the probabilities for the Chelsea at Reading game are shown in Figure 46-13. As we can see now, the draw has a higher probability compared to the two independent Poisson distributions. For what it is worth, that specific game during that season ended 2–2.

FROM POINT RATINGS TO PROBABILITIES

In Chapter 46 we learned how to calculate "power ratings," which allow us to estimate how many points one team is better than another. Even if we do not run our proposed rating systems, we can always look up power ratings at Jeff Sagarin's site on usatoday.com. In this chapter, we will show how to use power ratings to determine the probability that a team wins a game, covers a point spread bet, or covers a teaser bet. For NBA basketball, we will show how to use power ratings to determine the probability of each team winning a playoff series. We then show how power ratings can be used to compute the probability of each team winning the NCAA basketball tournament. We also show how to simulate the NFL playoffs.

Recall from Chapter 46 that based on power ratings we could predict that the average amount by which a home team will win a game is given by home edge + home team rating – away team rating. The home edge is considered to be three points for the NFL, college football, and the NBA. For college basketball the home edge is four points. Hal Stern (1991) showed that the probability distribution of the final margin of victory for the home team can be well approximated by a normal random variable **Margin** with mean = home edge + home team rating − away team rating, and standard deviation around 14 points. For NBA basketball, NCAA basketball, and college football, Jeff Sagarin has found that the historical standard deviation of game results about a prediction from a rating system is given by 12, 10, and 16 points, respectively. You can estimate this for your set of ratings by essentially estimating the standard error of the corresponding predictions.

CALCULATING NFL WIN AND GAMBLING PROBABILITIES

Let's now focus on NFL football. A normal random variable can assume fraction values but the final margin of victory in a game must be an integer. Therefore, we estimate that the probability that the home team wins by between a and b points (including a and b, where a < b) is probability(**Margin** is between a − .5 and b + .5). The Excel function NORM.DIST(x,mean,sigma,True) gives us the probability that a normal random variable with the given mean and sigma is ≤ x. With the help of this great function we can determine the probability that a team covers the spread, wins a game, or beats a teaser bet.

> SUPPOSE OUR POWER RATINGS INDICATE THE COLTS SHOULD WIN BY SEVEN POINTS. THEY ARE ONLY A FOUR-POINT FAVORITE. WHAT IS THE PROBABILITY THAT THE COLTS WILL COVER THE POINT SPREAD?

If **Margin** is ≥ 4.5, the Colts will cover the spread. If **Margin** is between 3.5 and 4.5, the game will be a push. The probability the Colts cover spread can be computed as 1 − NORM.DIST (4.5,7,14,TRUE) = .571.

$$\text{Probability of push} = \text{NORMDIST}(4.5,7,14,\text{TRUE}) - \text{NORMDIST}(3.5,7,14,\text{TRUE}) = .028.$$

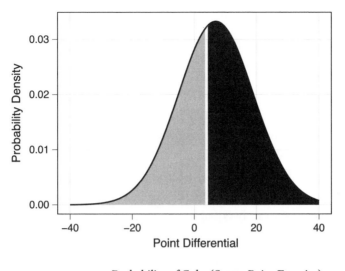

FIGURE 47.1   Probability of Colts (Seven-Point Favorite)
Covering a Four-Point Spread.

Therefore if we throw out the pushes (because no money changes hands) we would expect a bet on the Colts to have a .571/ $(1-.028)=58.7\%$ chance of covering the spread and a 41.3% chance of losing against the spread. Figure 47-1 visualizes these events, where the right side shaded area corresponds to the probability of the Colts covering the spread, the small unshaded area around the spread line is the probability of a push, and the left side shaded area corresponds to the probability of the Colts not covering the spread.

THE PATRIOTS WERE A THREE-POINT FAVORITE IN THE 2017 SUPER BOWL. WHAT IS THE PROBABILITY THAT THEY WILL WIN THE GAME?

Here we assume the point spread equals the mean outcome of the game (as we saw, this might not be the case for betting spreads but should or will be the case for our own power rating spread). The Patriots can win with a final margin of ≥ 1 point or win with say a .5 probability if regulation time ends in a tie. The probability that the Patriots win by 1 or more $=1-\text{NORMDIST}(0.5,3,14,\text{TRUE})=.571$.

Probability regulation ends in a tie $=$ NORMDIST

$(.5,3,14,$TRUE$) -$ NORMDIST$(-.5,7,14,$TRUE$) = .028.$

Therefore, we estimate the Patriots' chance of winning the 2017 Super Bowl to be $.571 + .5 * (.028) = .585$. Recall that the money line on this game was Patriots -150 and Falcons $+130$.[1] With this money line you should bet the Patriots to win if you think the Patriots had at least a 60% chance of winning and you should bet the Falcons to win if you think the Patriots have a chance of winning less than or equal to 56.5%. The average of these two probabilities is 58.3%, which is almost exactly equal to the chance of the Patriots winning, calculated based on the point spread (58.5%).

THE COLTS ARE AN EIGHT-POINT FAVORITE AND WE HAVE BET A SEVEN-POINT TEASER ON THE COLTS. WHAT FRACTION OF THE TIME CAN WE PREDICT THE COLTS WILL COVER OUR NEW SPREAD (COLTS BY 8 – 7 = 1 POINT)?

If the Colts win by ≥ 2 points we win the teaser. If the Colts win by one point, the teaser is a push. If regulation ends in a tie and the game goes overtime we will win the teaser if and only if the Colts win in overtime. We assume the Colts' chance of winning in overtime to be .5.

Probability Colts win

by $\geq 2 = 1 -$ NORM.DIST$(1.5,8,14,$TRUE$) = .679$

Probability of push $=$ NORM.DIST$(1.5,8,14,$TRUE$) -$

NORM.DIST$(.5,8,14,$TRUE$) = .025$

Probability game goes to OT $=$ NORM.DIST$(0.5,8,14,$TRUE$) -$

NORM.DIST$(-0.5,8,14,$TRUE$) = .024$

---

1. For a team with money line ML ≥ 100, the marginal win probability for the team is $p = 100/(100 + ML)$, while if the ML ≤ 100, the marginal win probability for the team is $p = |ML|/(100 + |ML|)$. For a given game, the marginal win probabilities of the two teams will not add up to 1, since the vig will not allow that, but one can easily normalize the win probability of team i (against team j) as $p_i/(p_i + p_j)$.

Therefore, we estimate the fraction of non-push games in which the Colts will cover the teaser to be $(.679 + .5 * .024)/(1 - .025) = 70.9\%$.

As we pointed out in Chapter 44, 70.6% of all seven-point teasers covered the teaser spread. This compares quite well with our estimate of 70.9% probability of the Colts covering this teaser. Note that for this teaser, a regulation tie gives us a chance to cover. If we were betting on a 15-point favorite with a seven-point teaser, a tie in regulation would be a sure loss.

WHAT IS THE CHANCE THAT THE WARRIORS WILL BEAT THE CAVS IN THE 2017 NBA FINALS?

We now turn our attention to trying to translate power ratings into an estimate of an NBA team winning a playoff series. The first team to win four games in an NBA playoff series is the winner. Even though in this specific example we could get a closed form solution to this question, we will turn to Monte Carlo simulations (see Chapter 4). In particular, we will "play out" or simulate a seven-game series thousands of times and track the number of times each team wins the series. The home point margin of victory in an NBA game will follow a normal distribution with mean = home edge + home rating − away rating and sigma = 12. Again, the sigma of 12 points is the historical standard deviation of actual game scores about a prediction from a ranking system.

The Excel formula NORM.INV(rand( ),mean,sigma) can generate a sample value from a normal random variable with a given mean and sigma. RAND( ) creates a random number equally likely to assume any value between 0 and 1. Suppose rand( ) = p. Then our formula generates the percentile from the given normal random variable. Thus RAND( ) = .5 yields the 50th percentile of the normal random variable, RAND( ) = .9 yields the 90th percentile, etc.

To illustrate how to determine the chance of a team winning a playoff series let's look at the 2017 NBA finals between the Warriors and the Cavs. Running our power rankings as described in the previous chapter we found the following ratings (using only regular season results). Warriors = +11.35 and Cavs = +2.89. Therefore, when the Warriors are at home they are favored to win by $11.35 - 2.89 + 3 = 11.46$ points, and when the Warriors play in Cleveland, they are favored to win by $11.35 - 2.89 - 3 = 5.46$ points. Therefore we can simulate the final margin (from the Warriors viewpoint) of the Warriors' home games (games 1, 2, 5, and 7) with the Excel function = NORM.INV((RAND( ),11.46,12). When Cleveland is at home (games 3, 4, and 6) we simulate the final Cavs margin with the formula = NORMINV(rand( ),5.46,12). Although the series may not go seven games, we have Excel simulate the result of seven games and declare the Warriors the winner if they win at least four games.

	C	D	E	F	G	H
1						
2						
3	Warriors	11.35				
4	Cavs	2.89				
5						
6						
7			Game	Home or Away	Warriors Favored by	RAND
8			1	H	11.46	0.779
9			2	H	11.46	0.823
10			3	A	5.46	0.882
11			4	A	5.46	0.169
12			5	A	5.46	0.251
13			6	H	11.46	0.722
14			7	H	11.46	0.08
15			Games Won by Warriors			
16			4	=COUNTIF(I8:I14,">0")		
17						
18						
19			Did Warriors Win Series?			
20			1	=IF(E16>=4,1,0)		
21						
22			Chance Warriors Win			
23			0.948	=AVERAGE(J21:J1020)		

FIGURE 47.2 A Sample Iteration of Warriors–Cavs Finals.

We used EXCEL's nice Data Table feature (see file NBAfinals2017.xlsx and the chapter appendix) to play out the series 1,000 times. A sample iteration is shown in Figure 47-2.

In this iteration, the Warriors would have won in seven games. We found the Warriors won 95% of the time. If we had included the playoff games in our ratings, we would have obtained a smaller chance of a Warriors series victory. In reality, the Warriors won the series in five games.

ESTIMATING NCAA TOURNAMENT PROBABILITIES

We can use a similar methodology and "play out" the NCAA tournament several thousand times. We recommend using the Sagarin power ratings from *USA Today*. The file NCAA2017.xlsx shows how we simulated the 2017 NCAA tournament.

To begin, we must simulate the crazy "play-in games." Figure 47-3 shows how we simulate them.

	BP	BQ	BR	BS	BT	BU	BV	BW	BX
1		Team	Rating	Winner					
2	65	Mt. St. Mary's	67.55	66	=IF(NORM.INV(RAND(),BR2-BR3,10)>0,BP2,BP3)				
3	66	NO	68.16						
4									
5	67	K. State	84.15	68	=IF(NORM.INV(RAND(),BR5-BR6,10)>0,BP5,BP6)				
6	68	Wake Forest	83.87						
7									
8	69	NC Central	78.44	69	=IF(NORM.INV(RAND(),BR8-BR9,10)>0,BP8,BP9)				
9	70	UC Davis	68.81						
10									
11	71	Providence	81.27	71	=IF(NORM.INV(RAND(),BR11-BR12,10)>0,BP11,BP12)				
12	72	USC	82.45						

FIGURE 47.3 Simulating the Play-In Games.

In Column BR we enter the Sagarin ratings for the play-in teams. Then the formulas in Column BS simulate the winner of each play-in game from the standpoint of the first team listed. For example, to determine the winner of the Providence–USC game we simulate Providence's margin as a normal random variable with mean −1.18 points and standard deviation of 10 points. In the iterations shown NO, Wake Forest, NC Central, and Providence advanced to the field of 64.

Next, in row 2 (Figure 47-4) we type in the names of the teams, and in row 4 we type in each team's Sagarin rating (printed in *USA Today* the Monday before the tournament starts). The teams listed in the first 16 columns can be from any region. The next 16 teams listed must be from the region that plays the first listed region in the semi-final games. Then the next two regions may be listed in any order. Within each region the teams should be listed in the order in which they appear in any bracket. Listing the ratings in this order ensures that in the later rounds (including the Final Four) the winners of the earlier games play the correct opponents. Most brackets list the teams according to their seeds in the following order: 1 seed, 16 seed, 8 seed, 9 seed, 4 seed, 13 seed, 5 seed, 12 seed, 3 seed, 14 seed, 6 seed, 11 seed, 7 seed, 10 seed, 2 seed, and 15 seed. **The code numbers listed in Row 1 are in ascending order 1–60. The only exception is that for the columns corresponding to play-in winners, we should point to the cell containing the winner of the relevant play-in game.** For example, as shown in Figure 47-4 in cell B1, we entered the Formula = BS2 to advance the winner of the play-in game to the tournament. In cell B2 the formula = VLOOKUP(BS2,BP2:BQ3,2) picks off the name of the team that won the relevant play-in game. In cell B3 the formula = B1 reenters the code number for the winner of the play-in game. In the play-in game columns a VLOOKUP formula is needed to "look up" the ratings of the teams that won the simulated play-in games.

A sample iteration of part of the first round of the 2017 East Region is shown in Figure 47-5.

Villanova (Team 1) plays against play-in winner NO. The final margin of the game (from the standpoint of the team listed directly above the game outcome) was simulated in cell A7 with the formula

	A	B	C	D	E	F	G	H	I	J
2	Villanova	NO	Wisconsin	Va Tech	UVA	Wilmington	Florida	ETSU	SMU	Wake Forest
3	1	66	2	3	4	5	6	7	8	68
4	93.93	68.16	89.01	82.93	91.23	79.91	91	78.32	88.73	83.87

FIGURE 47.4 Entering Team Data.

	A	B	C	D	E	F	G	H	I	J
2	Villanova	NO	Wisconsin	Va Tech	UVA	Wilmington	Florida	ETSU	SMU	Wake Forest
3	1	66	2	3	4	5	6	7	8	68
4	93.93	68.16	89.01	82.93	91.23	79.91	91	78.32	88.73	83.87
5	East									
6	1		66		2		3		4	
7	24.08659	1			6.135478	2			−8.57541	5
8										
9	1		2		5		6		8	
10	2.409695	1			−19.6427	6			−11.8399	9
11										
12		1		6			9	13		
13		9.030932912	1			5.820578407	9			
14										
15			1		9					
16			2.907665599	1						

FIGURE 47.5 Simulating the Eastern Region.

$$=NORM.INV(RAND(\ ),HLOOKUP(A6,Ratings,2,FALSE) -$$
$$HLOOKUP(C6,Ratings,2,FALSE),10).$$

Note that there is no home court advantage in the NCAA tournament. If a team plays in its own or in a neighboring city, you might want to give it a two-point home edge. We found Villanova won by around 24 points. Therefore, our spreadsheet puts a "1" in A9 to allow Villanova to advance to its next game.

The following formula simulates the outcome of Villanova's next game:

$$=NORM.INV(RAND(\ ),HLOOKUP(A9,Ratings,2,FALSE) -$$
$$HLOOKUP(C9,Ratings,2,FALSE),10).$$

The range A3:AL4 are named ratings. The HLOOKUP function works just like the VLOOKUP function, except that the third argument refers to the row of the range ratings (in this case the second row of the range, which is row 4) where the value of the HLOOKUP is found. In cell D16 we find that in this iteration Villanova won the Eastern Region.

Each region is played out in this fashion. Figure 47-6 displays the simulation of the Final Four. In row 59 Excel plays out the NCAA semifinals. The spreadsheet "knows" to pick off the winner of each region. We find that on this iteration Villanova and Louisville won the semifinal games. Then in row 64 we play out the final game. Here Louisville won by 10 points.

We then used Excel's Data Table feature to play out the tournament 1,000 times. The COUNTIF function is then used to tabulate

	A	B	C	D	E	F	G	H
57		Midwest		West		East		South
58		1		19		44		54
59		4.622219	1			26.22644165	44	
60			winner				winner	
61			Villanova				Louisville	
62	Finals							
63		1		44				
64		−4.645032161	44					
65								
66			Winner					
67			Louisville					

FIGURE 47.6   Simulating the Final Four.

	BI	BJ
13	Teams	Chance of winning
14	Gonzaga	0.151
15	UNC	0.117
16	Villanova	0.101
17	UK	0.075
18	WVU	0.069
19	Kansas	0.07
20	Louisville	0.051
21	Duke	0.061
22	Oregon	0.041
23	Baylor	0.029

FIGURE 47.7   Top 10 Teams for
the 2017 NCAA Tournament.

the fraction of the time each team won the tournament. Figure 47-7 shows the probabilities of winning the tournament for the 10 teams that our model gave the best chance to win the tournament. Our top two teams, Gonzaga and UNC, played in the championship game with UNC taking home the title. Before the championship game, brackets based on this model ranked in the 99th percentile on ESPN's Bracket Challenge.

FILLING OUT YOUR GAME-BY-GAME POOL BRACKET

Each year many readers fill out a game-by-game NCAA pool entry. To determine which team should be entered in your pool as the winner of each game, simply track the outcome of each game and choose the team that wins that game most often during our simulation. For example, to pick your pool entry for the East Regional champions you would track the outcome of the cell, which contains the winner of the East Region (that is, cell D16). Whichever team occurs in cell D16 during the most iterations should be your pick in the pool for that game. You will find that Villanova wins the East Region much more often than anyone else, so in your pool you will pick Villanova as the winner of the East Region.

SIMULATING THE NFL PLAYOFFS

The file NFLplayoffsimjan1__2017.xlsx contains a template that can be used to simulate the NFL playoffs. The logic of the NFL playoff simulation is similar to the logic used in our NCAA simulation, with the major wrinkle being that the spreadsheet reseeds the team each round so the higher seeded team is playing at home. As shown in Figure 47-8, in Column C you simply enter the NFC teams in seeded order followed by the AFC teams in seeded order. Then enter your ratings (we used our least squares ratings) in Column D. We lowered the Raiders' rating seven points because they had to play their third string QB in the playoffs after their starting QB David Carr and his backup, Matt McGloin, both missed the playoffs with injuries. After

	B	C	D	E	F	G
1			Rating	NFC	AFC	Super Bowl
2	1	DALL	6.97	0.41	0	0.19
3	2	ATL	8.48	0.42	0	0.22
4	3	SEA	2.13	0.06	0	0.02
5	4	GB	2.83	0.06	0	0.02
6	5	NYG	2.13	0.04	0	0.01
7	6	DET	−1.4	0.02	0	0.00
8	1	NE	9.29	0.00	0.61	0.35
9	2	KC	5.6	0.00	0.24	0.12
10	3	PITT	4.74	0.00	0.11	0.05
11	4	HOUS	−2.63	0.00	0.02	0.01
12	5	OAK	−3.74	0.00	0.01	0.00
13	6	MIA	−2.4	0.00	0.01	0.00

FIGURE 47.8   2017 NFL Playoff Predictions.

our hitting the F9 key, a Data Table replays the playoffs 10,000 times. Columns E–G give the chance of each team winning its Conference Championship and the Super Bowl. Due to the weak AFC competition, we had the Patriots with a 61% chance to win the AFC and a 35% chance to win the Super Bowl. Of course, the Patriots did win the Super Bowl.

WIN PROBABILITY CALIBRATION AND PLATT SCALING

In this chapter we saw how we can use team ratings to obtain win probabilities. A question that often arises is how to measure the accuracy of a set of win probabilities. To measure accuracy, we project the probabilities to a binary event, win/lose, and simply get the fraction of times that our projection matched the real outcome of the game. For example, the prediction tracker (http://www .thepredictiontracker.com) keeps tabs on the accuracy performance of various projections—including regression-based. However, accuracy might not be the best metric, depending on what one wants to do with the win probability output. For example, if you are interested in using this information for betting purposes, you do not only want to know the accuracy but also the *exact* probability of a team covering the spread, for instance. This will ensure that you are hedging your bets correctly. Therefore, it is important to evaluate your models based on the **probability calibration** by constructing the model's *validation curve*. While we have explored the validation curve concept in previous chapters (e.g., Chapter 39), we will see here how we can improve the calibration of a model.

Typically, logistic regression models tend to have well-calibrated output probabilities, since they have a genuine probabilistic interpretation. However, several classifiers do not, and they simply output a *classification score* that needs to be mapped to a (well-calibrated) probability. There are various reasons that can lead to an ill-calibrated probability model, such as flawed model assumptions or omitted variables from the model. The latter can be a problem for win probabilities obtained from regression-based team ratings, since there can be many important variables omitted. Luckily, there are methods that can be used as a post-processing step to calibrate the probabilities better. The most used and popular method for post-calibration is **Platt scaling**.[2] Platt scaling fits a logistic regression model to the same binary variable as the original (non-calibrated) model, using as its single independent variable the probability outcome of the original model. More specifically, if f($\mathbf{x}$) is the original model's classifier scores (i.e., probabilities), Platt scaling produces new probability estimates as

$$Pr(y=1|x) = \frac{1}{1 + \exp(Af(x) + B)}$$

where A and B are the coefficients to be learned.

To showcase the impact of Platt scaling, we use the team ratings for the 2016 NFL season to obtain win probabilities. For every week between weeks 4 and 16 for the regular season, we calculate the team ratings based on the square error regression described in

2. J. Platt, "Probabilistic outputs for support vector machines and comparisons to regularized likelihood methods," in *Advances in Large Margin Classifiers*, 10(3), 1999, 61–74.

Chapter 46 and consequently obtain win probabilities for the home team for the matchups of the following week as described above. We then calculate the reliability curve for the model as well as the log-loss of the model. The log-loss is an evaluation metric that, even though it is related with the accuracy of the classifier, also considers the calibration of the model. The log-loss is defined by:

$$\log - \text{loss} = -\frac{1}{N}\sum_{i=1}^{N}(y_i \cdot \log(p_i) + (1 - y_i) \cdot \log(1 - p_i))$$

where y_i is 1 (0) if the home team in game i won (lost)[3] and p_i is the probability assigned by our model to the home team winning game i. Smaller log-loss translates to a better model. We then use Platt scaling to calibrate the output of this original model and calculate the reliability curve and the log-loss for the calibrated model.

Figure 47-9 presents the results. The top part of the figure presents the validation curves for the original and the calibrated models. Focusing on the original model we can see that it exhibits fairly good calibration (the curve is around the $y = x$ line), especially for probabilities higher than 30%. The log-loss was calculated equal to 0.67. Platt scaling calibration has a slightly better log-loss, equal to 0.64. The fitted values for the scaling are $A = -1.17$ and $B = 2.74$. Figure 47-10 visualizes the sigmoid function learned from Platt scaling and that is used for the probability calibration. The reliability curve is right around the $y = x$ line. However, as we can see from the bottom part of Figure 47-9, Platt scaling reduced the confidence of the predictions (i.e., the probability coverage). The calibrated model provided probabilities in the range [0.27, 0.8], while the original—but *worse*-calibrated model—provided probabilities in the range [0.08, 0.92]. This is also visible from the sigmoid function in Figure 47-10. Given that the input in Platt scaling is already a probability (i.e., between 0 and 1), the sigmoid transformation learned maps the probabilities to the smaller range with the black lines at the vertical axis.

---

3. Given that the probability of a tie in an NFL game is less than 1%, we ignore the games that ended with a tie for simplicity.

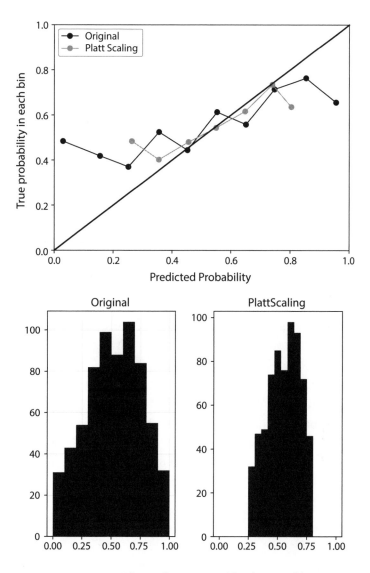

FIGURE 47.9 Platt scaling can provide a better calibration for the win probabilities (top figure), but at the expense of reduced confidence (bottom part).

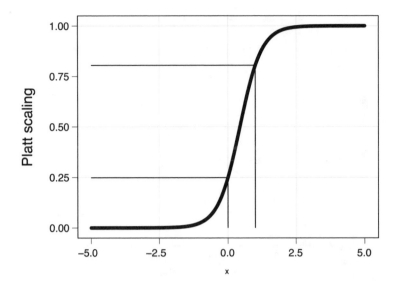

FIGURE 47.10   The sigmoid function used to calibrate the
win probability model. The confidence of the calibrated model is smaller
compared to the original model.

Platt scaling is just one of several ways that we can calibrate a
probability model. For example, isotonic regression can be applied
to the output of an ill-calibrated model to improve its probability
calibration. The interested reader can learn more in Kuhn and Johnson's book, *Applied Predictive Modeling*.[4]

4. M. Kuhn and K. Johnson, *Applied Predictive Modeling*, Springer, 2016.

CHAPTER 47 APPENDIX: USING DATA TABLES TO PERFORM SIMULATIONS

Our Chapter 1 Appendix reviewed the basics of Excel data tables. In the file NBAfinals2017.xlsx we replayed the Warriors Cavs series 1,000 times. Cell E16 calculates how many games the Warriors would win if the series went the full seven games. Then cell E20 records a 1 if the Warriors win the series and a 0 otherwise.

We make cell E20 the output cell for our one-way data table. When we set up the data table (the table range is J16:J1015) we make the column input cell any blank cell in the spreadsheet. Then Excel places a 1 in the selected column input cell and computes cell E20. Excel recalculates the RAND( ) functions and the result is one iteration of the Warriors Cavs series. Then Excel places a 2 in the blank column input cell. Then all our = RAND( ) functions recalculate and we get another "play" or iteration of the Warriors Cavs series. Therefore, our data table replays the series 1,000 times, and we find that the Warriors win around 95% of the time.

Data tables often slow down a spreadsheet. For this reason, when using a data table for simulation, we usually go to the Formulas tab under Calculation Options, and choose Automatic Except for Data Tables. With this option, data tables only recalculate when you hit the F9 (sometimes the Fn+9) key(s).

CHAPTER 48

THE NCAA EVALUATION TOOL (NET)

During the college basketball season, hoop fans anxiously anticipate the selection and seeding of teams for the NCAA tournament. The NCAA selection committee wants an accurate view of the teams' relative abilities. For almost 40 years (from 1981 to 2018) the NCAA tournament selection committee wanted to use only teams' win-loss records and not the scores of their games to rank teams. The NCAA believes that including game scores in the ranking and seeding process would cause the top teams to try and run up the score on lesser opponents. In Chapter 46 we explained how to use a logistic regression based ranking system to rank NFL teams. An identical system would do an excellent job of ranking college basketball teams based solely on win percentages. Defying logic, however, the NCAA used the complex and flawed RPI (Rating Percentage Index) to rank college basketball teams for 37 years. In the previous version of this book, we were hopeful that the NCAA would eliminate this eyesore from the beautiful landscape of college basketball, and gladly for us and all of college basketball fans, it did so in 2018.

The descendant of RPI is NET (NCAA Evaluation Tool). NET takes into consideration five different factors:

- Team value index
- Net efficiency

- Winning percentage
- Adjusted winning percentage
- Scoring margin.

Let's examine each one of these factors separately.

Team value index: TVI is a team rating that is based on the results of the game, adjusting for opponent and location of the game (home/away/neutral). Any final point margin is capped to 10 points and games that end up in overtime are capped to a one-point margin. The day of the game is not taken into consideration (i.e., the first game of the season is weighted equally with the conference championship game). While NCAA does not provide details on how this TVI is computed, we believe that it is very similar to the regression-based ratings we described in previous chapters.

Net efficiency: This is the difference between the points scored and allowed by a team per 100 possessions. In order to estimate the number of offensive possessions during a game for a team, NET uses the following formula: $N(\text{offensive possessions}) = FGA + TO - OREB + 0.475FTA$. Similarly, the number of defensive possessions for the team is: $N(\text{defensive possessions}) = OppFGA + OppTO - OppOREB + 0.475OppFTA$. For example, Figure 48-1 shows the box score information from the Georgia Tech–Pitt game in 2020. In that game the net efficiency for Pitt was +30.32, which means that Pitt outscored GT by 30 points per 100 possessions. Similarly the net efficiency for GT was −30.32. We can calculate the season net efficiency of a team by combining the box score information from all the games. The problem with this is that it defies the purpose of capping the win margin to 10 points while calculating TVI. A team might still try to run the score up, since this will lead to much higher net efficiency.

Winning percentage: This is the typical win percentage of a team, i.e., games won/total games.

Adjusted win-percentage: In the adjusted win percentage the location of the win-loss matters. More specifically, a road win is equivalent to 1.4 wins, a home win is equivalent to 0.6 wins, while a neutral site win is equivalent to one win. Similarly, a road loss is

2/8/20	Points	FGA		TO	OREB	FTA	Possessions
Pitt	73		57	8	11	15	61.13
Georgia Tech	64		50	22	13	27	71.83
	Pitt	Offensive Efficiency	119.43				
		Defensive Efficiency	89.11				
		Net Efficiency	30.32				

FIGURE 48.1   Estimating Net Efficiency from Box Score.

equivalent to 0.6 losses, a home loss is equivalent to 1.4 losses, while a loss in a neutral site is equivalent to one loss. So let's assume that Pitt has won two games at home, has lost one game on the road, and has lost one game on a neutral site. Then Pitt's adjusted win percentage is 1.2 − 1.6, or .428 (compared to a .500 winning percentage).

Scoring margin: This is the total points scored by the team minus the total points allowed by the team (with the individual games being capped at 10 points margin—and one point for OT games).

NCAA did not provide any details on how these five quantities combine to obtain the final ranking. However, it mentions that its objective is to provide good predictions for future games. Therefore, it is reasonable to assume that using data from previous seasons, NCAA cross-validated a predictive model for college team matchups.

How does NET compare to RPI? Of course, every ranking metric has its pros and cons, so here we want to just simply compare the rankings obtained by the two metrics. For example, if the two metrics give the same ranking, why bother with NET? We collected the NET and RPI rankings for the 2018–2019 season (the first year NET was introduced), and we estimated the Spearman rank correlation coefficient between the two. Our results gave a surprising .95 correlation! However, digging a little deeper we found that this is partly because both metrics are good at ranking bad teams low (and the majority of college teams are bad). If we focus on the rankings at the top 10, top 25, etc. the two metrics are still correlated, but to a lesser degree. Figure 48-2 presents the correlation results.

Top-k	Spearman
10	0.45
25	0.72
50	0.86
100	0.87
200	0.88
300	0.93
353	0.95

FIGURE 48.2 Rank Correlation Coefficient (at the Top k) for RPI and NET Rankings.

Only time will tell whether fans, media, and teams will call for the *death* of NET (if we had to . . . bet, we would put our money on "yes," but don't ask us to predict the over and under in terms of years that this happens).

CHAPTER 49

OPTIMAL MONEY MANAGEMENT

The Kelley Growth Criterion

Suppose we believe we have an almost sure bet on Colts –12. We believe the Colts have a 90% chance of covering the spread. This would probably never happen, but humor us and assume that such a bet really exists! What fraction of our capital should we allocate to this bet? If we bet many times all our money on bets with a 90% chance of winning, eventually we will be wiped out when we first lose a bet. Therefore, no matter how good the odds, we must be conservative in determining the optimal fraction of our capital to bet.

Edward Kelley (1956) determined the optimal fraction of our capital to bet on a gamble. The story of Kelley's work is wonderfully told in William Poundstone's *Fortune's Formula* (2005). Kelley assumes our goal is to maximize the expected **long run** percentage growth of our portfolio, measured on a per gamble basis. We will soon see, for example, if we can pick 60% winners against the spread, on each bet we should bet 14.55% of our bankroll, and in the long run our capital will grow by an average of 1.8% per bet. Kelley's solution to determining the optimal bet fraction is as follows. Assume we start with $1. Simply choose the fraction to invest which maximizes the expected value of the natural logarithm of your bankroll after the bet. The file Kelley.xlsx contains an Excel Solver model to solve for the optimal bet fraction given the following parameters:

- WINMULT = The profit we make per $1 bet on a winning bet.
- LOSEMULT = Our loss per $1 bet on a losing bet.
- PROBWIN = Probability we win bet.
- PROBLOSE = Probability we lose bet.

For a typical football point spread bet, WINMULT = 1 and LOSEMULT = 1.1. For a Super Bowl money line bet on the Colts –240, WINMULT = 100/240 = .417 and LOSEMULT = 1. For a Super Bowl money line bet on the Bears +220, WINMULT = 220/100 = 2.2 and LOSEMULT = 1. Kelley tells us to maximize the expected value of the logarithm of our final asset position Given a probability p of winning the bet we should choose f = our fraction of capital to bet to maximize:

$$p \cdot \ln(1 + \text{WINMULT} * f) + (1 - p) * \ln(1 - \text{LOSEMULT} * f) \ (1). \quad (1)$$

We find the optimal value for f by setting the derivative of the above expression to 0. This derivative is:

$$\frac{p * \text{WINMULT}}{(1 + \text{WINMULT} * f)} - \frac{(1 - p) * \text{LOSEMULT}}{(1 - \text{LOSEMULT} * f)}.$$

and will be equal to 0 if:

$$f = \frac{p\text{WINMULT} - (1 - p)\text{LOSEMULT}}{\text{WINMULT} * \text{LOSEMULT}}.$$

The numerator of the equation for f is our expected profit on a gamble per dollar bet (often called the "edge"). Our equation shows that the optimal bet fraction is a linear function of the probability of winning a bet, a really elegant result!

	M	N	P
			Expected growth per
14	Prob Win	Fraction	gamble
15	0.54	0.0309	0.053%
16	0.55	0.0500	0.138%
17	0.56	0.0691	0.264%
18	0.57	0.0882	0.430%
19	0.58	0.1073	0.639%
20	0.59	0.1264	0.889%
21	0.60	0.1455	1.181%
22	0.61	0.1645	1.516%
23	0.62	0.1836	1.896%
24	0.63	0.2027	2.320%
25	0.64	0.2218	2.790%
26	0.65	0.2409	3.307%
27	0.66	0.2600	3.873%
28	0.67	0.2791	4.488%
29	0.68	0.2982	5.154%
30	0.69	0.3173	5.873%
31	0.70	0.3364	6.647%
32	0.71	0.3555	7.478%
33	0.72	0.3745	8.368%
34	0.73	0.3936	9.319%
35	0.74	0.4127	10.335%
36	0.75	0.4318	11.418%
37	0.76	0.4509	12.572%
38	0.77	0.4700	13.800%
39	0.78	0.4891	15.106%
40	0.79	0.5082	16.496%
41	0.80	0.5273	17.973%
42	0.81	0.5464	19.544%
43	0.82	0.5655	21.214%
44	0.83	0.5845	22.991%
45	0.84	0.6036	24.883%
46	0.85	0.6227	26.898%
47	0.86	0.6418	29.047%
48	0.87	0.6609	31.342%
49	0.88	0.6800	33.795%
50	0.89	0.6991	36.424%
51	0.90	0.7182	39.246%

FIGURE 49.1 Kelley Growth Strategy and Average Growth Rate of Bankroll as Function of Win Probability.

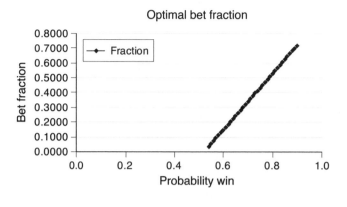

FIGURE 49.2 Optimal Bet Fraction as Function of Win Probability.

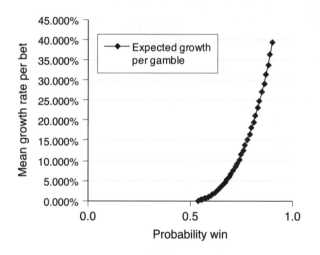

FIGURE 49.3 Average Wealth Growth per Period as a Function of Win Probability.

Kelley also showed that in the long run betting a fraction f of your bankroll each time leads to a long-term growth rate per gamble of $e^{\text{Expected LN final wealth}}$, where Expected Ln Final Wealth is given by equation (1).

Simplifying our expression for f we may rewrite our equation for f as:

$$f = \frac{p}{\text{LOSEMULT}} - \frac{q}{\text{WINMULT}}.$$

This equation makes it clear that an increase in the probability of a winning bet or an increase in WINMULT will increase our bet. Also, an increase in the probability of losing or an increase in LOSEMULT will decrease our bet.

As an example, let us compute our optimal bet fraction for an NFL point spread bet with a 60% chance of winning. We find that

$$f = \frac{.6(1) - .4(1.1)}{1(1.1)} = .145.$$

Figure 49-1 summarizes the optimal bet fraction and expected percentage growth per gamble if we use the Kelley growth criterion.

Figures 49-2 and 49-3 summarize the dependence of the optimal bet fraction and expected long-term growth rate on our win probability (of the bet).

As stated earlier, the optimal bet fraction is a linear function of our (bet) win probability, but our average capital growth rate per gamble increases at a faster rate as our win probability increases.

	K	L
		average
2	f	growth rate
3	0.05	0.67%
4	0.10	1.06%
5	0.15	1.18%
6	0.20	1.01%
7	0.25	0.53%
8	0.30	−0.28%
9	0.35	−1.43%
10	0.40	−2.96%
11	0.45	−4.91%
12	0.50	−7.33%

FIGURE 49.4   Average Long-Term
Growth Rate vs. Fraction Bet.

To show the importance of the optimal bet fraction, suppose that we can win 60% of our football point spread bets. Figure 49-4 shows how our long-term average growth rate per bet varies as a function of the fraction bet on each game.

Note that if we bet 30% or more of our money on each game, in the long run our capital will decline even though we win 60% of our bets!

CHAPTER 50

CALCUTTAS

In a Calcutta auction, you bid with other players for teams (in the NCAA basketball tournament) or golfers (in a Masters Calcutta). In this chapter, we will focus on NCAA basketball Calcuttas. Bidding for each team begins usually in random order, and every team is "bought" by the highest bidder. The owner will then get a payoff based on how the team(s) he bought performed in the tournament. In particular, after the tournament is over, the total amount bid is divided up on a percentage basis. Let's assume a team's payout (as a percentage of the pot) is as follows:

- Teams winning exactly one game split 2% of the pot.
- Teams winning exactly two games split 5% of the pot.
- Teams winning exactly three games split 15% of the pot.
- Teams winning exactly four games split 20% of the pot.
- Teams winning exactly five games split 25% of the pot.
- Champion gets 33% of the pot.

Sometimes a team that loses by the most gets say, 5%, of the pot to stimulate betting on bad teams.

Clearly a Calcutta auction is a zero-sum game, so if we pay less for a team than the team is expected to pay out, we should on average

	E	F	G	H	I	J	K	L	M
6			16	8	4	2	1	1	1
7		fractionofpottoeachteam	0.00125	0.00625	0.0375	0.1	0.25	0.33	
8		fraction	2	5	15	20	25	33	
9		howmanywinthismoney	16	8	4	2	1	1	
10	Team	0	1	2	3	4	5	6	Expectedfraction
11	KENT	0.0024	0.0532	0.0668	0.1132	0.1954	0.1446	0.4244	0.200471
12	WISC	0.0156	0.1112	0.209	0.2476	0.2392	0.0706	0.1068	0.08754425
13	UVA	0.0164	0.1956	0.1842	0.2398	0.145	0.1266	0.0924	0.08703025
14	DUKE	0.0166	0.1358	0.2096	0.23	0.1792	0.1386	0.0902	0.09244075
15	VILL	0.0118	0.1338	0.196	0.2656	0.1656	0.1414	0.0858	0.09157625
16	ARI	0.0076	0.2036	0.1846	0.2638	0.2116	0.055	0.0738	0.07056475
17	ZAGS	0.0206	0.2206	0.231	0.2412	0.142	0.093	0.0516	0.0652425
18	UNC	0.1214	0.2472	0.4036	0.125	0.0774	0.0146	0.0108	0.022473
19	ND	0.0774	0.3484	0.2506	0.2516	0.0472	0.0152	0.0096	0.02312475
20	KANS	0.096	0.3174	0.2628	0.259	0.04	0.0154	0.0094	0.02270375
21	IOWA ST	0.0552	0.2588	0.385	0.1746	0.0774	0.0396	0.0094	0.03001925
22	OKLA	0.0792	0.2822	0.4156	0.1344	0.0552	0.0248	0.0086	0.02254825
23	UTAH	0.24	0.2404	0.3212	0.1102	0.0532	0.0272	0.0078	0.0211345
24	VILLE	0.1306	0.3052	0.3704	0.124	0.048	0.0174	0.0044	0.0179485
25	BAYLOR	0.159	0.2722	0.3686	0.1328	0.0526	0.012	0.0028	0.016808
26	WVA	0.2702	0.2696	0.4066	0.0288	0.0188	0.004	0.002	0.00749825
27	WICH	0.3324	0.3688	0.1616	0.1156	0.0158	0.0042	0.0016	0.008964
28	MD	0.2866	0.36	0.317	0.0218	0.0106	0.0028	0.0012	0.00540475
29	MSU	0.3312	0.502	0.0764	0.0604	0.0216	0.0074	0.001	0.00771
30	ARK	0.2108	0.4872	0.2386	0.0444	0.0166	0.0016	0.0008	0.00608925
31	N IOWA	0.1918	0.44	0.2772	0.0642	0.0222	0.0038	0.0008	0.008124
32	IOWA	0.4216	0.4276	0.085	0.0454	0.0144	0.0052	0.0008	0.00577225

FIGURE 50.1 NCAA 2015 Expected Calcutta Payouts.

win money by purchasing that team. The key to success in Calcuttas is obtaining at any given time a good forecast for the total pot size.

To begin, we figure out the expected fraction of the pot that each team will get. We can run a simulation like our Chapter 47 NCAA simulation that estimates the probability that each team wins each possible number of games. This allows us to compute the fraction of the pot each team is expected to win. Our work for the 2015 NCAA basketball tournament is in the file Calcutta.xlsx and is shown in Figure 50-1.

For example, Kentucky was expected to win 20.04% of the pot, Wisconsin 8.8%, etc.

Before there is any bidding, we need to come up with our estimate of the total size of the pool. This can be tough, but a good start would be to assume that the pool size will equal last year's pool size. Suppose your estimate of pool size is $50,000. Suppose Kansas is the first team to go up for bids. Based on its chances of advancing through each round, Kansas, on average, will receive 2.27% of the pot. If you can get Kansas for a lot less than $.0227 * (50,000) = \$1,135$, then that purchase will have positive expected value and you should take Kansas. Suppose Kansas goes for \$2,000. Since \$2,000 appears to be worth 2.27% of the pot, you would now estimate the total size of the pot as $[(1/.0227)] * 2,000 = \$88,091$. Now assume Notre Dame (ND) comes up next. If you can purchase ND for less than $.023 * (88,091) = \$2,026$ you should take it. Suppose ND goes for \$2,500. Now my estimate of the pool is $[1/(.023 + .0227)] * (4500) = \$98,468$.

And as Kurt Vonnegut would say, "and so it goes."

PART VI

METHODS AND MISCELLANEOUS

CHAPTER 51

HOW TO WORK WITH DATA SOURCES

Collecting and Visualizing Data

Sports analytics are powered by data! Do you need a dataset with the 2016–2017 season's NBA game results? Fire up your web browser, point it to google.com, and search for "2016–2017 NBA game results." The first hit is from www.basketball-reference.com, from where you can obtain the data you were looking for easily. Other times, people have done the hard work for you and have already compiled datasets with exactly what you are looking for. However, this is not always the case. And most certainly it will not be the case when you are trying to build something novel that most probably will require a piece of data that has not been extensively used in the past. Therefore, a modern *computational sports (data) scientist* needs to be capable of collecting the datasets she needs from the web or other online databases.

There are in general two possible options for accessing sports datasets. In the first case, the entity that possesses the data is interested in opening its datasets (at least parts of them) to the public and it provides an Application Programming Interface (API). An API is loosely a set of hooks that allows your code to access the parts of the data that its owner has opened to the public. Typically, APIs might come with *rate limits*, that is, there is a limit on the number of times that one can invoke these hooks within a predetermined amount of time (usually a day). In the second case, there is no available API, but the data reside on a web server, typically stored within an HTML file (or possibly more than one HTML file). In this case, you must download and process the HTML file(s) to extract the required information. Of course, these steps cannot be done manually! We will provide you with some examples to get you started, but we urge the interested reader to explore resources dedicated to automated web data collection for more details.

Let us start with the setting where an API is provided by the data owner. One such example is the NHL, which provides an API to access its stats pages. Before looking at the NHL API, let's see what extended URLs are. If you want to search on Google on a specific keyword (e.g., "NBA"), you point your web browser to Google's URL, that is, www.google.com, and then you type in the search box the query you are searching for. You can achieve the same result if you point your browser to a slightly different URL. In particular, if you point your browser to https://www.google.com/search?q=nba, you are going to get the same result without having to type in the query in the search box. This is called an extended URL. Extended URLs include arguments in their string that follow a specific format that allows the user to specify inputs that dictate the page she gets in return. In the example above, the format identifies that we are looking for the "search" functionality, and the query is specified with the argument "q." An API for a data provider is very similar to these extended URLs. Even though there are different ways to create these "hooks" (e.g., there are different interface description languages that describe an API in different ways or formats), the main idea is that there is a structure in the URL that allows the user to communicate with the data provider and obtain the data she needs. The key of course is to identify the format of these URLs, which most of the time the provider is documenting. An API has several *endpoints* that provide you a path to different parts of the data, and each endpoint has specific *input arguments* that allow you to further filter the data. As an example, let's see NHL's API that uses the OpenAPI 3.0.[1] The URL that one has to create to obtain the required data consists of three parts: (i) the *base* URL, which is the part of the URL that essentially specifies the server that needs to be visited for obtaining the data, (ii) the *endpoint* path, and (iii) the *input arguments.* For example, the base URL for the NHL API is https://statsapi.web.nhl.com/api/v1. There are several endpoints that extend this URL and allow someone to obtain specific types of information. For example, there is an endpoint, /game/{id}/feed/live. This endpoint provides information from any game played. So if you extend the base URL to https://statsapi.web.nhl.com/api/v1/game/{id}/feed/live, you will be able to get detailed game data. However, we are not done yet. As you see, the endpoint includes some information in brackets. These are the input arguments that we need to specify in order for the API to return the data we need. In this case, the only input argument needed is the game ID. Let's assume that we know the game ID we are looking for (in fact there is another endpoint that you can use to get this game ID), and it is 2019020110.[2] If you fire up your browser and point it to https://statsapi.web.nhl.com/api/v1/game/2019020110/feed/live, you will get a JSON file with detailed information about the game, including shots, locations, penalties, etc. APIs allow you to systematically collect data for your analysis, but you need to make sure that you are following the format of the corresponding API. You can imagine that the base URL is the address of an apartment building, while the endpoint is the key to a specific apartment, and the input arguments are the keyless door code for a specific room in the apartment. If you use the wrong key or code, you will not be able to open the apartment or room. However, locks and codes change over time, and so do APIs. So you need to make sure that the data provider has not

---

1. https://swagger.io/docs/specification/about/

2. The game ID has a specific format as well. The first four digits specify the season (e.g., 2019 is the 2019–2020 season), the next two digits specify the type of the game (e.g., 02 is a regular season game), and the last four digits are the number of the game (e.g., 0110 is the 110th game of the season).

changed the version of the API, which means that the endpoints have been deprecated, changed, etc.

Let us now turn to the case where no API exists for accessing the data. In this case you will need to use libraries that deal directly with the web page that has the data you are looking for. Returning to our original example, the results from the 2016–2017 NBA season can be found at the URL https://www.basketball-reference.com /leagues/NBA_2017_games-{month}.html, where {month} can be substituted by the corresponding month (e.g., "october," "november," etc.). This URL is not an API URL. An API URL would return the data in a *machine-readable* format, pretty much ready for analysis (e.g., JSON, XML, CSV, etc.). In this case, the result returned is a web page with the data integrated in the HTML file specifying the web elements and markups of the page. There is good and bad news in this. The good news is that the data are there, in the HTML file! The bad news is that you need to do some extra work to extract them. Once you download the web page(s), the HTML file(s), you will notice, as mentioned earlier, that the downloaded data include HTML markup, that is, tags that inform the web browser how to display the content of the web page. Therefore, you will need to use appropriate parsing libraries to process this markup (e.g., Beautiful-Soup in Python) and extract the required data.[3] However, similarly to the APIs, the structure of a web page can change, and hence you might need to rewrite your scrapers.

When performing HTML scraping, the web server may be tracking your IP address to make sure that the requests you are making are not excessive. If the server considers you to be an automated scraping process (which you are!), it can block your IP address from any further requests. One of the reasons for this is that a high volume of requests can degrade the experience of other users who try to access the website at the same time. The resources of the server are

---

3. On the book's GitHub page we have included a detailed example on using Python to extract the four factors of teams in different games from basketball -reference.com.

not unlimited, and hence, if your software performs many requests at a high rate, it will utilize a large fraction of these resources, possibly not allowing other users to use them. Therefore, it is always a good practice to be considerate and reduce the rate of your requests to the extent possible. This most easily can be done by putting your software to sleep every so many requests and/or seconds, and giving time to the web server to serve other users and requests.

Many times, when there are popular APIs, people might have done the hard work for you and developed wrapper libraries that allow you to use these APIs, or scrape web pages, without your need of knowing their detailed structure. You can think of a wrapper library as a key tag for allowing you to find the key to the door faster. Typically, wrapper libraries have a much more intuitive structure that allows the programmer to access the third-party API of interest without the need to know the exact structure of the latter. For example, the popular nflscrapR developed by Max Horowitz, Ronald Yurko, and Samuel Ventura—that literally gave the keys to public analytics for the NFL—was a wrapper library in R for accessing the NFL's API.

EFFECTIVE VISUALIZATION

Being able to communicate the results of your analysis is important for every (data) analyst. Even more so in sports analytics, where the decision makers (i.e., general managers, coaching staff, etc.) are typically not well versed in the technical details of modeling and evaluation. However, they need to be able to grasp the conclusion and actionable information through an effective visualization. While whole books could be written on the "dos and do nots" for a visualization—and in fact they have been written—we will just provide a set of basic principles to be followed when visualizing results. We will also discuss tools that can help us effectively build visualization applications to present the results of our analysis to the ultimate decision makers, allowing them to also interact with the results.

Visualizations can be used for two purposes: (i) to understand a story (i.e., data exploration), and (ii) to tell a story (i.e., explanation).

Regardless of its purpose, when developing a visualization, one needs to consider human cognition and how humans perceive, learn, read, and attend to information. For example, color choice is crucial when the color lightness represents a ranking of sorts. A bad color choice can lead to difficulty in reading and understanding the presenting information. When we create a visualization we want to:

- Make it easy for the consumer of the information to spot differences and understand the relationships presented.
- Keep the visualization lean and avoid overloading it: "less is more."
- Not (visually) distort the data.
- Always remember that our audience might be color blind.

One of the most common problems with visualization (either intentional or unintentional) is the distortion of the information being presented. Many times, the reason for this is that the so-called *principle of proportional ink* is violated. According to this principle, if a shaded region is used to represent some value, then the area of this shaded region needs to be proportional to this value. This means that when we present some data using bar charts, we should never truncate the vertical axis (or, in general, the axis representing the values). Doing so will lead to a violation of this principle and mislead the audience of the graph. Figure 51-1 provides an example of this violation (PFF does great job, but its visualizations can certainly be improved).

Today there are several tools (both commercial and open source) that allow analysts to try a variety of charts and visualizations. One of the aspects that is important for a sports analyst is to allow the ultimate decision makers to interact with the results of her analysis. She needs to provide the decision maker with the ability to see different *views* of the data. Two open source softwares (among several other options) that allow for this and are extremely helpful are the shiny library for R and the bokeh library for Python.[4] These libraries

---

4. We provide examples on the book's GitHub page as well.

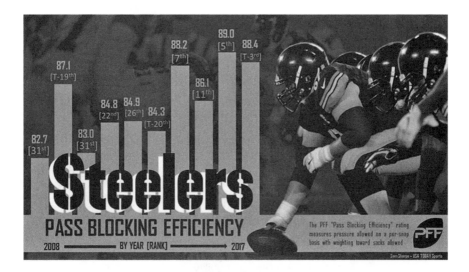

FIGURE 51.1 The distortion of the vertical axis is violating the proportional ink principle. The second bar is almost twice as tall as the third bar and represents a 4.1 percentage points difference (source: PFF).

allow an analyst to build web interfaces that can help visualize her analysis and allow the user to interact with the results. These applications can be deployed on a web server, which means that the end user does not need to have access to any particular software—other than a web browser. Deploying the application on a web server owned by the analyst gives more flexibility on who has access, the resources available etc. However, for a *casual* analyst, maintaining such a server might not be cost-efficient. In this case, one can freely deploy applications in various cloud services that provide free hosting. For example, shinyapps.io and heroku.com provide such free services.

Figure 51-2 visualizes the output of a shiny app that we built for a fourth down decision bot. The decision engine uses the analysis in Chapter 21. The user of the app can provide the input parameters, namely, the yards needed for a first down, the distance from the goal line, and the time remaining in the half. The application will generate a plot that provides the *optimal* decision for different FG success probabilities and fourth down conversion probability. The application also pinpoints (circle) the league-average values for

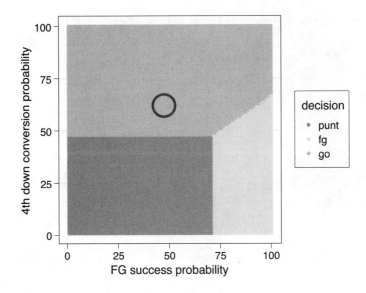

FIGURE 51.2   Fourth Down Decision Chart from a Shiny App
(the application can be accessed at https://athlytics.shinyapps.io
/4thDown/).

these probabilities. In this example, we faced a fourth and two, at the opponent's 36-yard line. As we can see, the league-average decision should be to go for it. Any changes in the input will automatically trigger the app to recalculate and output the new decision chart. Now this is a fairly straightforward example, but these tools allow an analyst to develop more complicated interfaces that allow the user to explore data, models, and results. However, keep in mind that when it comes to visualizations, most of the times (if not all of the times), less is more! And always remember: "You don't know it if you cannot effectively communicate it"!

TEXT MINING IN EXCEL

One of the misconceptions about Excel is that it is not very friendly with text analysis and mining. While obviously scripting programming languages (like Python) are very powerful in analyzing text, you can still perform basic text mining with Excel. For example, let's consider the website pro-football-reference.com (PFR) and its Play

	M	N	O	P	Q	R
1		How Many	Avg. Pts. Gained	Stdev	Downside Risk	
2	Passes	167	0.044	1.605	−0.486	
3	Runs	285	−0.142	0.725	−0.303	
4						
5						
6	EPA	Pts Gained	Pass	Sack	Pass or Run	Downside
7	7	5.27	18		pass	0
8	7	3.69			run	0
9	3.64	3.36			run	0
10	4.31	3.24	18		pass	0
11	6.74	3.03	18		pass	0

FIGURE 51.3   Analyzing 2014 Texans First and 10 Plays.

Index Game Play Finder. This tool allows you to slice and dice NFL plays and download as a text or Excel file information about any subset of plays. The information contains the starting line of scrimmage, a play description, down, yards to go, and the point value of the play (PFR does not provide any details on how this value is calculated). As an example of the type of analysis a team or armchair fan can conduct consider the 2014 Houston Texans. We can get information (see file Texansfinal.xlsx) on points gained on each Texan's 2014 first and 10 play from scrimmage. In row 7 of Figure 51-3, for example, we find that a 58-yard TD pass gained 5.27 points. We want to figure out for runs and passes the average points gained per play, the standard deviation of the points gained per play, and how many runs and passes were called.

The problem is that in order to identify whether a play is a pass or a rush we need to process the text description of the play. For this we proceeded as follows:

1. We copied from O7 to O8:O458 the formula =IFERROR (FIND("pass",J7,1),"") to see if a play is a pass. If the word "pass" is contained in the play description, the FIND function

returns the number of the position in the cell with data that begins with the word "pass." Otherwise the formula returns a blank cell.

2. In a similar fashion, we copied from P7 to P8:P458 the formula =IFERROR(FIND("sack",J7,1),"") to see if the play is a sack. We copied from Q7 to Q8:Q458 the formula =IF(LEN(O7)+LEN(P7)>0,"pass","run") to see if play was a pass or a run. If anything is in row 7 of the pass or sack column, then the play is a pass or a sack. Then LEN(O7)+LEN(P7) is greater than 0, and the play is counted as a pass. Otherwise the play is counted as a run.

3. In N2 and N3 we count how many passes and runs with the formulas =COUNTIF(Pass_or_Run,"pass") and =COUNTIF(Pass_or_Run,"run"). The first formula counts the number of passes and the second formula counts the number or runs.

4. In cells O2 and O3 we find average points per play with the formulas =AVERAGEIF(Pass_or_Run,"pass",Pts_Gained) and =AVERAGEIF(Pass_or_Run,"run",Pts_Gained). The first formula averages the points per play on passes or sacks and the second formula averages the points per play on running plays.

5. In cells P2 and P3 we array entered the formulas =STDEV(IF(Pass_or_Run="pass",Pts_Gained,"")) and =STDEV(IF(Pass_or_Run="run",Pts_Gained,"")). To array enter a formula, we first select the range the formula will fill (it can be more than one cell), then type the formula and finally hold down the keystroke combination Control+Shift+Enter. If you have the newest Office 365 you do not need to hit Control+Shift+Enter. The first formula loops through each row of data and creates an array that contains the value of the play when the play is a pass, and a blank cell when the play is not a pass. Excel ignores text, so this formula will find the standard deviation of the points per passing attempt. Similarly, the second formula calculates the standard deviation of the points per running play.

We find that on first and 10 the Texans passed 37% of the time and ran 63% of the time. Despite running more, passing plays gained an average of 0.04 points per play and running plays lost an average of 0.14 points per play. It seems like the Texans (and every team that has a QB) should have passed more on first down. As measured by the standard deviation on passing and running plays, we find, however, that passing plays were more than twice as risky as running plays.

Many people in finance use standard deviation as a measure of portfolio or investment risk. This is not a good idea, because we should care only about downside risk, not risk on the upside. Similarly, in evaluating passing and running plays we should choose a "target points per play" (say 0) and track the average amount by which a play fails to meet the target. To accomplish this goal, in Column R we computed the amount by which a play fails to achieve 0 (0 for a play with positive value; otherwise simply the play's value). Using the AVERAGEIF function we find that passing plays have an average downside risk of 0.49 points per play and running plays have an average downside risk of 0.30 points per play.

One other misconception is that we cannot automate (or semi-automate) a data collection with Excel. However, this is not necessarily true. Rintaro Masuda,[5] a data engineer at Microsoft, has provided us with Excel spreadsheets that are able to collect NBA data with a single click! Masuda created the file 2020–2021_v1.4.xlsm, which we can use to obtain team-level data directly from the NBA stats website. Using the "Controller" sheet we select the team we want to download data for and with one click get in the "View" sheet several statistics ranging from basic box score stats to advanced and opponent stats.

5. https://github.com/rintaromasuda

CHAPTER 52

ASSESSING PLAYERS WITH LIMITED DATA

The Bayesian Approach

Many times in sports we need to assess players with limited information. Should we sign a QB for whom we have only observed 30 pass attempts? What is the probability distribution for his yards/attempt? What about a kicker that has only taken five field goals of 50+ yards? How can we decide on whether to sign him?

To answer similar questions, we can rely on one of the most important theorems in probability theory: the Bayes theorem. Bayes' theorem operates the same way the scientific method operates. We begin with a prior belief on a hypothesis and then as we collect more evidence, i.e., data, we update our belief for this hypothesis. This is the high-level operation of Bayesian inference, and the reader who is not interested in the mathematics behind Bayes' theorem can now skip to the next section.

Let us assume that we have a prior probabilistic belief for an event A (e.g., whether Ben Roethlisberger will be a first ballot Hall-of-Famer, whether Joe Burrow is an 8 yards/attempt QB, etc.), P(A). We then observe a new event B (e.g., whether Roethlisberger was elected in this year's Pro Bowl, a new pass attempt from Burrow, etc.) and we update our belief. Specifically, we calculate the following posterior probability (that is, the conditional probability of A given B):

$$P(A|B) = \frac{P(B|A)}{P(B)} \cdot P(A) \tag{1}$$

Once you understand conditional probability and the law of total probability, Bayes' theorem is easy to understand. In many situations, we are trying to estimate the probabilities of various states of the world. Then we receive information that we use to change our probability estimates. Let's begin with some non-sports examples. For instance, consider a 40-year-old woman with no risk factors for breast cancer. The states of the world can be defined by the events C = (event woman has cancer) and NC = (event woman does not have cancer). Given no other information, the probabilities of these events (known as *prior* or *a priori* probabilities) are given by P(C) = 0.004 and P(NC) = 0.996.

Now we receive more information (e.g., the results of a mammogram) that changes our estimates of prior probabilities. Suppose the mammogram yields a positive (+) test result. To update our probability estimates, we need to know the likelihood of a positive test result for each state of the world. The likelihoods for a positive test result are known to be P(+ | C) = 0.80 and P(+ | NC) = 0.10. Now we want to update our prior probability (0.004) of cancer after receiving the positive test result. This new probability (P(C | +) is called *posterior* or *a posteriori* probability. Applying Bayes' theorem and the law of total probability, we find the following:

$$P(C|+) = \frac{P(+|C)P(C)}{P(+)} \quad \frac{P(+|C)P(C)}{P(+|C)P(C) + P(+|NC)P(NC)}$$

$$= \frac{0.8 \cdot 0.004}{0.8 \cdot 0.004 + 0.996 \cdot 0.1} = 0.031$$

Perhaps surprisingly, even after a positive test result, there is (thankfully!) only a small chance that the woman has cancer. This is because most women do not have breast cancer, and so many of their mammograms will result in false positives. Another way to see this is to look at a representative sample of 10,000 women. Then a contingency table would show that the 10,000 women would be classified as follows:

TABLE 52.1

	Positive Test	Negative Test
Cancer	$10000*(.004)*(.8)=32$	$10000*(.004)*(1-.8)=8$
No Cancer	$10000*(.996)*(.1)=996$	$10000*(.996)*(.8)=8964$

Given a positive test result, we are working with the 1,028 women in the first column. Therefore, after a positive test result, the chance the woman has cancer is $32/1,028=0.031$.

As a final (non-sport) example of Bayes' theorem, consider the classic Let's Make a Deal problem popularized by Marilyn Vos Savant in her "Ask Marilyn" column in *Parade Magazine*. A car is behind one of three doors, and there is a goat behind the other two doors. I choose a door (let's say door 1). Now Monty Hall (the host of *Let's Make a Deal*) chooses to open a door (door 2 or door 3) and reveals a goat. You are now allowed to switch doors. Should you?

Let's assume Monty opens door 2. The following are the relevant events:

- The states of the world are D_i, $i \in \{1, 2, 3\}$, i.e., the event that the car is behind door i.
- Define S_i, $i \in \{1, 2, 3\}$, as the event that Monty opens door i.
- We know $P(D_1)=P(D_2)=P(D_3)=1/3$ are the prior probabilities.
- The likelihoods are $P(S_3 | D_1)=1/2$, $P(S_2 | D_1)=1/2$, $P(S_3 | D_2)=1$, and $P(S_2 | D_3)=1$ (and of course $P(S_i | D_i)=0$).

When Monty opens door 2, we know that the car is behind either door 3 or door 1. Using Bayes' theorem, we calculate the following:

$$P(D_3|S_2) = \frac{P(S_2|D_3) \cdot P(D_3)}{P(S_2|D_3) \cdot P(D_3) + P(S_2|D_2) \cdot P(D_2) + P(S_2|D_1) \cdot P(D_1)}$$

$$= \frac{1 \cdot \frac{1}{3}}{1 \cdot \frac{1}{3} + 0 \cdot \frac{1}{3} + \frac{1}{2} \cdot \frac{1}{3}} = \frac{2}{3}$$

Therefore, $P(D_1|S_2) = 1 - (2/3) = 1/3$. Thus, we should switch our guess to door 3!!

Now let's return to using Bayes' theorem for estimating how new information changes our estimate that Ben Roethlisberger will make the Hall of Fame. In this context, we have a prior belief on Roethlisberger being a first ballot Hall-of-Famer, $P(A)$; and then, with the new evidence we have that he was elected to the Pro Bowl at the age of 38, we update our estimate. The term $\dfrac{P(B|A)}{P(B)}$ **is the support that event B provides for event A.** So, the Bayes equation above can be thought as: posterior probability = prior probability x support new evidence provides.

Equation (1) expresses the Bayes' theorem for point estimate probabilities, that is, single value probabilities that represent the "best guess" for the probability we are looking for. However, many times we are interested in the uncertainty around this point estimate, and therefore we are working with probability distributions, rather than point estimates. The Bayes' theorem can still be applied in the following form:

$$\pi(A|B) = \frac{f(B|A)}{f(B)} \cdot \pi(A) \tag{2}$$

In this equation, $\pi(A)$ is the prior probability (density) function for event A and $\pi(A|B)$ is the posterior probability function given the evidence collected by observing event B (i.e., Roethlisberger's recent Pro Bowl selection). $f(B|A)$ is the likelihood of observing B, i.e., observing the data related to event B, while $f(B)$ is the total probability of these data. f(B) can be calculated using the law of total probability. With some abuse of the notation, $f(B) = \int f(B \mid A)\pi(A)dA$.

To further explain Bayes' theorem for probability distributions, let us consider an event A that has a prior probability distribution that follows a normal distribution with mean μ and standard deviation τ, $N(\mu, \tau^2)$. For example, our best estimate for the average yards/attempt θ for Burrow's first year is what we have seen from rookie QBs, which follows a normal distribution with mean μ and standard deviation τ (computable from past data). Once we get new observations on Burrow, e.g., a new pass attempt from Burrow, we can update our belief at the distribution of his yards/attempt. The corresponding likelihood $f(B \mid A)$ in this case is also normally distributed $N(\theta, \sigma^2)$, where σ^2 is a known variance (e.g., it is realistic to assume that the variance in his performance is the same as that of the other rookie QBs we have observed in the past, or we can simply compute the variance from his limited pass attempts) and θ is the unknown average yards/attempt for Burrow. Hence, we can now compute the right-hand part of equation (2), and thus obtain the posterior probability distribution for θ. Note that in this case the posterior will also follow a normal distribution. For specific priors, such as normal distribution, there is a closed-form solution for the likelihood function, and hence for the posterior probability distribution as well. Furthermore, in similar cases where the prior and posterior distributions are from the same family, they are called **conjugate distributions**.

In the following section, we will present a detailed example of how one can apply the "distribution version" of Bayes' theorem to evaluate NFL kickers.

EVALUATING KICKERS

Your team (or a brewery in Chicago)[1] has brought in the practice a new kicker for evaluation. The special team's coach has him attempt 20 50-yard field goals. He makes 16 of those! What can we say about

---

1. https://www.sbnation.com/nfl/2019/1/12/18180142/chicago-bears-fans-field-goal-challenge-fails

his success rate on 50-yard field goal attempts that can help the team decide on whether to sign the kicker or not?

We will rely on Bayes' theorem (equation (2)). The variable of interest is the kicker's success rate at 50-yard FGs, σ_{50} (or simply σ). First, we need to identify a relevant prior distribution for the success rate σ_{50} of the kicker. For this we can use the success rate of all the kickers in the NFL. On average, there is approximately a 70% success rate on 50-yard FGs. There are some kickers who are extremely good (e.g., Justin Tucker), but most of the kickers are *average* (i.e., with a success rate around 70%). Furthermore, the success rate can only take values between 0 and 1 (i.e., a normal distribution is not appropriate as a prior for the distribution of σ_{50}). This type of *behavior* can be modeled through a beta distribution:

$$f_{\text{Beta}}(\sigma;\alpha,\beta) = \frac{1}{B(\alpha,\beta)}\sigma^{\alpha-1}(1-\sigma)^{\beta-1}, B(\alpha,\beta)$$

$$= \frac{(\alpha-1)!(\beta-1)!}{(\alpha+\beta-1)!}, 0 \leq \sigma \leq 1.$$

The parameters α and β control the mean of the distribution, $E[\sigma] = \frac{\alpha}{\alpha+\beta}$. For our prior distribution, we chose $\alpha=5$ and $\beta=2$, since this gives us an average success rate of approximately 70%. The left part of Figure 52-1 presents this distribution.

We now need to estimate the likelihood of observing 16 successful field goals in 20 attempts given the success rate σ. We can consider these 20 attempts as a series of Bernoulli trials,[2] and hence, $f(\text{"16-of-20"}|\sigma) = \binom{20}{16}\sigma^{16}(1-\sigma)^{4}$. Finally, we calculate the total probability of the observed data as:

$$f(\text{"16-of-20"}) = \int_{0}^{1}\binom{20}{16}\sigma^{16}(1-\sigma)^{4}\frac{1}{B(\alpha,\beta)}\sigma^{\alpha-1}(1-\sigma)^{\beta-1}d\sigma$$

---

2. In fact, the beta distribution is the conjugate prior for the Bernoulli likelihood.

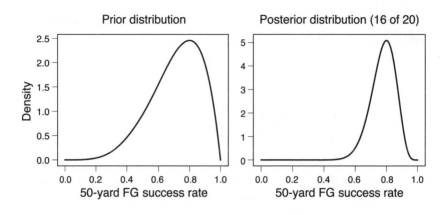

FIGURE 52.1 The Prior and Posterior Distribution for the 50-Yard FG
Success Rate of the Kicker Being Evaluated.

Putting everything together we obtain the right part of Figure 52-1
as the posterior distribution $\pi(\sigma|$"16-of-20"$)$ for the 50-yard FG suc-
cess rate of the kicker.

Using this posterior distribution, we can now estimate the
probability that the kicker our team is evaluating has a success
rate of 50-yard FGs better than 80%. This probability is equal
to: $\Pr[\text{kicker} \geq 80\%] = \int_{0.8}^{1} \pi(\sigma|$"16-of-20"$)\,d\sigma$, i.e., equal to the area
under the posterior probability curve between $\theta = 0.8$ and $\theta = 1$.
This corresponds to a 42% probability. A *generic* kicker, i.e., a kicker
who is drawn from the prior probability distribution (left part at
Figure 52-1), has a smaller chance of having an 80% or better success
rate kicker for 50-yard FGs, i.e., 34%. If we have the kicker attempt
another 20 FGs and he again scores 16 of them, we can further update our posterior probability and obtain a 44% probability that this
kicker is an 80% or better 50-yard FG kicker.

Bayes' theorem is a powerful analytical tool and can provide us
with a lot of insights.

HOW TO CHOOSE THE PRIOR

One of the challenges when taking a Bayesian approach is the choice of prior. To that end, without getting into the debate between *Bayesians* and *frequentists*, we would like to make the following suggestions and points.

When data is available, we can use data-driven priors. For example, in the case described above, the NFL kickers' performance itself drives the prior distribution for the kicker we evaluate. While the choice of a beta distribution might appear arbitrarily, one can take a more principled approach. Various distributions can be fit on the available data through a maximum likelihood estimation, and the one that fits the data better can be chosen as a prior. With experience, one can quickly eliminate candidate distributions almost immediately (e.g., as we described above, the normal distribution is not appropriate for modeling the FG success rate).

The choice of prior is clearly important, and when data are not available one might have to make a *best guess*. However, to some extent the choice of prior probability becomes irrelevant when *enough* and *informative* data and observations become available. For instance, in the case of the cancer example described above, and assuming the woman indeed has cancer, a second positive mammogram increases the probability of the woman examined having cancer to 20%, while a third positive mammogram increases the probability of cancer to 66.6%. As we can see, the accumulation of new, informative evidence moves the probability toward its correct value. Of course, this is not to say that making a good choice of prior is not important and should be ignored. One of the properties of Bayes' theorem is that if we start with two completely different priors, our posterior beliefs should move toward one another—and both toward the *truth*—as we obtain more evidence over time. Nate Silver in his book *The Signal and the Noise* (2015) illustrates this property with an interesting example. Three investors are trying to determine whether we are in a bull (i.e., price-rising) or bear (i.e., recession) market. They have different prior beliefs: 90%, 50%, and 10%, respectively, of being in a bull market. They observe the market daily on whether it goes up or down. Assuming that the market increases 60% of the time in the long run, eventually all investors agree that they are in a bull market with high probability.

DID 49ERS MAKE A GOOD DECISION IN TRADING FOR GAROPPOLO?

Of course, in hindsight he took the 49ers to the Super Bowl, and people will certainly be biased in answering yes. However, as objective analysts what one should be interested in evaluating is whether **at the time of decision** that was the right move. ESPN's Brian Burke used a similar approach to the one described above to put all the information we had about Jimmy Garoppolo at the moment of the trade into perspective.[3] Garoppolo had two starts with the Patriots during Brady's 2016 suspension and he attempted 63 passes. This is a small sample size, so the question is, what information can we learn from these 63 passes? Burke used Bayes' theorem to estimate whether a team should attempt to get Garoppolo instead of a first-round QB. His analysis supports that Garoppolo has a 64% chance to be better than a generic-first round QB. So, the 49ers appear to have made a well-informed trade when they traded their 2018 second round draft pick to the Patriots for Garoppolo.

3. http://www.espn.com/nfl/story/_/id/18741560/jimmy-garoppolo-edge-rookie-qbs

CHAPTER 53

FINDING LATENT PATTERNS THROUGH MATRIX FACTORIZATION

Often, sports data can be represented by a matrix, i.e., a rectangular array of numbers arranged in rows and columns. The rows of the matrix typically represent an *entity*, while its columns represent a *unit* of observation (e.g., a feature). For example, the shooting tendencies of NBA players (or teams) can be described through a matrix **S**, where rows represent players (or teams) and columns represent locations on the court. Each element S_{ij} corresponds to the number of shots player (or team) i took from location j. The location needs to be discretized, and one can use a grid over the court (and consequently each column represents one of the grid cells), or the court zones (e.g., midrange slot, restricted area, right corner 3, etc.). Player tracking data can also be described through matrices, where rows can represent individual snapshots and columns court locations as above. The (i, j) entry of the matrix then corresponds to the number of players in location j during snapshot i. The benefit of describing data through matrices is the availability of many analytical tools for analyzing matrices and identifying *latent patterns*. The latter typically allow us to express the objects of interest (e.g., shooting patterns) in a more compact—and at times meaningful—representation as compared to the original data. One such fundamental tool is **matrix factorization**. Matrix factorization (or decomposition) aims at expressing a given matrix as the product of two (or more) matrices, e.g., $S = WH$, where the matrices W and H are the factors. The rows (columns) of these matrices represent a *transformation* of the original data in such a way that they can capture latent patterns for the original data in S. This will become more clear in what follows.

Before providing specific examples for sports, let's understand better the notion of matrix factorization. While there are many different ways to factorize a matrix, we will focus on nonnegative matrix factorization (NMF), which deals with matrices that have nonnegative elements (e.g., matrices that represent counts—such as the one described above). With S being the original data matrix, the objective of NMF is to find matrices W and H, such that $S = WH$ (in reality, $S \approx WH$), with the additional constraint that all elements of W and H are nonnegative. In order to learn these matrices, we can solve the following optimization problem:

$$\min_{W,H} \|S - WH\|_F^2$$

$$\text{s.t.} \quad \begin{aligned} W_{ij} &\geq 0, \forall i, j \\ H_{ij} &\geq 0, \forall i, j \end{aligned} \qquad (1)$$

The above optimization essentially aims at identifying matrices W and H that minimize the distance (Frobenius norm) between S and WH. Depending on the data, it might be beneficial to add a regularization term for one or both matrices W and H, similarly to the discussion in Chapter 30 for regularizing the adjusted plus/minus of the players. Assuming matrix S has n rows and m columns (i.e., is $n \times m$), matrix W must have n rows as well, while matrix H needs to have m columns. Furthermore, it is obvious that the number of columns for W needs to be the same as the number of rows for H, and this is a parameter (say k) that we need to decide beforehand.

For example, let us consider that

$$
S = \begin{pmatrix}
5 & 5 & 0 & 0 & 0 & 0 & 0 \\
4 & 4 & 0 & 0 & 0 & 0 & 0 \\
5 & 5 & 0 & 0 & 0 & 0 & 0 \\
0 & 0 & 4 & 4 & 0 & 0 & 0 \\
0 & 0 & 3 & 3 & 0 & 0 & 0 \\
0 & 0 & 4 & 4 & 0 & 0 & 1 \\
0 & 0 & 0 & 0 & 5 & 5 & 0 \\
0 & 0 & 0 & 0 & 4 & 4 & 0
\end{pmatrix}
$$

In this case, the nonnegative matrix factorization for $k = 4$ gives:

$$
W = \begin{pmatrix}
2.7 & 0 & 0 & 0 \\
2.1 & 0 & 0 & 0 \\
2.7 & 0 & 0 & 0 \\
0 & 0 & 0.02 & 4.8 \\
0 & 0 & 0.01 & 3.6 \\
0 & 0 & 1.4 & 0 \\
0 & 4.8 & 0 & 0 \\
0 & 3.8 & 0 & 0
\end{pmatrix}
$$

$$
H = \begin{pmatrix}
1.9 & 1.9 & 0 & 0 & 0 & 0 & 0 \\
0 & 0 & 0 & 0 & 1.04 & 1.04 & 0 \\
0 & 0 & 2.9 & 2.9 & 0 & 0 & 0.7 \\
0 & 0 & 0.8 & 0.8 & 0 & 0 & 0
\end{pmatrix}
$$

What do matrices W and H mean? While this will be more evident through the analysis of NBA shooting data, you can imagine that the columns of matrix H represent *concepts* (*latent patterns*) that are linear combinations of the original columns of matrix S. For example, the first row of H describes a *concept* that is a linear combination with equal weights of the first two columns of matrix S, while the rest of the columns are not represented at all in this concept.

Matrix **W** corresponds to the transformation of the rows of matrix **S** to these concepts. For example, the first row of matrix **W** (i.e., first row of matrix **S**) includes only the first concept (columns 2, 3, and 4 are zero). The structure of matrix **S** in this example is such that one can identify the concepts or patterns even with a quick visual inspection (in fact, **WH** is almost equal to matrix **S**). In a real-life dataset the latent patterns will not be this easily identifiable through visual inspection. As we will see next, NMF can help simplify and obtain a better understanding of the data and patterns.

SHOOTING PATTERNS IN THE NBA

The shot selection process in basketball can be thought of as one of the dimensions—a small one—that characterizes the (offensive) identity of a team and/or player and is usually analyzed through shot charts. Despite the fact that shot selection by itself does not tell us anything about how these shots were created, being able to obtain a good understanding of these tendencies could be important for player and team comparisons as well as for pregame scouting. Typically shot charts are analyzed in a heuristic—and most of the times purely visual—manner, not following a principled approach to learn the prototype patterns hidden in the data. NMF can be used to achieve just this. To show how, we used a dataset from all NBA shots from the 2014–2015 NBA season. This dataset includes 184,209 shots from 348 players. It includes several pieces of information such as the game in which the shot was taken, the time on the game clock, the location from where the shot was taken, and, of course, the outcome of the shot. Using this information one can create shot charts such as the one presented in Figure 53-1, where we depict the shots taken by Isaiah Thomas during that season after he was traded to the Boston Celtics.

In particular, Thomas's chart can be represented through a 12-dimensional vector, each element of which corresponds to one of the 12 court zones (we eliminate the backcourt since there are very few shots taken from there in general—less than 0.5%). The value of

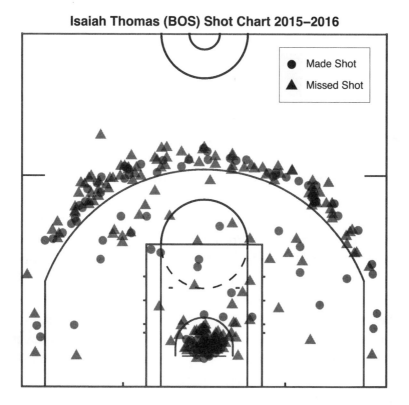

FIGURE 53.1 An Example Shot Chart.

the corresponding element is the number of shots taken by Thomas from that zone. Therefore, we can represent the whole dataset using a matrix, whose rows correspond to the different players while the columns correspond to the court zones. Therefore, our matrix S is a 288-by-12 matrix (we removed players with less than 50 shots during the season), where element S_{ij} corresponds to the total number of shots player i took from zone j. We can then perform NMF on matrix S. Matrix H will then represent k latent shooting patterns. The latter will correspond to combinations of shot frequency from different areas of the court that appear in the data. Matrix W will then express every player in the dataset as a linear combination of these k concepts or patterns.

The first question we need to answer is how many concepts should we look for, i.e., what is the value of k? The answer to this question is not trivial and it is one of those things that it is more of an art than science. There are some basic *constraints* for the value of k that need to be met for the obtained model to be meaningful. In particular, we should set $k \leq \min(m, n)$, i.e., the number of latent patterns should be smaller than the smallest dimension of the original matrix. After all, as mentioned earlier, one of the reasons for using matrix factorization is to represent the players' shooting patterns in a more compact way as compared to the original data. For instance, in the example above, Thomas is described through a 12-dimensional vector. If we choose a value greater than 12, there is no benefit in this new description. However, this constraint provides only an upper bound to the value of k. How can we choose the exact value of k? We could start looking at the quality of the approximation $S \approx WH$ for different values of k. However, one problem is that as we increase the value of k this approximation monotonically improves, even though the additional factors might not provide any useful information. To understand the last point, imagine the 288×12 matrix above. If we set $k = 1$, we are essentially trying to represent 12-dimensional information with a single number. The approximation error will be large. As we increase k, though, this approximation will improve, and as we go beyond $k = 12$, this approximation error will be extremely close to 0. However, this does not provide us with any new information or more compact representation! To solve this problem, one could possibly resample the raw data several times and choose a value of k based on the stability of the obtained patterns. Another approach would be to estimate the approximation error for increasing values of k. Consequently, we can choose the value of k based on the point where the improvement starts to slow down, thus exhibiting diminishing returns. We present this in Figure 53-2, where we can see that the error is reducing as we increase k. Obviously for $k > 12$ the error is practically 0, while we can see that around $k = 10$ (or even for a bit smaller value of k) the improvement in the approximation is getting smaller. Furthermore, the application at hand can provide further

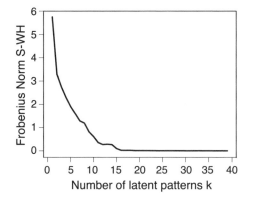

FIGURE 53.2 The approximation of the original matrix improves as we increase the value of k, but we should choose a value that balances the quality of the approximation and the compactness of the new representation.

information for the value of k or a way to obtain an educated guess. Choosing the value of k is a very similar problem to choosing the number of clusters in a clustering problem (Chapter 37).

Figure 53-3 presents the results for k=10. The court images at the top of the table represent the 10 latent patterns identified (i.e., the rows of matrix **H**). Each one of the court areas corresponds to one of the columns of **H**, while the color/shade represents the corresponding value learned. The table itself corresponds to the rows of matrix **W**, which are the new representation of the corresponding player. For instance, Steph Curry's (first row) most dominant—i.e., with the largest coefficient—latent pattern is pattern 4, which corresponds to wing 3s, while players like Chris Paul (last row) and Lamarcus Aldrige (fourth row) exhibit high coefficients for the midrange shooting patterns. On the contrary, LeBron James (third row) is taking advantage of his physicality and his dominant component includes shots from the paint and the restricted area. These coefficients are also proportional to the total number of shots taken by the players (James, Paul, Aldridge, and Curry took many more shots than Ariza during the 2014–2015 season, and therefore, overall, their coefficients are larger). Another interesting thing to note is that in this case, where we chose k=10, some of

	Component 1	Component 2	Component 3	Component 4	Component 5
S. Curry	2.8	2.9	5.7	12.6	0.1
T. Ariza	0.9	10.1	0.4	12.4	2.1
L. James	8.5	1.2	5.3	5.4	0.0
L. Aldridge	28.1	1.4	5.4	0.0	1.3
C. Paul	0.5	1.3	5.9	2.5	9.7

FIGURE 53.3   The 10 Components Obtained from the NMF and the Coefficients of Five Players.

the components identified exhibit some similarity. For example, components 3 and 8 are fairly similar (mainly shots from the paint), and if we had chosen a smaller number for the latent patterns, it is probable these components would be merged together. This could potentially be another way to decide on the value of k, i.e., checking the inter-component similarity obtained for different values of k.

In the above analysis, we have used as the spatial granularity the court zones. This certainly removes some of the details of the shot chart, since all shots within a zone are considered equally. More detailed patterns can be obtained if we consider a finer grid over the court. For example, instead of using the court zones as the columns of our matrix, we can apply NMF by overlaying a grid over the (half) court with 1×1 feet cells. This essentially creates $47 * 50 = 2,350$ possible locations (instead of 12 when using the court zones) that will be our columns. This spatial granularity can provide much more detailed information for the distribution of the shots within a court zone. In this case we will also feel the benefits of NMF in terms of dimensionality reduction and data representation. Without NMF, to describe the shooting of a player we would need a vector of 2,350 elements (as

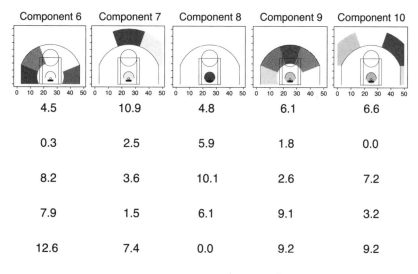

Component 6	Component 7	Component 8	Component 9	Component 10
4.5	10.9	4.8	6.1	6.6
0.3	2.5	5.9	1.8	0.0
8.2	3.6	10.1	2.6	7.2
7.9	1.5	6.1	9.1	3.2
12.6	7.4	0.0	9.2	9.2

FIGURE 53.3 (*continued*)

many as the grid cells). However, after applying NMF on the data, we can describe a player in a significantly reduced dimensionality (equal to the number of latent patterns chosen). Each of these dimensions is not a simple grid cell but corresponds to one of the patterns. Miller, Bornn, Adams, and Goldsberry[1] have further shown that even better results can be obtained if the NMF is applied on intensity surfaces obtained from the shooting points rather than the raw shooting counts. This essentially smooths the shooting charts, removing potential noise from the sparsity of the data due to the finer spatial granularity.

Finally, while we have introduced and used for our examples NMF, there are other matrix factorization techniques that can be used and possibly with better results. Gohberg, Kaashoek, and Spitkovsky[2] provide a nice overview of matrix factorization techniques.

1. A. Miller, L. Bornn, R. Adams, and K. Goldsberry, "Factorized point process intensities: A spatial analysis of professional basketball," in *International Conference on Machine Learning*, January 2014, 235–243.

2. Israel Gohberg, Marinus A. Kaashoek, and Ilya M. Spitkovsky, "An overview of matrix factorization theory and operator applications," *Factorization and Integrable Systems*, Birkhäuser, Basel, 2003, 1–102.

CHAPTER 54

NETWORK ANALYSIS IN SPORTS

One of the crucial elements in team sports like basketball and soccer is the interactions between players on the court or pitch and the underlying relationships that these interactions create. For example, players pass the ball to each other creating a complex system of passing combinations that the team hopes to ultimately capitalize on and take a good quality shot. Teams also have win-loss relationships with each other based on their game outcomes. To study similar relationship patterns, we can rely on network science, which provides a set of tools that can be helpful in sports analytics tasks. A **network** is described by **two** different **sets**; a set of *objects* and a set of *relationships* between these objects. In our case the objects (termed as *nodes* or *actors* or *vertices*) can be the players (e.g., when studying the passing behavior of a team) or teams (e.g., when we study win-loss relationships). Two nodes are connected in the network if the relationship (termed as *edge* or *link*) that the network represents exists between these nodes. In the simplest case the relationship is reciprocal (e.g., Joe is friends with Nick—and vice versa). However, in many cases there is a direction in this relationship (e.g., LeBron passes to Anthony Davis). Furthermore, the edges can be annotated with a *weight* that captures the intensity of the relationship (e.g., how many times LeBron did pass the ball to Davis).

PASSING NETWORKS

One of the most straightforward ways to apply network science in sports is by analyzing passing networks, i.e., structures that capture information about who passes to whom. Fewell and his colleagues (2012) used data from the first round of the 2010 NBA playoffs and created passing networks for all the 16 playoff teams, whose edges are weighted based on the number of interactions or passes observed between the two nodes. The nodes in the network represent the players (more specifically the position of every player). They also added starting and ending nodes that capture the way a possession started (e.g., inbound pass, rebound) and ended (e.g., turnover, made three-point FG, etc.). One of the issues that appears when analyzing similar *small* networks (i.e., networks with a small number of nodes) is that over a large number of observations (which is the case when we deal with basketball possessions), the network will be fully connected, since there will be almost certainly at least one pass between every two players. To deal with this, the typical approach is to set a threshold on the weight of the edges and eliminate edges with a weight below that threshold. Fewell et al. (2012), instead of using an absolute threshold, first ranked the interactions from most to least frequent, and kept the minimal set of transitions that represent a particular percentile of all ball movements (they used the 60th percentile). These networks can be used to study things like the *unpredictability* of the offense based on the connectivity patterns observed, or they can reveal roles in the team. In the data analyzed by Fewell et al. (2012) the Bulls and the Celtics exhibit a *centralized* structure, with the point guard position being the central one, getting 60% of the inbound passes and then distributing the majority of the rest of the passes. In these networks, removing the point guard node will lead to a disconnected network. On the contrary, the networks for the Cavaliers and the Lakers deviate from this centralized structure, which matches leadership roles within these two teams (Lebron James and Kobe Bryant). But what do these networks reveal?

Centralized networks have their benefits and drawbacks. As the authors argue, on the one hand, a centralized network is a sign of a team with clearly defined roles. On the other hand, in such a structure some players are less involved and therefore the opponent's defense could exploit that (e.g., by adding more double teams involving the defenders responsible for guarding these less involved offensive players).

One of the most promising ways to use networks in the analysis of sports is for quantifying *intangibles*. The value of a network does not necessarily lie with the nodes, but rather with the connection patterns between them. This means that for the same set of nodes, one connectivity pattern might lead to a system with a different total value as compared to a different connectivity pattern. This is true not only for sports networks but for any kind of network. For example, let us consider the case of a pencil and a diamond.[1] They are both systems whose core is carbon atoms (the nodes). However, the value of each of these systems is vastly different, and the reason behind this is the way that the carbon atoms are interconnected! The same is true for a team network. You hear people all the time talking about a team that has *great chemistry,* and that it performs much better than expected by considering its individual units. In other words, the team has a value that is larger than the sum of the individual parts of the team. We have seen in Chapter 32 that we can use the adjusted +/− ratings of individual players and the lineup ratings to quantify how much better a lineup is performing compared to the expectation from simply adding the contributions of each player. Network science tools can further allow us to quantify the connectivity of the underlying player (passing) network. One such tool is the *algebraic connectivity*. Before discussing this let us introduce some network science and linear algebra terminology.

---

1. This example taken from the excellent TED talk of Nicholas Christakis—a pioneer in network science—on the hidden influence of networks: https://www.ted .com/talks/nicholas_christakis_the_hidden_influence_of_social_networks

Every network can be represented through its *adjacency matrix* that captures the underlying connections. This adjacency matrix A has n rows and n columns (where n is the number of nodes in the network) and its element A_{ij} represents whether there is a connection between nodes i and j or not. We can also define the degree matrix D, which also has n rows and n columns and is a diagonal matrix. The element in the ith diagonal place is the degree of node i, that is, the total number of edges attached to node i. With these two matrices, we can then define the Laplacian matrix for the network as $L = D - A$. Every matrix is associated with a set of *eigenvalues* and *eigenvectors*. The intuitive way to understand what these represent is to think of a matrix M as a linear transformation that takes as input vector x and returns a vector $y = Mx$. The eigenvectors of matrix M are vectors whose **direction** does not change under the linear transformation of matrix M. The eigenvalues are the corresponding scaling factors. For a matrix M, the eigenvalues λ and eigenvectors x are the solutions to the equation $Ax = \lambda x$, where $x \neq 0$. The eigenvalues and eigenvectors of a matrix have found several applications in a large spectrum of fields ranging from control theory and stability analysis to facial recognition.

In networks, the smallest eigenvalue of the Laplacian matrix is always 0, while the second smallest eigenvalue λ_2 reveals information about the network connectivity. If this is also 0, then the network is disconnected. In general, the higher the value of λ_2 the better connected the network is. Figure 54-1 presents a toy example of a network with seven nodes and the corresponding adjacency matrix A and degree matrix D. The eigenvalues for the Laplacian matrix are 0, 0.37, 1.57, 2, 3, 3.47, and 4.6. If we remove the edge between nodes 3 and 6, the network is disconnected. In this case, the second smallest eigenvalue of the Laplacian matrix is also 0 and the corresponding eigenvalues are 0, 0, 1.38, 2, 2, 3.62, and 4. If we further remove edges 3–4 and 3–1 then we have three disconnected components in the network and the corresponding eigenvalues are 0, 0, 0, 1, 2, 3, and 3.

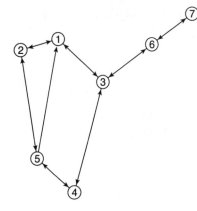

$$A = \begin{bmatrix} 0 & 1 & 1 & 0 & 1 & 0 & 0 \\ 1 & 0 & 0 & 0 & 1 & 0 & 0 \\ 1 & 0 & 0 & 1 & 0 & 1 & 0 \\ 0 & 0 & 1 & 0 & 1 & 0 & 0 \\ 1 & 1 & 0 & 1 & 0 & 0 & 0 \\ 0 & 0 & 1 & 0 & 0 & 0 & 1 \\ 0 & 0 & 0 & 0 & 0 & 1 & 0 \end{bmatrix} \quad D = \begin{bmatrix} 3 & 0 & 0 & 0 & 0 & 0 & 0 \\ 0 & 2 & 0 & 0 & 0 & 0 & 0 \\ 0 & 0 & 3 & 0 & 0 & 0 & 0 \\ 0 & 0 & 0 & 2 & 0 & 0 & 0 \\ 0 & 0 & 0 & 0 & 2 & 0 & 0 \\ 0 & 0 & 0 & 0 & 0 & 2 & 0 \\ 0 & 0 & 0 & 0 & 0 & 0 & 1 \end{bmatrix}$$

FIGURE 54.1   A Network Toy Example with the Corresponding
Adjacency and Degree Matrices.

At the MIT Sloan Sports Analytics Conference ESPN Hackathon
in 2017, we showcased the use of the passing network's algebraic
connectivity, which we termed as the *team chemistry eigenvalue*, as a
way to quantify intangibles in the game. Using the (admittedly small)
sample of games provided during the event, we found that the game
point differential exhibits a medium correlation with the difference
in the team chemistry eigenvalues of the two teams ($r = 0.59$); but
this correlation is not statistically significant. Now this can be either
because there is indeed no relationship between the two or because
the dataset is small and there was not enough statistical power in
the corresponding test. The takeaway is not that we should use the
Laplacian eigenvalues to quantify team chemistry, but rather that
networks offer a wide range of powerful tools that can be adopted
for quantifying intangibles in sports and are currently underutilized.

Similar network analysis has been performed in soccer studies.
Statistical properties of soccer passing networks have been analyzed and found to exhibit *small-world* behavior, which means that
there is a small average shortest path between any pair of nodes and
that there are more connected triangles (triplets of players who are
all connected to each other) in the network compared to what is

expected by chance.[2] Gyarmati, Kwak, and Rodriguez (2014)[3] went deeper in the network analysis by analyzing motifs of passes (i.e., beyond a single pass). They particularly analyzed the sub-possessions that are three passes long and examined their prevalence in the game of each team in the Spanish, Italian, English, French, and German first division teams for the 2012–2013 season. They consequently compared the motifs extracted with a randomized version of the passing networks and obtained the z-score for each motif. Their results show that flow motif analysis can reveal unique strategies used by teams. FC Barcelona, known for its *tiki-taka* game that includes quick, short passes, stands out when the game is analyzed with this method (other methods using established statistics such as number of passes, number of goals, timing of passes, etc. have failed). In particular, Barcelona applies the motif ABAC (i.e., player A passes to a player B, who passes back to A and then A passes to C) significantly more often than other teams—at least a 2.5 standard deviations difference. Also, FC Barcelona applies the motif ABCD significantly less frequently as compared to the rest of the teams. Using the z-scores for the five different motif types examined,[4] one can cluster teams based on their passing signatures. The latter is simply a five-motif-long feature vector. Figure 54-2 presents the clusters obtained from k-means for the Spanish league, where for visualization purposes the five-dimensional features are projected to their first two principal components (PC1 and PC2). Principal component analysis (PCA) is a dimensionality reduction method (similar to the NMF discussed in Chapter 53) that allows us to project a high-dimensional vector to a lower dimensionality, while preserving a pre-determined amount of the variance in the data. As we see for the data, Barcelona's cluster does not contain any other teams!

2. Takuma Narizuka, Ken Yamamoto, and Yoshihiro Yamazaki, "Statistical properties of position-dependent ball-passing networks in football games," *Physica A: Statistical Mechanics and Its Applications*, 412, 2014, 157–168.

3. Laszlo Gyarmati, Haewoon Kwak, and Pablo Rodriguez, "Searching for a unique style in soccer," in SIGKDD workshop on Large Scale Sports Analytics, 2014.

4. The five motifs explored are: ABAB, ABAC, ABCA, ABCB, and ABCD.

PRINCIPAL COMPONENT ANALYSIS

The general idea of PCA is straightforward. Assuming our data is organized in an m × n matrix **X**, we first calculate its covariance matrix C_x. The principal components of the original data are then simply the eigenvectors of the covariance matrix C_x. We can then reduce the dimensionality of our original data by choosing only some of the eigenvectors of C_x to represent our data. A typical rule of thumb on the choice of the number of principal components r includes the corresponding eigenvalues λ_i; r is the minimum value such that $\sum_{i=1}^{r} \lambda_i^2 / \sum_{i=1}^{n} \lambda_i^2 > 0.95$, which translates to the r first principal components capturing 95% of the variance in the original data. Obviously, the threshold of 0.95 can be different for different applications. For the interested reader, a nice tutorial on PCA is provided by Jonathon Shlens. Going back to our Spanish league team clustering, when the goal of applying PCA is to visualize the data, the value of r is limited to 1, 2, or 3, with the most typical value being r = 2 as in Figure 54-2.

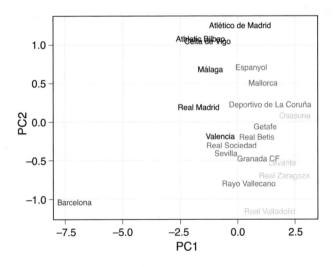

FIGURE 54.2 Barcelona stands out, defining one cluster by itself, when considering motifs of three-pass long sequences (figure obtained from Gyarmati et al., 2014).

THE EWING THEORY

Dave Cirilli was convinced that Patrick Ewing's teams (Georgetown and NY Knicks) played inexplicably better when Ewing was missing games—or even missing extended times in a single game. Cirilli and Simmons further termed a similar phenomenon, that is, a team plays better without its superstar(s), as the "Ewing theory." Skinner (2010)[5] used network theory and an analogy from the well-known Braess's paradox[6] for transportation networks to show how the Ewing theory is plausible. In short, the Braess's paradox states that if you have a transportation network and add a new link, i.e., street, with the goal of improving the traffic—by increasing the total capacity of the network—it might actually have the reverse effect. As an example, of Braess's paradox the New York City authorities found that when Times Square was closed to cars and trucks, congestion lessened, rather than increased as many expected. The main reason behind this paradox is the fact that individuals are myopic to socially optimal solutions and try to selfishly optimize their own utility. While the final state that the system reaches is an equilibrium, unfortunately it is not socially optimal.

Before getting into the details of the toy example from Skinner that clearly shows that the Ewing theory is **plausible**, a critical part of the theory is the notion of *skill curves* introduced by Dean Oliver in his seminal book *Basketball on Paper* (2004). The idea behind the skill curves is that the efficiency of a player is not stable, but changes with his utilization. In fact, the efficiency of a player $f(x)$, is a function of his utilization x. The utilization is captured through the fraction of the team shots that this player takes (or a player's usage rate, which is the fraction of plays or possessions *used* by the player). The efficiency can be captured from a variety of metrics such as effective

5. B. Skinner, "The price of anarchy in basketball," *Journal of Quantitative Analysis in Sports*, 6(1), 2010.

6. D. Braess, A. Nagurney, and T. Wakolbinger, "On a paradox of traffic planning," in *Transportation Science*, 39(4), November 2005.

field goal (eFG), true shooting percentage (TS%), etc. For example, Figure 54-3 shows the skill curve for Ray Allen obtained from Skinner's paper. One of the issues with skill curves is that the players' variability in terms of utilization is very small. Simply put, LeBron James is never going to take only 5% of the shots of his team, while on the other hand a role player will never take 50%+ of his team's shots. This creates a problem in the estimation of the skill curves. However, it is widely accepted that these skill curves are decreasing functions of the player's utilization. This is similar to what happens in a transportation network, where a road with higher utilization can exhibit lower efficiency, i.e., larger delays.

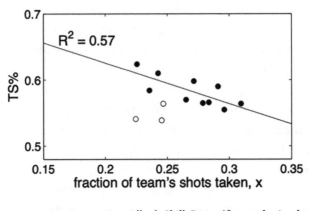

FIGURE 54.3 Ray Allen's Skill Curve (figure obtained from Skinner, 2010).

Having this concept of skill curves in mind let's see two toy examples of offensive schemes (Figure 54-4). The left part of the figure shows a possession that can have players 1 or 2 dribble the ball to the basket and then either take a shot or layup or pass the ball to player 5 to take the shot. Each edge is annotated with the efficiency of the play. For example, player 1 sees a reduced efficiency the more he is utilized (variable x) for driving the ball in the paint. However, this option is always better than having player 2 drive the ball, who has a constant efficiency of 0.5. Therefore, the first observation is that in the Nash equilibrium solution, player 1 always drives the ball to the basket. He then has two options: (i) pass the ball to player 5, or (ii) take the ball to the basket on his own. If player 5 receives the ball, he has the option to finish to the rim or to pass the ball to 1 or 2 for a lay-up or shot. It should be clear that the efficiency of player 1 finishing the shot is always smaller than or equal to that of player 2 and 5, and therefore there is no reason for player 1 to ever finish the possession. So we know that in the Nash equilibrium player 1 drives the ball to the low post and passes to 5, and then the only remaining question is whether player 5 passes to 2 or takes it to the basket. In the Nash equilibrium situation, to identify what fraction of possessions player 5 should take the shot and what fraction of possessions player 5 should pass the ball to player 2 for him to take the shot, we need to equate their respective efficiencies. The result then is that player 5 should take the shot 1/3 of the times, while player 2 should take the shot 2/3 of the time (with an expected TS% of 2/3). Given that the drive by player 1 has an efficiency of 0.5, the overall efficiency of the possession is $0.5 \cdot 2/3 = 0.33$.

Let's now examine a similar possession but without the presence of player 5. This is depicted in the right part of Figure 54-4. In this case, the two players are equally good options, so they will split the shots 50–50. So, the overall efficiency is $0.5 \cdot (1 - 0.5 \cdot 0.5) = 0.375$, which is greater than the efficiency when player 5 was in the game!

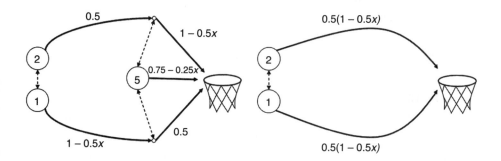

FIGURE 54.4 The offensive scheme on the right provides overall higher efficiency for the team, even though player 5, who is the most efficient player of the team, is not included in the scheme (figure obtained from Skinner, 2010).

To reiterate, the problem is that when trying to play according to the Nash equilibrium, a team is essentially optimizing every part of the possession individually. This will not lead to an overall optimal solution. In this case (left part of Figure 54-4), Skinner shows that the optimal solution is obtained by having player 1 drive the ball half of the time, and player 2 drive the ball the other half of the time. Then players 2 and 5 should split the shot attempts 50–50. This strategy will give an overall efficiency of 0.43, higher than both scenarios utilizing the Nash equilibrium. This boils down to the fact that if the team considers one possession at a time, the offensive efficiency will be only 33%, while the best possible strategy will give it an efficiency 43%. Skinner's takeaway message from this simple offensive set is that the apparent improvement of a team when losing its best player (i.e., the plausibility of the Ewing theory) is a result most of all of short-sightedness while this best player is on the court.

RATING TEAMS THROUGH NETWORKS

We have discussed in previous chapters how you can rate teams through simple regression models. Another possible way to rank teams is through networks. Central to this ranking system is a network of who wins whom. The nodes of this network are teams while the edges represent win-loss relationships. There is a direction in the edge from node i to node j if team i *lost* to team j. The edges can be further annotated with the point differential of the corresponding game.

There are several approaches that one can use to rank teams using this network. The simplest one is to use PageRank,[7] a network centrality metric that is used to identify *important* nodes in a network. PageRank was developed by Google's founders and is the core mechanism behind its search engine. The high-level idea behind PageRank (at the context of Google and a web page) is that a web page is important if (i) several other nodes link to it and/or

---

7. Lawrence Page, et al., *The PageRank Citation Ranking: Bringing Order to the Web*, Stanford InfoLab, 1999.

(ii) important web pages link to it, leading to a recursive definition. The way PageRank operates can be described through an infinite random walker on the network. At each step, the random walker picks with a probability $(1-\alpha)$ any node in the network to *teleport* uniformly at random, while with probability α it picks uniformly at random one of the outgoing edges from the current node at which the walker resides. If we let the random walk go on for an *infinite* amount of time, the PageRank of node j is then the fraction of times that our walker passed through node j. Simply put, the PageRank is the steady state solution π of the random walk described above. This can be expressed through the equation: $\pi = D(D - \alpha A)^{-1} 1$, where 1 is the unit vector, i.e., a vector with all elements equal to 1.

When applying the concept of PageRank to a sports team ranking, a team is *important* if it has won against (many) other teams— and even more if these teams are important themselves. This implicitly captures the strength of schedule aspect of team rankings. We have developed a simple ranking scheme as described above—termed SportsNetRank[8]—which we evaluated by predicting upcoming matchups. Before presenting the performance of SportsNetRank in the prediction task, let us present a simple toy example to understand the PageRank computations. Let us consider a small league with six teams where each one will play against each other team once, for a total of 15 games. Let us consider the network after the first six games, where the matchups in Table 54-1 have played out.

The adjacency matrix for this directed network is

$$
A = \begin{bmatrix}
0 & 0 & 0 & 0 & 1 & 0 \\
1 & 0 & 0 & 0 & 0 & 0 \\
1 & 1 & 0 & 0 & 0 & 0 \\
0 & 0 & 0 & 0 & 0 & 0 \\
0 & 0 & 0 & 1 & 0 & 1 \\
0 & 0 & 0 & 0 & 0 & 0
\end{bmatrix}
$$

8. K. Pelechrinis, E. Papalexakis, and C. Faloutsos, "SportsNetRank: Network-based sports team ranking," in ACM SIGKDD workshop on Large Scale Sports Analytics, 2016.

TABLE 54.1

A Toy Example on Matchups from a Six-Team League

Loser	T1	T1	T2	T4	T6	T5
Winner	T2	T3	T3	T5	T5	T1

while the degree matrix is:

$$D = \begin{bmatrix} 2 & 0 & 0 & 0 & 0 & 0 \\ 0 & 1 & 0 & 0 & 0 & 0 \\ 0 & 0 & 1 & 0 & 0 & 0 \\ 0 & 0 & 0 & 1 & 0 & 0 \\ 0 & 0 & 0 & 0 & 1 & 0 \\ 0 & 0 & 0 & 0 & 0 & 1 \end{bmatrix}$$

With $\alpha = 0.85$ (this is the typical value[9] used when calculating PageRank, i.e., 85% of the time the random walker is following the outgoing links from its current position), we have:

$$(D - \alpha A)^{-1} = \begin{bmatrix} 0.50 & 0.00 & 0.00 & 0.36 & 0.43 & 0.36 \\ 0.43 & 1.00 & 0.00 & 0.31 & 0.36 & 0.31 \\ 0.79 & 0.85 & 1.00 & 0.57 & 0.67 & 0.57 \\ 0.00 & 0.00 & 0.00 & 1.00 & 0.00 & 0.00 \\ 0.00 & 0.00 & 0.00 & 0.85 & 1.00 & 0.85 \\ 0.00 & 0.00 & 0.00 & 0.00 & 0.00 & 1.00 \end{bmatrix}$$

The final PageRank vector is: $\pi = (3.3, 2.4, 4.4, 1, 2.7, 1)^T$ or, normalized, $\pi = (0.22, 0.16, 0.3, 0.07, 0.18, 0.07)^T$. As we can see, the top-ranked team in this case is team T3, with a record 2–0. Team T1 is ranked second, even though its record is 1–2. The reason for this is the fact that its sole win has been against team T5, which itself had a record of 2–0 prior to their matchup.

---

9. For prediction tasks like the ones described in this section, the value of α can also be picked using a validation set (see Chapter 30).

Going back to the evaluation of SportsNetRank, we used the games up to week $\tau - 1$, to predict the matchups of week τ. The prediction algorithm is simple: if $\pi_i(\tau - 1) > \pi_j(\tau - 1)$, then team i is projected to win team j in week τ. Table 54-2 presents the results of these predictions for the NFL seasons 2009–2015. We also provide the accuracy of predictions of a similar algorithm where the winning percentage was used instead of the PageRank. SportsNetRank exhibits an overall accuracy of approximately 61% with a standard error of 1.2%, while the win-loss percentage can correctly predict future matchups with 52% accuracy. The performance of Sports-NetRank could be further improved with an appropriate selection of PageRank's dumping factor α (e.g., through cross-validation). However, even with the default value of $\alpha = 0.85$, SportsNetRank's performance on predicting future outcomes is on par with the performance of systems such as Microsoft's Cortana[10] and ESPN's FPI,[11] despite being (most probably) significantly less complicated. Sports-NetRank—the way described above—does not produce probabilistic predictions. It is simply a ranking model with a binary prediction based on which team has better ranking. However, one could use the actual PageRank values of the teams matching up and build a logistic regression model that would predict the probability of the home team winning the game.

We would have hoped to have seen more of similar approaches when seeding the NCAA conference tournaments during the COVID-19 pandemic. In particular, the pandemic and the health and safety protocols led to highly imbalanced schedules among teams within the same conference, which renders win percentage not a great proxy. For example, in the ACC late in the season, Geogia Tech was the seventh seed based on win percentage (it ended up moving to fourth with the last few games). However, at the same time, it had the second highest page rank in the conference. Georgia Tech

---

10. https://www.firstscribe.com/bing-predicts-looks-average-in-nfl-week-17-wildcard-weekend-preview/

11. http://www.espn.com/blog/statsinfo/post/_/id/123048/a-guide-to-nfl-fpi

TABLE 54.2

SportsNetRank has approximately 61% accuracy in
predicting NFL matchups

Year	SportsNetRank	Winning Percentage
2009	0.64	0.57
2010	0.56	0.5
2011	0.64	0.58
2012	0.65	0.56
2013	0.56	0.5
2014	0.66	0.56
2015	0.59	0.53

went on to win the ACC championship (the other finalist, Florida State, had the highest page rank, and was the second seed based on win percentages).

Network structures are extremely prevalent in sports and go beyond passing relationships. One can build substitution networks (who is subbing for whom), screen networks, handoff networks, or pick-n-roll networks in basketball, etc. Each one of these structures captures different relationships and can potentially reveal different information and help us understand team or player performance in more detail. State-of-the-art network science tools can help toward this goal. Particularly helpful can be *network embeddings*, that is, methods that allow us to learn a vector representation of network nodes based on the connectivity structure. These vector representations can then be used as input to traditional statistical learning models for downstream tasks like predictions. For instance, by finding vector representations of players from substitution networks, we could possibly identify similar players. While providing an in-depth treatment of networks is beyond the scope of this book, there are several resources for the interested reader, such as the excellent book by Easley and Kleinberg *Networks, Crowds, and Markets: Reasoning About a Highly Connected World* (2010), which is also available as an e-book from the authors' personal webpage.[12]

12. https://www.cs.cornell.edu/home/kleinber/networks-book/

CHAPTER 55

ELO RATINGS

Physicist Arpad Elo developed a rating system in the 1960s for rating chess players. The main idea behind the Elo rating system is the *exchange* of *rating points* between the players after their matchup. The winner adds x points to her rating, which are subtracted from the losing player's rating. Therefore, the Elo rating is a zero-sum rating. The question is then, how can we decide how many points to award to the winner (and equivalently how many points to deduct from the losing player)? The general idea is that if the win was *expected* the winner is rewarded less compared to the case where the win was not expected (i.e., an upset). This expectation is calculated from the ratings of the opponents prior to the matchup.

In more detail, if team A plays against team B, and their Elo ratings are R_A and R_B, respectively, the expectation that A **does not lose** to B is given by $E_A = \dfrac{1}{1 + 10^{((R_B - R_A)/400)}}$. If we want to incorporate home field advantage for team A (in the case of chess this corresponds to a player playing with the white), we can update the above equation as $E_A = \dfrac{1}{1 + 10^{((R_B - R_A - H_e)/400)}}$, where H_e is the home edge in Elo points. One way to identify the home edge H_e is to consider two teams of equal strength (i.e., $R_A = R_B$). In the NFL, home teams win (or don't lose) approximately 58% of the games, which means we should have $E_A = \dfrac{1}{1 + 10^{((-H_e)/400)}} = 0.58$. Solving this equation for H_e we get $H_e \approx 56$.

To reiterate, E_A is the expectation of A winning if there is no possibility of a draw (e.g., in basketball). If there is the possibility of a draw, E_A captures the probability of A winning or drawing. The Elo rating cannot further differentiate between these two possibilities. Elo treats draws as half-win and half-loss, and therefore E_A is the sum of the win probability as well as half the probability of a draw. If, for example, $E_A = 0.6$, this could be either a 60% probability of winning and a 0% chance of drawing, or a 40% chance of winning and a 40% chance of drawing. Now let us assume that player A beats player B. With S_i representing the event of player i *winning*, we have $S_A = 1$, while $S_B = 0$ (in the case of a draw, $S_A = S_B = 0.5$). The Elo rating of player A is then updated to: $R'_A = R_A + K(S_A - E_A)$. The parameter K is crucial for the performance of the rating system and controls how fast the system reacts to new results. A large value for K will react fast to new results—especially to unexpected results—while a smaller value for K will require a larger number of unexpected results for a player's rating to change dramatically. The Elo process resembles to some degree the Bayesian inference. As the system obtains more observations (wins, losses, draws), the ratings of the players or teams change in response to the new evidence. Depending on the support the evidence provides for the new rating, the change can be large or small.

The Elo rating system is used for several sports today and forms the basis for many of the predictions (at least the early ones) made on Nate Silver's FiveThirtyEight.com. There are several challenges associated with the implementation of an Elo rating system. Some of the most important ones are:

1. Choice of initial ratings for a team
2. Transfer of team ratings between seasons
3. Challenge of dealing with expansion teams
4. Choice of K.

The choice of the initial ratings will impact the projections of the system at least in the short run. In the long run the impact of these initial ratings will be negligible (similar to the impact of a prior). For example, FiveThirtyEight.com calculates the Elo rating of a team starting from the inception of each league. Therefore, the initial ratings chosen have little impact (if at all) on the projections for the current season. However, it might be preferable to build a rating system that is not based on the performance of the teams in past seasons, especially when teams undergo significant roster changes from one season to the next. Nevertheless, how can one pick the initial ratings in this case? A typical choice might be to start with every team having the same rating. However, it should not take a lot to convince you that this is not a good idea. The New England Patriots and the New York Jets were not of equal strength at the beginning of the 2017 NFL season (or any season for that matter). So how can we get these ratings?

We can turn to betting markets. Betting markets give the opportunity to people to bet before the beginning of the season on the total number of wins that a team will have. The market provides a line, which can be thought of as the expected number of wins for the team, and people can bet on whether the team will go over or under this expected number of wins. These lines encode several pieces of information that all boil down to the expected strength of the teams prior to the beginning of the season. One can use this information to obtain an initial Elo rating for the teams at the beginning of the season. More specifically, we can obtain a set of ratings for the teams that minimize the (squared) error between the expected number of wins according to the betting market and the expected number of wins as obtained using these initial ratings. We further constrain these ratings to have a prespecified average that can be chosen arbitrarily (e.g., 1,500 is what FiveThirtyEight uses as the average rating).

More specifically, assuming team i plays against teams H_s at home and against teams V_s on the road, the expected number of wins for team i for a given set of ratings $\mathbf{R}$ is:

$$\mathbb{E}[W_i | R, H_e] = \sum_{j \in H_s} \frac{1}{1 + 10^{((R_j - R_i - H_e)/400)}} + \sum_{j \in V_s} \frac{1}{1 + 10^{((R_i - R_j - H_e)/400)}}$$

With l_i being the betting wins total for team i, we can then learn the set of ratings **R** by solving the following optimization problem:

$$\min_R \sum_{i=1}^{32} (l_i - \mathbb{E}[W_i | R, H_e])^2$$

under the constraint that $\sum_i R_i = 1,500$ (or any other value we want the rating of an average team to be). Table 55-1 presents the initial Elo ratings obtained for the 2017 NFL season, along with the total regular season wins predicted by the oddsmakers. We can see that some teams are expected to have eight wins, which represents essentially an average team. However, their Elo rating is fairly different from 1,500, which is our Elo rating average (e.g., Arizona). One of the reasons is the fact that teams have an uneven schedule in the NFL. A team that is below average might still be expected to have eight wins because, for example, it has several easy games (with other below-average teams). The same is true for a team that is above average but is expected to have eight wins because it has a tough schedule to play. Note here that if we do not want to use the value of $H_e = 56$ that we estimated earlier for the home edge, we can include this as an additional variable to be learned from the optimization problem. It is worth noting that we can use a similar approach to obtain pre-season regression ratings (Chapter 46).

As mentioned above, FiveThirtyEight.com takes a different approach and calculates historic Elo ratings for the teams from the beginning. In this case, the choice of initial ratings will have a minimal impact on the current projections. However, this approach comes with a different challenge, namely, how does one transfer the end-of-season Elo rating of a team to its beginning-of-season Elo rating for the following year? Someone could argue that we can leave it intact. However, as mentioned earlier, this might not be a good idea. The Denver Broncos won the Super Bowl in 2016, the last season Payton Manning played. Clearly, they were not the same team at the

TABLE 55.1

Initial Elo Ratings for the 2017 NFL Season Based on
Betting Total Win Lines

Team	Rating	Over/Under Line
Arizona Cardinals	1466	8
Atlanta Falcons	1578	9.5
Baltimore Ravens	1532	9.5
Buffalo Bills	1431	6
Carolina Panthers	1518	8.5
Chicago Bears	1383	5.5
Cincinnati Bengals	1504	8.5
Cleveland Browns	1323	4.5
Dallas Cowboys	1578	9.5
Denver Broncos	1549	8.5
Detroit Lions	1479	7.5
Green Bay Packers	1589	10
Houston Texans	1502	8.5
Indianapolis Colts	1506	9
Jacksonville Jaguars	1373	6
Kansas City Chiefs	1575	9
Los Angeles Chargers	1491	7.5
Los Angeles Rams	1360	5.5
Miami Dolphins	1495	7.5
Minnesota Vikings	1510	8.5
New England Patriots	1726	12.5
New Orleans Saints	1499	8
New York Giants	1549	9
New York Jets	1341	4.5
Oakland Raiders	1612	10
Philadelphia Eagles	1502	8
Pittsburgh Steelers	1607	10.5
San Francisco 49ers	1307	4.5
Seattle Seahawks	1580	10.5
Tampa Bay Buccaneers	1529	8.5
Tennessee Titans	1492	8.5
Washington Redskins	1488	7.5

beginning of the following season (and beyond that). The approach that FiveThirtyEight.com takes is to regress the rating of teams to the mean. In particular using a mean for the Elo ratings of 1,500, the beginning-of-season rating for team A is $R_{A,start} = 0.75 \cdot R_{A,end} + 0.25 \cdot 1505$, where $R_{A,end}$ is the Elo rating of the team at the end of the previous season. Notice this weighted average consists of 75% of the past year's ending rating and 25% of an "average" team's rating. However, the average team's rating is taken to be 1,505 and not 1,500, which is the actual mean of the Elo ratings used by the site. The reason for this has to do with another challenge that these historic Elo ratings have to deal with, which is the expansion teams. Each franchise begins with a rating of 1,300; therefore, this can bring the average Elo rating a lot below 1,500. To avoid this in the long run, FiveThirtyEight.com gives each team a rating only slightly higher than 1,500 to counteract the presence of expansion teams.

One of the things that remains to tackle in applying Elo ratings to our favorite league is the choice of K. To reiterate, K controls how fast the ratings react to new results. A large value of K will lead to a more noisy series of Elo ratings for a team, while a smaller value for K will lead to a more stable Elo rating over the season. On the contrary, a small value of K might end up biasing the rating toward the initial Elo of a team, which if not set properly will have cascading effects on the prediction performance. Therefore, we need to have a principled way of choosing the value of K. Different sports will have different *optimal* values for K. For example, FiveThirtyEight.com uses a value of K around 20 for the NBA and the NFL, while a smaller value is used for MLB. Given that K is essentially a hyper-parameter of the model, we can choose the value of K for a league by using historic matchups and a validation set. Assuming for simplicity that there is not a possibility of a tie, different values of K will provide different win probabilities for a matchup in the validation set. We can then choose the value of K that provides the best probabilistic predictions, i.e., the one that minimizes the following metric: $\frac{1}{N}\Sigma_{i=1}^{N}(S_i - E_i)^2$, where S_i is 1 if the home team of matchup i won the game and 0 otherwise, while E_i is the win probability for the home team of matchup i according to the Elo ratings.

GLICKO RATINGS

Since Elo ratings were developed to rate chess players, it would be a blunder of ours not to discuss its improvement, the Glicko rating, introduced by the great Mark Glickman. The main improvement is the calculation of a rating deviation *RD*. RD measures the uncertainty around the player's (team's) rating and is equal to one standard deviation. A player with a rating of 1,600 and an RD of 30 will have a true rating anywhere between 1,510 and 1,690 (i.e., within three standard deviations) with 99.7% confidence. RD is used in updating the rating of a player after a game. For example, if a player's RD is low, it translates to low uncertainty about her true rating, and hence the change after a game should be smaller. RD changes itself with more games played, where typically more games result in lower RD. However, it can also slowly increase with absolute time if the player does not play any games during a (long) time period. Mark Glickman has provided detailed explanations and implementations of Glicko that the interested user can look at.

CHAPTER 56

COMPARING PLAYERS FROM DIFFERENT ERAS

In Chapter 15 we tried to figure out whether it would be likely that Ted Williams would hit .406 if he played in 2019. Our analysis required that we compare the pitching and fielding abilities of players from different eras. In Chapter 15 we used a simplistic approach and found that it was unlikely that Ted Williams would hit .406 in 2019. In this chapter, we use our WINVAL ratings to determine whether the players in the NBA have improved or declined in quality since the year 2000. Then we summarize the results of Berry, Reese, and Larkey (BRL 1999), who analyzed the change in player quality over time for Major League Baseball, professional hockey, and professional golf. We close by discussing a relatively simple approach of Peterson, Penner, and Stanley (PPS 2011) that can be used to compare baseball players from different eras.

ANALYZING CHANGE IN NBA PLAYER QUALITY 2000–2007

We used WINVAL player ratings for all NBA players during the 2000–2007 seasons to estimate change in player quality during those years. For example, if Dirk Nowitzki had a +10 rating for the

2004–2005 season, that would indicate that, per 48 minutes, if Dirk played instead of **an average 2004–2005 player,** then our best estimate is that the team would improve its performance by 10 points per 48 minutes. We can use our WINVAL ratings to estimate the relative level of player abilities during the 2000–2007 seasons. Let's arbitrarily assign the 2006–2007 season a strength level of 0. If our model estimates, for example, that the players who played during the 2003–2004 season had a strength level of +4, that would mean that, on average, players in 2003–2004 were four points better than players in 2006–2007. Each "data point" for us is a player's WINVAL rating for a given season. We restricted our analysis to players who played at least 1,000 minutes during a season. We used the Excel Solver to estimate the following:

- For each player we estimated an overall ability level relative to 2006–2007 players. For example, if we come up with a +10 estimate for Dirk Nowitzki, we are estimating that Dirk, during our years of data, averaged playing 10 points better per 48 minutes than an average **2006–2007 player.**
- An overall ability level for each season. For example, if we obtain an estimate of −3 for the 2002–2003 season, we would estimate that per 48 minutes, players during the 2002–2003 season were three points worse than the 2006–2007 players.

Let's continue using Dirk Nowitzki as our example. We would hope that during each year x,

$$\text{(Dirk's ability relative to the 2006–2007 season)} = \text{(Dirk's rating during year x)} + \text{(Year x strength relative to 2006–2007)}. \quad (1)$$

For example, if Dirk had a +10 rating during 2002–2003 and players during the 2002–2003 season averaged five points better in ability than players during the 2006–2007 season, then we would estimate that relative to the 2006–2007 players Dirk was 15 points better (per

48 minutes) than an average 2006–2007 player. Rearranging equation (1) we can obtain

(Dirk's rating during year x) = (Dirk's ability relative to
the 2006–2007 season)—(Year x strength relative to the
2006–2007 season).								(2)

We know the left side of (2), but we do not know either of the quantities on the right-hand side of (2). Following our approach to rating teams in Chapter 46, we use the Excel Solver to choose each player's rating relative to 2006–2007 and each season's average player rating relative to 2006–2007 to minimize the sum over all players and seasons:

$$[(\text{Player rating during year x}) - \{(\text{Player's ability}$$
$$\text{relative to the 2006–2007 season}) -$$
$$(\text{Year x strength relative to the 2006–2007 season})\}]^2.$$

After running this optimization, Table 56-1 shows the following estimates of player strength for each season relative to 2006–2007.

TABLE 56.1
Season Strengths Relative to 2006–2007

Season	Relative Player Strength
1999–2000	− 0.32
2000–2001	− 0.78
2001–2002	− 0.56
2002–2003	− 0.69
2003–2004	− 0.17
2004–2005	− 0.53
2005–2006	− 0.02
2006–2007	0

For example, we found the average level of player ability in the 2000–2001 season was 0.78 points worse than that of an average 2006–2007 player. As another example, we found that the 2005–2006 season player strength was virtually indistinguishable from that of the 2006–2007 season.

BRIDGING ERAS IN SPORTS: A MORE SOPHISTICATED APPROACH

In a landmark paper (BRL), Berry, Reese, and Larkey (1999) analyzed the changes over time in abilities of Major League Baseball players, golfers, and National Hockey League players. Their major goal was to compare the abilities of players from different eras. They used 1996 as their "base" season and developed equations like (1) and (2) for their analysis, but also included for each player some additional terms to adjust for the influence of a player's age on the player's performance. In what follows we summarize their results.

AGING IN HOCKEY, GOLF, AND BASEBALL

BRL found that hockey players improve steadily in their ability to score points (points = assists + goals) until age 27 and then experience a sharp decline in ability. They found that golfers improve until ages 30–34 with there being little difference in a golfer's ability in the 30–34 age range. For baseball, BRL found that home run hitting ability increases until age 29 and then drops off. Batting ability peaks at age 27 and then drops off, but not as steeply as hockey ability drops off post-peak.

COMPARING THE ALL-TIME GREATS

BRL found that Mario Lemieux and Wayne Gretsky were the two greatest hockey players (non-goalies) of all time. BRL estimated that at their physical peak Mario Lemieux would be a 187-point player and Wayne Gretsky would be an 181-point player. The best "old timer" was found to be Gordie Howe, who was estimated to be a 119-point player. Again, these numbers are predictions for how the player would perform in the base year (1996).

For golf, BRL found Jack Nicklaus, Tom Watson, and Ben Hogan to be the three best golfers of all time. For example, BRL projected that in a Grand Slam tournament in 1996 Nicklaus (at his peak) would average 70.42 strokes per round, Watson (at his peak) would average 70.72 shots per round, and Hogan (at his peak) would average 71.12 shots per round. Tiger Woods was just beginning his legendary career, and BRL estimated that Woods in his peak would average 71.77 strokes per round in a 1996 Grand Slam tournament. In hindsight, it is clear that this underestimated Woods's abilities.

BRL estimated that the best all-time hitter was the legendary Ty Cobb. They estimated that if he were at his peak in 1996, he would have hit for a .368 batting average. Second on the list was Hall of Famer Tony Gwynn. If at his peak in 1996, Gwynn was estimated to be a .363 hitter. Finally, Mark McGwire was estimated to be the best home run hitter of all time. BRL estimated that if McGwire were at his peak in 1996, he would have averaged .104 home runs per at bat. Second on the list was Texas Ranger Juan Gonzales, who was projected to hit .098 home runs per at bat during 1996 if he had been at his peak. The legendary Babe Ruth was estimated to hit .094 home runs per at bat in 1996 if he were at his peak. BRL did their study before Barry Bonds's home run hitting really took off, so they estimated that, at his peak, Bonds would have hit only .079 home runs per at bat in 1996.

We strongly urge the reader to read BRL's wonderful paper. Their work shows how a sophisticated mathematical model can answer age-old questions such as, who is the greatest home run hitter of all time?

COMPARING ERAS: A SIMPLER APPROACH

Peterson, Penner, and Stanley (PPS 2011) utilize a simpler approach to compare baseball players. We all know the US Consumer Price Index (CPI) is based on comparing prices at various times to a base period (currently 1982–1984). In a similar fashion, PPS normalize a player's statistics using a base level equal to the average MLB performance during the years 1890–2010.

Their approach can best be illustrated with a simple example. Barry Bonds in 2001 hit the most HRs (73) in a season. In 1920 Babe Ruth hit 54 home runs. Which performance was more outstanding? PPS claim that Babe Ruth's 1920 HRs was the best performance ever. After detrending, Ruth's 1920 performance was equivalent to 128 HRs, while Bonds's 73 HRs in 2001 was equivalent (after detrending) to 47 HRs. Basically, PPS adjust a player's HR total for a season by multiplying actual home runs by:

Average HRs per plate appearance over MLB history (1890–2010)/ (Average HRs per plate appearance during the relevant season).

To illustrate the idea, suppose that average HRs per PA 1890–2010 = 0.019, and average HRs per PA during the relevant season is 0.038. In this case, we would halve the player's HRs because he played during a hitter-friendly season. PPS ignored players with fewer than 100 at bats (such as pitcher's) so our calculations below do not exactly agree with their results.

In 2001 HRs/PA in the National League was 0.03, and in the 1920 American League HRs/PA was 0.008. Also, during the years 1890–2010, average HRs/PA were 0.019. Therefore,

Adjusted Babe Ruth 1920 HRs = 54 * (0.019)/(0.008) ≈ 128 HRs
Adjusted Barry Bonds 2001 HRs = 73 * (0.019)/(0.03) ≈ 46 HRs.

Barry Bonds's 73 HRs do not even make PPS's top 50 performance, while Babe Ruth 1920 HRs is #1. PPS implicitly assume that players during different eras are of relatively equal ability, and this seems a pretty strong assumption. Daniel Eck argues that most ratings of MLB players over baseball's entire history tend to overrate players from earlier eras. We tend to agree with Eck on this point.

In closing, we note that even if we side with the PPS approach for baseball, this approach is probably not valid for the NFL and NBA due to rule changes that have greatly impacted the game. For example, in the NBA the three-point shot was introduced during the 1979–1980 season and in 2008 hand checking was limited (among other rule changes that greatly benefited the offense). In 1978 the NFL rule changes permitted a defender to maintain contact with a receiver within five yards of the line of scrimmage, but restricted contact beyond that point. The pass-blocking rule was interpreted to permit the extending of arms and open hands. These changes certainly made life easier for NFL passing offenses. In fact, the average yards passing per team jumped from 1,976 during the 1977 season to 2,541 yards during the 1978 season! Changing the kickoff yard line, moving the goalposts back, introducing the two-point conversion, and other changes have also had a major impact on the game. Many recent rule changes in the NFL, while being introduced for protecting players (e.g., changes in penalties for roughing the passer), have as a result made *life easier* for offenses. All these impact the performance we observe and can use to compare players.

CHAPTER 57

DOES FATIGUE MAKE COWARDS OF US ALL?

The Case of NBA Back-to-Back Games and NFL Bye Weeks

"Fatigue makes cowards of us all" is a famous anonymous quote popularized by the late, great Green Bay Packer coach Vince Lombardi. The idea, of course, is that if you are tired you cannot perform at peak performance level. In this chapter we use the following two different types of game situations to show that, contrary to what we may think, fatigue does not have a significantly deleterious impact on team performance:

- We show that NBA teams that play back-to-back games do perform significantly worse than expected during the second game of the back-to-back.
- Additionally, we show that the week after an NFL team has a bye or an open date, it does not perform significantly better than expected.

NBA BACK-TO-BACK GAMES

Typically, an NBA team will play between 10 and 20 games for which it also has played the night before. These games are called "back to backs." Usually back-to-back games are played in different cities. For example, the Mavericks might play Friday in Minnesota and Saturday have a home game in Dallas. The NBA never schedules a team to play three consecutive nights. The NBA extended the season by one week beginning in 2017–2018, so teams no longer must play four games in five days, as they had before.

One would think the combination of travel and having played a game the night before would lead to fatigue, causing inferior performance. Using the game scores for the 2018–2019 NBA regular season, we investigate the effects of back-to-back performances. Our work is in the file Chapter57.xlsx—Back to Back. For each game in which **only one of the teams** played a back-to-back game we predicted the outcome of the game based on Sagarin ratings and home court edge as:

Back-to-back margin prediction = (Rating of back-to-back team) – (Rating of non-back-to-back team) + (Home edge of +3 points if back-to-back team is home or –3 points if back-to-back team is away).

Then we define for each game:

Back-to-back residual = Actual margin for back-to-back team – Back-to-back prediction.

Historically, the standard deviation of NBA game margins about the predictions made using season-ending Sagarin ratings is around 12 points and the actual game margins follow a normal distribution. If back-to-back hindered the performance of the team, we would expect the average of the back-to-back residuals to be significantly negative. We can essentially perform a hypothesis test to see whether our evidence supports our null hypothesis (see Chapter 11):

H_0: back-to-back has no effect on performance

H_1: back-to-back has effect on performance.

Under the null hypothesis, each back-to-back game residual is normally distributed, with a mean of 0 and a standard deviation of 12. The average of n identically distributed random variables has a mean equal to the individual random variables and a standard deviation equal to (standard deviation of individual random variable)/$\sqrt{n}$. Hence, if there are n back-to-back games and back-to-back games have no effect, then the average of these residuals should follow a normal random variable with mean 0 and standard deviation $\dfrac{12}{\sqrt{n}}$.

During the 2018–2019 season there were 291 back-to-back games in which only one team faced a back-to-back. We found Sum of back-to-back residuals $=-551.1$. Thus, the teams facing a back-to-back in this sample performed on average $551.1/291 = 1.89$ points worse than expected. This corresponds to a z-score of $\dfrac{-1.89}{12/\sqrt{291}} = -2.69$. Hence, we can reject the null hypothesis that back-to-backs have no effect on the performance of the team. This result shows that it is abnormal for a team to perform worse than expected by 1.89 points in a back-to-back game.

DOES A BYE WEEK HELP OR HURT AN NFL TEAM?

During the NFL regular season, each team gets a week off, or a "bye" week. Public opinion has it that teams coming off of a bye week have an advantage over their opponent because the extra week of rest enables injured players to heal and gives the coaches an extra week to prepare for their opponent. We will see that this is more of an urban legend than a fact and we can investigate this question with a similar approach to the one used for NBA back-to-backs. Our data consists of every post-bye week NFL game during the 2014–2018 regular seasons. Of course, we eliminated games in which both teams were coming off a bye week. This left us (see file Chapter57.xlsx—NFL

Bye Week) with 136 games in which only one team had a bye. As before, we computed

$$\text{Bye team margin prediction} = (\text{Rating of bye team}) -$$
$$(\text{Rating of non-bye team}) + (\text{Home edge of} +3 \text{ if bye team is}$$
$$\text{home or} -3 \text{ points if bye team is away}).$$

Then we define for each game:

Bye team residual = Bye team actual margin − Bye team prediction.

Over the 136 games, the average bye team residual was 0.35 points, i.e., a team coming off of a bye week played 0.35 better than expected. We used again the end-of-season Sagarin ratings, which for NFL exhibit a standard deviation of 14 points around the actual point margin. Hence, bye week teams played $\dfrac{0.35}{14/\sqrt{136}} = 0.29$ standard deviations better than we would have expected if the bye week had no effect. This deviation is not outside of the norm. Therefore, we cannot conclude that after a bye week NFL teams do play significantly better than expected.

Another way to look at this is by looking at how coming off a bye week correlates with the win probability of the team. For every game and team, we can have an indicator variable on whether the team is coming off a bye week and what is the predicted point

TABLE 57.1

Coming off a bye week is not correlated
with win probability

Variable	Coefficient	p-value
Intercept	− 0.02	0.62
Bye (1)	− 0.02	0.89
Rating	0.13	< 0.0001

differential according to Sagarin ratings. We can input these in a logistic regression and see whether the "bye week" indicator variable is a significant predictor for the team that wins. Using data from the 2009 to 2016 seasons we obtained the results presented in Table 57-1. As we can see, this further supports our previous findings, since having a bye week does not correlate with winning once we adjust for the teams' Sagarin ratings and home field.

CHAPTER 58

THE COLLEGE FOOTBALL PLAYOFF

If you are a college football fan, you know that in 2014 the NCAA switched from relying on the two-team Bowl Championship Series (BCS) system to a four-team College Football Playoff. This chapter will discuss the former BCS system, the current CFP system, issues with the current CFP system, and finally which factors make a good team for playoff selection.

A BRIEF HISTORY OF THE BCS

Starting in 1997 teams were ranked using the following four factors:[1]

- subjective polls,
- computer rankings,
- strength of schedule, and
- team record.

In the first BCS championship game, on January 4, 1998, Tennessee defeated Florida State 23–16. In 2001 a "quality wins" factor that

---

1. For a complete history of BCS you can look at: https://en.wikipedia.org/wiki /Bowl_Championship_Series#History_leading_to_the_creation_and_dissolution _of_the_BCS

gave teams credit for defeating one of the top 15 ranked teams was added to the mix of factors. During the 2001 season Nebraska made the championship game and was clobbered 37–14 by Miami. Most observers felt that Nebraska undeservedly made the championship game because many lopsided Cornhusker wins against weak teams "padded" its computer rankings. Therefore, beginning with the 2002 season, BCS computer ranking systems were changed to exclude margin of victory from their algorithms. In 2004 the team records, strength of schedule, and quality wins were eliminated from the rankings, because the BCS believed the computer rankings already included these factors. So let's take a look at the current ranking system.

CFP RANKINGS

The rankings of the CFP are now created by a college selection committee. The CFP administration is responsible for appointing and managing the members of the selection committee. This administration is made up of one representative from each of the FBS conferences, and one representative for Notre Dame.

No "artificial intelligence" is involved, and the appointed selection committee determines the rankings based on subjective ("human intelligence") evaluations, rather than the statistics utilized by the BCS. At the end of the season, the top four teams in the CFP ranking are invited to compete in a playoff to determine the national champion. The playoff consists of two games. In the first round, the first and fourth ranked team play in one game and the second and third ranked teams play in another. The winners of these two games play in one final national championship game the following week to determine the official national champion.

ISSUES WITH THE CFP

When the CFP replaced the BCS, it solved many of the complaints that fans and teams had. However—as it happens every time when a new system is in place to solve a problem—with time, fans and teams are growing increasingly impatient with the current four-team playoff. In the 2017–2018 season, both the Big Ten champion (Ohio State) and the only undefeated team (University of Central Florida) were left out of the four-team playoff. The CFP committee instead selected two SEC teams (which eventually met in the championship). In the 2018–2019 season, the Big Ten champion (Ohio State) and the undefeated University of Central Florida were again left out of the four-team playoff selection. Additionally, many argued that a two-loss Georgia team was deserving of a chance at the championship due to its difficult strength of schedule. While the CFP has not officially discussed expansion of the playoff, many fans, coaches, and experts have suggested an eight-team playoff (of course, with time this will also cause similar discussions of team X being left out).

WHAT DETERMINES A GOOD TEAM?

Now that the selection for the playoff is up to a subjective group of voters rather than a computerized system incorporating the factors associated with the BCS, an important question for both the members of the committee as well as those in charge of determining the proper size for the CFP is, "What determines if a team is good?"

To better evaluate teams and identify which statistics determined if a team is good in college football, we ran a regression with numerous team statistics and the final Associated Press (AP) rank of the team. The College Football Playoff has been around only since 2014, and not enough data exists to perform a worthwhile regression. For this reason, we utilized the final AP rank (which has been around since 1968) and data from 2008–2016 from sports-reference.com.

We first ran three separate regressions, for defense, offense, and special teams, to determine relevant factors. We evaluated seven defensive statistics: tackles for loss per game, defensive RedZone touchdown percentage, defensive RedZone field goal percentage, points allowed per game, turnovers created, yards allowed per play, yards allowed per game. The only defensive factors that were significant (at the .05 level) were defensive RedZone touchdown percentage and points allowed. Next, we ran a regression to determine if five key special team statistics had impact on the AP final rank: field goal percentage, average kick return, average punt return, average kick return allowed, and average punt return allowed. No special team factors had significant predictive value. For the offensive regression, five key statistics were used: points scored per game, RedZone touchdown percentage, RedZone field goal percentage, time of possession per game, and turnovers lost. Points scored per game, time of possession per game, and turnovers lost all had significant value on final AP rank.

In order to form a holistic picture of how the relevant factors contribute to success, we performed a final regression that combines both offensive and defensive data. Additionally, a factor of whether or not a team was in a Power Five conference was added to account for difficulty of opponents (Notre Dame was included in a Power Five conference due to its association with the ACC). The results of this final regression are shown in Figure 58-1.

Complete Data Regression

Regression Statistics	
Multiple R	0.655687099
R Square	0.429925572
Adjusted R Square	0.41423545
Standard Error	5.531341397
Observations	225

ANOVA

	df	SS	MS	F	Significance F
Regression	6	5030.129193	838.3548655	27.40103459	2.88058E–24
Residual	218	6669.870807	30.59573765		
Total	224	11700			

	Coefficients	Standard Error	t Stat	P-value	Lower 95%	Upper 95%	Lower 95.0%	Upper 95.0%
Intercept	41.87687461	7.870361899	5.320832148	2.5579E–07	26.36513426	57.38861496	26.36513426	57.38861496
PPG	–0.635696423	0.065833943	–9.656058779	1.36113E–18	–0.76544891	–0.505943936	–0.76544891	–0.505943936
TOP/G	–704.7751082	281.8915326	–2.500164165	0.01315016	–1260.356709	–149.1935077	–1260.356709	–149.1935077
Redzone TD Defense %	–0.04922546	0.049899853	–0.986485077	0.324988746	–0.14757336	0.049122439	–0.14757336	0.049122439
Points Allowed/Game	0.645234289	0.09954453	6.481865826	5.96096E–10	0.449041416	0.841427161	0.449041416	0.841427161
Power Five?	–4.948638115	0.975112708	–5.074939619	8.29325E–07	–6.870493212	–3.026783017	–6.870493212	–3.026783017
Turnovers Lost	0.108833794	0.078490946	1.386577677	0.166986899	–0.045864452	0.263532039	–0.045864452	0.263532039

FIGURE 58.1 Linear Regression for College Football AP Rankings.

ORDINAL LOGISTIC REGRESSION

In the above example the dependent variable, i.e., the AP ranking of a team, while it can be considered to be numerical, in principle is an ordinal variable. Therefore, a more suitable model is to build an ordinal logistic regression (OLR). OLR can be thought of as a special case of multiclass logistic regression. In multiclass logistic regression, the dependent variable essentially takes values from a predefined set of labels. These labels do not have a specific ordering. In OLR we have again a set of *labels* but now there is an explicit ordering in the categories. For example, in the case of AP ranking, label "1" corresponds to the best team, while label "25" to the worst team (from those ranked), and hence, there is an explicit ordering that "1" > "25." In multiclass logistic regression, the log of odds that a specific label is observed is modeled as a linear combination of the independent variables. However, in this case the ordering of labels does not come into play when learning the model. OLR overcomes this limitation by using cumulative events according to the label ordering for the logs of the odds.

Similarly to any other regression problem, we have a set of features $\mathbf{x}$ and the learning algorithm will learn a coefficient vector $\mathbf{w}$ for each of the variables. However, the model will also learn a set of $K-1$ thresholds or intercepts (with K being the total number of ordered labels) such that $\theta_1 < \theta_2 < \cdots < \theta_{K-1}$. These intercepts provide a division of the linear response to K levels. The OLR model is then $\Pr(y \leq i \,|\, x) = \dfrac{1}{1 + e^{-(\theta_i - \mathbf{w} \cdot \mathbf{x})}}$. If we want to now get the probability of our variable getting a specific label, we can simply use $\Pr(y = i \,|\, x) = \Pr(y \leq i \,|\, x) - \Pr(y \leq i-1 \,|\, x)$.

So we can now build our OLR using the combined offensive/defensive data above. Figure 58-3 at the end of the chapter presents the full model. There are two tables; the first one includes the coefficients for the dependent variables, while the second one includes the 24 thresholds or intercepts. This model was trained using 75% of the data and evaluated on the remaining 25%. Similar multiclass problems are evaluated using the *top* N metric. Essentially, if the true class is within the top N-ranked probabilities output from the model, then this prediction is marked as correct. It should be obvious that the larger the value of N, the better the performance (appears to be). With $N = 3$, 14% of the test set samples are correctly predicted. If 14% seems low, keep in mind that there are 25 labels, which means a random assignment would be correct 4% of the time. With $N = 10$, the correct rank is included within the top N predictions in 72% of the test. Another way to evaluate the quality of the model is to obtain the expected rank for each test sample by taking the expectation of the corresponding probability distribution from the OLR. For example, if the predicted probability distribution for sample i is $p_{i,r}$ for rank $r \in \{1, 2, 3, \ldots, 25\}$, then the expected rank for this sample is $\sum_{r=1}^{25} r \cdot p_{ir}$. Figure 58-2 presents the expected rank as obtained from the model and the actual rank. As we can see, we cannot reject the hypothesis that the linear trend of the results is equal to the $y = x$ line, which means that the predicted rank is close to the actual rank. Of

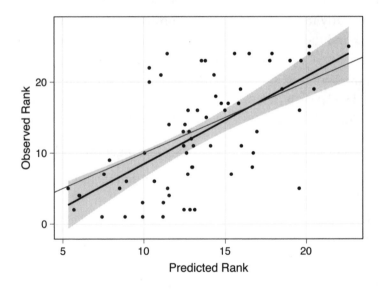

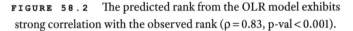

FIGURE 58.2 The predicted rank from the OLR model exhibits strong correlation with the observed rank ($\rho = 0.83$, p-val < 0.001).

A	Dependent Variable
	Rank
PPG	−0.179***
	(0.026)
TOP/G	−0.04
	(0002)
Redzone TD	
Defense %	−0.013
	(0.018)
PtsAllowed/G	0.213***
	(0.038)
Power5	−1.608***
	(0.404)
TOs Lost	0.053*
	(0.030)
Obserations	168

Note: *p<0.1; **p<0.05; ***p<0.001

B	Dependent Variable
	Rank
1|2	−7.66***
	(1.44)
2|3	−6.56***
	(1.40)
3|4	−6.05***
	(1.39)
4|5	−5.52***
	(1.37)
5|6	−5.24***
	(0.404)
6|7	−4.95***
	(0.03)
7|8	−4.74***
	(1.35)
8|9	−4.41***
	(1.34)
9|10	−4.22**
	(1.34)
10|11	−3.91***
	(1.35)
11|12	−3.75**
	(1.34)
12|13	−3.51**
	(1.33)
13|14	−3.26**
	(1.33)
14|15	−3.05**
	(1.34)
15|16	−2.87**
	(1.33)
16|17	−2.63**
	(1.32)
17|18	−2.32*
	(1.32)
18|19	−2.11
	(1.33)
19|20	−1.90
	(1.33)
20|21	−1.63
	(1.33)
21|22	−1.29
	(1.32)
22|23	−0.85
	(1.31)
23|24	−0.21
	(1.32)
24|25	0.44
	(1.35)
Obserations	168

Note: *p<0.1; **p<0.05; ***p<0.001

FIGURE 58.3 Model Summary for the AP Rank OLR.

course, there is nonnegligible variability, which can be attributed to several factors ranging from data issues (e.g., small sample size) to modeling issues (e.g., omitted variables) to underlying mechanism issues (e.g., AP committee is not consistent from year to year in its rankings).

CHAPTER 59

QUANTIFYING SPORTS COLLAPSES

Fans and media are always fascinated with big team collapses (or historic comebacks—if you remember that there are always two teams in a game). Some of these collapses come at crucial games and they are of course amplified. With 17 games left in the 2007 MLB season, the New York Mets held a seemingly comfortable seven-game lead in the NL East over the second-place Philadelphia Phillies. The Mets collapsed and the Phillies won the division. This collapse inspired Todd Behrendt to write an article ranking "all-time great sports collapses." The Warriors blew a 3–1 lead to the Cavaliers in the 2015–2016 finals. The Falcons blew a 28–3 lead to the Patriots in Super Bowl LI. The Mavericks blew a 30-point lead with 14 minutes left to the Raptors during the 2019–2020 NBA season. In this chapter, we will make some simple assumptions and then use basic probability to try and determine the probability of each collapse occurring. The "greatest collapse" can then emerge itself as the one with the smallest probability of occurring.

WHAT WAS THE PROBABILITY THAT THE WARRIORS WILL BLOW A 3–1 LEAD IN THE 2016 NBA FINALS?

In 2016 the Warriors set the record for most wins during the NBA regular season with 73! They were the heavy favorite for the back-to-back championship. They were also leading the finals series with

TABLE 59.1

Win Probability for the Warriors during the 2015– 2016
NBA Finals

Game	Betting Line	Warriors Win Probability
5 (home)	8	0.75
6 (away)	2.5	0.58
7 (home)	5	0.66

3–1. But then they lost the ring. What was the chance of the Cavaliers making a comeback? According to betting markets, the Warriors were favored in games 5, 6, and 7 by 8, 2.5, and 5 points, respectively. As we discussed in Chapter 47, the final point margin of a game follows a normal distribution with mean equal to the predicted point margin (the betting line) and a standard deviation equal to 12 points. Hence, we can calculate for every game the win probability for the Warriors according to the betting markets (Table 59-1).

Under the assumption that the results of the games are independent, the probability that the Warriors lose the finals series is equal to

$$1 - \Pr[\text{Cavs win series}] = 1 - ((1 - 0.75) * (1 - 0.58) * \\ (1 - 0.66)) = 0.035.$$

So the probability of the Warriors losing the championship was small but not astronomically small, as it might have "felt" to people, possibly due to the dominant season they had that year.

HOW PROBABLE WAS THE PATRIOTS' COMEBACK AT SUPER BOWL LI?

In Super Bowl LI the Falcons had a 28–3 lead with 8:31 left in the third quarter (22 minutes and 31 seconds for the game). It seemed all but certain that the Falcons would win their first Super Bowl, until they didn't. What was the probability of this happening? As stated in

Chapter 47, the outcome of an NFL game follows a normal distribution with mean the predicted point margin and standard deviation 14 points. Betting markets had New England a three-point favorite during that game. Since we are interested in a smaller portion of the game (and not the whole game), the standard deviation of the expected point margin during that time period will be different. In particular, the standard deviation during an n-minute portion of a game is given by:

$$\frac{\text{game stDev of margin}}{\sqrt{\text{game duration}/\text{n}}}. \tag{1}$$

Thus, for 22 minutes and 30 seconds in an NFL game, the standard deviation of the point margin is $\dfrac{14}{\sqrt{60/22.5}} = 8.5$. Similarly, the Patriots were a three-point favorite over 60 minutes, which translates to a $3 * (22.5/60) = 1.13$ points favorite for the 22.5 minutes left. Then the chance that the Patriots would win this game is given by:

NORMDIST(−25.5, −1.13,8.5, TRUE) + 0.5 * (NORMDIST (−24.5, −1.13,8.5, TRUE) − NORMDIST(−25.5, −1.13,8.5, TRUE)).

The first term gives the probability that the Falcons are outscored in the rest of the game by 25.5 points or more, and hence they lose the game in regulation. The second term provides the probability that the Falcons are outscored by 25 points during the rest of the game, and hence they have a 50% chance of losing the game in OT. Calculating the above probability gives a 0.15% probability or a 1 in 666!

HOW DID WE GET EQUATION (1)?

We can express the final point of margin in a game as the sum of the points of margin during each minute in the game. Assuming a 60-minute game (NFL), we have:

Final margin = margin for minute 1 + margin for minute 2 + · · · · + margin for minute 60.

> Under independence assumption (possibly a strong assumption), the variance of the sum is the sum of the individual variances. So we have: 60 $*$ (variance for a minute) $= 14^2$, or the standard deviation for one minute is $\dfrac{14}{\sqrt{60}}$. Then the variance of point margin for an n-minute period is: $n \cdot \left(\dfrac{14}{\sqrt{60}}\right)^2$. Finally, this gives the standard deviation of the margin for n minutes: $\dfrac{14}{\sqrt{60/n}}$.

As a bonus let's see what happened in Super Bowl LIV. Kyle Shanahan was the offensive coordinator for the Falcons in Super Bowl LI, and after that game he moved on to be the head coach for the 49ers. In Super Bowl LIV, with 8:33 minutes left in the game, Shanahan seemed poised to finally get the Lombardi trophy; the 49ers were up 10 points before punting the ball back to the Chiefs. However, they lost. The chance of this happening based on the above method was (the Chiefs were a 1.5-point favorite):

$$\text{NORMDIST}(-10.5, -0.21, 5.2, \text{TRUE}) + 0.5 *$$
$$(\text{NORMDIST } (-9.5, -0.21, 5.2, \text{TRUE}) -$$
$$\text{NORMDIST}(-10.5, -0.21, 5.2, \text{TRUE})) = 3\%.$$

Not quite the Super Bowl LI collapse—but quite the same as the Warriors blowing a 3–1 lead!

HOW BIG WAS THE MAVERICKS' COLLAPSE IN TORONTO IN THE 2019–2020 SEASON?

In the 2019–2020 season the Mavericks were visiting the Raptors in Toronto and they held a 30-point lead with 14-and-a-half minutes left in the game. This game will stay forever in the history of the Mavericks as the biggest lead blown by the team (until the time of writing at least). Before the game, the Raptors were a 1.5-points favorite.[1] Using

---

1. 1.5 $*$ (14.5/48) $= 0.45$ points for the remaining 14.5 minutes.

the same approach as above, we can find that the standard deviation for the point margin for the final 14:30 minutes is $\frac{12}{\sqrt{48/14.5}}=6.6$. Hence, the probability of the Mavericks losing this game is:

$$NORMDIST(-30.5, -0.45, 6.6, TRUE) + 0.5 *$$
$$(NORMD\ IST(-29.5, -0.45, 6.6, TRUE)$$
$$-NORMDIST(-30.5, -0.45, 6.6, TRUE)) = 0.0004\%, \text{ or } 1 \text{ in } 2{,}500!$$

IN-GAME WIN EXPECTANCY

Another way to evaluate sports collapses or comebacks is through in-game win probability models. These are models similar to the win expectancy models for baseball that we introduced in Chapter 8. For example, Table 59-2 provides the odds from various in-game win probability models during Super Bowl LI.

TABLE 59.2
Chance of a Comeback during Super Bowl LI

Model	Chance of Comeback
ESPN	1 in 500
nsflscrapR	1 in 100
PFR	1 in 1,000
Gambletron	1 in 20
iWinRNFL[1]	1 in 50

1. K. Pelechrinis, "iWinRNFL: A simple, interpretable & well-calibrated in-game win probability model for NFL," arXiv preprint arXiv:1704.00197, 2017

These models can account for various things that can impact the win probability, such as personnel available for the team, timeouts left, etc. Furthermore, they can be a bit more adaptive to changes in the game (assuming they are trained frequently)—e.g., big leads in the NBA are not as safe anymore, as they were a decade ago, due to

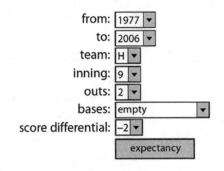

from: 1977 ▼
to: 2006 ▼
team: H ▼
inning: 9 ▼
outs: 2 ▼
bases: empty ▼
score differential: –2 ▼
expectancy

games: 3033
home won: 41
expectancy: 0.014

FIGURE 59.1   The Chance the Mets Come Back
to Win Game 6 of the 1986 World Series.

changes in the pace and the long ball. One thing, though, that we can
see is that these models can give vastly different predictions. How
do we know which one is the *correct* one? We cannot be certain.
After all, "all models are wrong but some are useful." Mike Lopez,
the director of data and analytics in the NFL, has discussed this topic
extensively in an article on his website[2] (from which we also got the
numbers in Table 59-2). Using the win expectancy model we introduced in Chapter 8, let's see some of baseball's collapses.

WHAT WERE THE CHANCES THE METS WOULD WIN GAME 6
OF THE 1986 WORLD SERIES BEFORE BUCKNER'S ERROR?

Figure 59-1 shows that by using the Win Expectancy Finder, with
two outs and nobody on in the bottom of the ninth inning, the Mets
had about a 1.4% chance of winning the game. We used the bottom

2. https://statsbylopez.com/2017/03/08/all-win-probability-models-are
-wrong-some-are-useful/

of the ninth inning even though the Mets were batting in the bottom of the 10th inning, because a team batting in the bottom of the ninth has the same chance of winning as a team batting in the bottom of the tenth.

WHAT WAS THE PROBABILITY THAT THE ATHLETICS WOULD COME BACK FROM BEING DOWN 8–0 TO BEAT THE CUBS DURING GAME 4 OF THE 1929 WORLD SERIES?

Again we consult the Win Expectancy Finder. Figure 59-2 shows that with a seven-run deficit at the bottom of the seventh inning, there are roughly 6 chances in 1,000 of winning the game. Of course, an eight-run deficit would have a slightly smaller chance of being erased. The Win Expectancy Finder found 415 games with the same situation and an eight-run deficit. No team came back to win the game. Can we say that the probability of a comeback for an eight-run deficit is 0? Even with large sample sizes that imply a zero-probability event, we cannot be sure that the event is impossible to happen. For

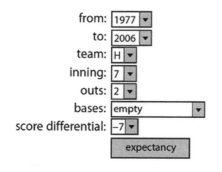

from: 1977 ▼
to: 2006 ▼
team: H ▼
inning: 7 ▼
outs: 2 ▼
bases: empty ▼
score differential: –7 ▼
expectancy

games: 3053
home won: 17
expectancy: 0.006

FIGURE 59.2 The Chance the Cubs Blow an 8–0 Lead in Game 4 of the 1929 World Series.

this reason, Laplace suggested that we can smooth the probability by adding a pseudocount of events. In this case a comeback from an eight-run deficit has a chance of $(0+1)/(415+2) = 0.002$ or 2 in 1,000! (Remember Faber College from Chapter 46?)

LAPLACE SMOOTHING

This technique is attributed to Laplace, who allegedly came up with it when he was trying to estimate the probability that the sun will rise the next day. His rationale was that even with a large sample size, we cannot be completely sure that the sun will rise again tomorrow. To solve this problem he suggested the use of a **pseudocount**. In general, if we have observed N instances of a binary event (Yes/No), with y positive instances and n negative instances, the empirical probability of a positive instance is y/N. If we want to perform Laplace smoothing on this estimate, we will have $(y+\alpha)/(N+2\alpha)$. If we choose $\alpha=1$, we get the **add-one smoothing**, $(y+1)/(N+2)$. This equation resembles the Bayesian average (see Chapter 32), where the *prior* is two observations with one success.

CHAPTER 60

DAILY FANTASY SPORTS

Daily fantasy sports (DFS) is a relatively recent development within the broader set of fantasy sports games, gaining popularity over the past decade with the advent of sites like DraftKings and FanDuel. Those who play DFS games compete with many others, or go head-to-head with one opponent by building a team of professional athletes, while staying under a salary cap. Each player has a cost or "salary," with more popular and better performing athletes requiring greater expenditure. The measure of success is fantasy points (FPs), in which desirable player actions or outcomes (e.g., rushing yards, pass completions, and touchdowns in football) receive positive FPs, while undesirable results (like interceptions or fumbles, sticking with football as an example) receive negative FPs. Scoring systems vary across sports, but generally follow this concept. A DFS contest may focus on a single day, or it could include all games played in a single week of competition. One key differentiator of traditional fantasy sports game like rotisserie baseball is that each time you enter a contest, it is for a fixed period much shorter than the entire season.

HOW CAN WE BUILD THE BEST
(OR OPTIMAL) FANTASY SPORTS TEAM?

Let's say that we are interested in building a fantasy basketball team. The contest rules specify that a team must consist of a total of eight basketball players, as specified below:

- One point guard (PG)
- One shooting guard (SG)
- One small forward (SF)
- One power forward (PF)
- One center (C)
- One guard (G) [can be either a PG or an SG]
- One forward (F) [can be either an SF or a PF]
- One utility (U) [can be any of the five real positions—PG, SG, SF, PF, C].

Note that the last three listed positions are "synthetic" or "flex" positions, allowing the player some flexibility in team composition, but within reasonable limits. Additionally, the roster must include players from at least two different NBA teams (you can't just pick the starters and key bench players from your favorite team!) and must represent at least two real-life NBA games.

If we think about this from an optimization (or prescriptive analytics) perspective, this problem is very similar to the knapsack problem, one of the first models taught to those learning mathematical programming. The general idea is to get as much value from the objects you put in a knapsack (or backpack), without exceeding its carrying capacity in terms of weight. In this case, we want to put the right number of basketball players (meeting the criteria described above) on our team, to give us the highest expected FPs, while sticking within the salary cap or budget.

For each player i, we have a cost (salary) c_i and an FP projection p_i. Our decision is relatively straightforward—which eight players do we include on our team? But it is better to think of it as a separate but related (yes or no) decision for each individual player, as this is the way we will model the decisions. We will create a binary decision variable x_i for each player, which will take on the value 1 if player i is selected for our team and 0 if he is not. Recall that our goal (or objective) is to maximize the total number of FPs with the players we select, which we represent as Max $\sum_i p_i x_i$. The projected FPs for a player only get added to the total if that player is selected for our team. We know that we cannot spend more than our budget B, so this means that the sum of all the costs for our selected players must be less than this, or $\sum_i c_i x_i \leq B$. Next, we must make sure that we select the right number of players. We start by making sure that the total number of players is exactly eight, meaning $\sum_i x_i = 8$. The first five are all relatively easy to figure out, and can be expressed as: $\sum_{i \in PG} x_i = 1, \sum_{i \in SG} x_i = 1, \sum_{i \in SF} x_i = 1, \sum_{i \in PF} x_i = 1, \sum_{i \in C} x_i = 1$. That's a good start, but now we must think about the synthetic positions, which are a little trickier. First, for the guard (G), the player can be either a PG or an SG. So, we will add two binary decision variables, g_{PG} and g_{SG}. Each will be equal to 1 if we select a player at that position to be the additional guard. But clearly, we can only select one of them, so we must add the constraint $g_{PG} + g_{SG} = 1$. Similarly, we add binary decision variables f_{SF} and f_{PF} and the constraint $f_{SF} + f_{PF} = 1$ for the forward (F) position. Likewise, for the utility (U) position, we can add five binary decision variables u_{PG}, u_{SG}, u_{SF}, u_{PF}, and u_C and the constraint $u_{PG} + u_{SG} + u_{SF} + u_{PF} + u_C = 1$. We're almost there, except for those earlier constraints where we limited selections to one PG, one SG, one SF, one PF, and one C. Clearly, we need to modify those to allow three additional players to be selected. So, these constraints become $\sum_{i \in PG} x_i = 1 + g_{PG} + u_{PG}, \sum_{i \in SG} x_i = 1 + g_{SG} + u_{SG}, \sum_{i \in SF} x_i = 1 + g_{SF} + u_{SF}, \sum_{i \in PF} x_i = 1 + g_{PF} + u_{PF}, \sum_{i \in C} x_i = 1 + u_C$. There are alternate formulations that are a bit more concise, but this is likely the most straightforward expression for those not well versed in optimization. That's enough to get us started. We can deal with the restrictions related to teams and games later, if we need to do so.

The overall formulation thus appears as follows.

Objective function

$$\text{Max} \sum_i p_i x_i$$

subject to constraints

$$\sum_i c_i x_i \leq B \text{ (budget or salary cap)}$$
$$\sum_i x_i = 8 \text{ (total players)}$$
$$\sum_{i \in PG} x_i = 1 + g_{PG} + u_{PG} \text{ (point guard position)}$$
$$\sum_{i \in SG} x_i = 1 + g_{SG} + u_{SG} \text{ (shooting guard position)}$$
$$\sum_{i \in SF} x_i = 1 + g_{SF} + u_{SF} \text{ (strong forward position)}$$
$$\sum_{i \in PF} x_i = 1 + g_{PF} + u_{PF} \text{ (power forward position)}$$
$$\sum_{i \in C} x_i = 1 + u_C \text{ (center position)}$$
$$g_{PG} + g_{SG} = 1 \text{ (extra guard position)}$$
$$f_{SF} + f_{PF} = 1 \text{ (extra forward position)}$$
$$u_{PG} + u_{SG} + u_{SF} + u_{PF} + u_C = 1 \text{ (utility position)}$$

with binary (0–1) variables:

$$x_i, \forall i \text{ (one for each player)}$$

$g_{PG}, g_{SG}, f_{SF}, f_{PF}, u_{PG}, u_{SG}, u_{SF}, u_{PF}, u_C$ (for the synthetic positions).

SO HOW DO WE DO THIS WITH SOME DATA?

On a typical day in the NBA, there are between three and 12 games. On February 1, 2019, there were five games; the dataset we use here includes 97 players from the 10 teams that played on that date. The data available for the first 10 athletes on the Data_01FEB2019 tab in Chapter60.xlsx is shown here (Figure 60-1).

You can see that, in addition to the salary for each athlete, you can find out which team the athletes play for (and who on their team is playing), the position(s) at which they are eligible to be selected,

	A	B	C	D	E	F	G	H
1	PlayerName	Salary($K)	Team	Position	Opp	SalaryChange	MinPerGame	ProjectedPoints
2	Mario Hezonja	4.2	NYK	SF	BOS	−200	18.9	27.57
3	Kyle Anderson	4.4	MEM	PG	@ CHA	−1200	29.8	27.99
4	Kadeem Allen	3.3	NYK	PG	BOS	−200	17.9	20.64
5	Paul Millsap	5.0	DEN	PF	HOU	−400	25.7	30.19
6	Bruno Caboclo	3.4	MEM	SF	@ CHA	−500	22.9	20.4
7	Wayne Ellington	3.6	MIA	SG	OKC	−700	22	20.87
8	Monte Morris	5.1	DEN	PG	HOU	600	24.4	29.49
9	Tyler Johnson	4.2	MIA	PG	OKC	−1100	25.6	23.9
10	Gary Harris	5.3	DEN	SG	HOU	−200	29.9	29.14
11	Justin Holiday	4.1	MEM	SG	@ CHA	−200	32.5	22.4

FIGURE 60.1 Data for Daily Fantasy Basketball Contest on February 1, 2019.

and the projected (expected) number of FPs they are expected to score in this game, as computed by some predictive model. Note that in this simple example, each player is only eligible for one position; in real DFS competitions, many players are eligible at two positions, which can complicate things a bit. More advanced players may develop their own model to determine the projected FPs; beginners usually download projections from one of numerous websites that provide them. Note that players with higher projected points (column H) generally command a higher salary (column B), but this is not always the case. If you scroll through the data file, you will see that the player with the highest salary is James Harden ($12.9K), who is projected to score 60.46 FPs.

So now we have to build the optimization model discussed earlier. Shown on the Model1 tab is an implementation of this set of decision variables, constraints, and an objective function. Columns A through F just contain the data as before, except we have renamed column B from Salary to c_i, representing the cost c_i. Likewise, we have renamed column F from ProjectedFP to p_i representing the projection p_i. Column G has the binary decision variables x_i, one for each player. We need a way to represent the position at which each player is eligible to be selected; columns H through L contain a 1 if the player is eligible to play that position, and 0 otherwise. Cells H2 through L98 were initially filled in using a simple IF function like $= \mathrm{IF}(\$D3 = H\$2,1,0)$ in the case of cell H3. However, to avoid issues with Excel Solver, we then issued a "Copy, Paste Special, Values" command over this entire range, leaving just the appropriate numeric value.

Moving on across the spreadsheet, columns M through Q indicate when a player has been selected at a particular position. For example, as shown in Figure 60-2, a 1 in cell O3 would indicate that Mario Hezonja was selected for our team as an SF. This was computed by multiplying the binary variable x_i in cell G3 by the indicator in cell J3, using the formula $= \$G3 * J3$. Column R similarly computes the projected FPs for selected players by multiplying p_i times x_i with $= F3 * G3$. For example, if Hezonja is selected, he would add 27.57 FPs to the team.

	A	B	C	D	E	F	G	H	I	J	K	L	M	N	O	P	Q	R
1								Eligible					Selected					
2	PlayerName	c_i	Team	Position	Opp	p_i	x_i	PG	SG	SF	PF	C	PG	SG	SF	PF	C	FPs
3	Mario Hezonja	4.2	NYK	SF	BOS	27.57	1	0	0	1	0	0	0	0	1	0	0	27.57
4	Kyle Anderson	4.4	MEM	PG	@ CHA	27.99	0	1	0	0	0	0	0	0	0	0	0	0
5	Kadeem Allen	3.3	NYK	PG	BOS	20.64	0	1	0	0	0	0	0	0	0	0	0	0
6	Paul Millsap	5	DEN	PF	HOU	30.19	0	0	0	0	1	0	0	0	0	0	0	0
7	Bruno Caboclo	3.4	MEM	SF	@ CHA	20.4	0	0	0	1	0	0	0	0	0	0	0	0

FIGURE 60.2   Computed Data Fields for Daily Fantasy Basketball
Contest on February 1, 2019.

Moving down to the bottom of the sheet, in rows 100 through 107, as illustrated in Figure 60-3, we find the rest of the model. Row 100 contains the total number of players selected at each of the five real positions, the expected FPs (our objective that we want to maximize), and information about the salary cap. Row 101 limits the total number of players selected to eight, by setting cell G101, which is SUM(G3:G99), equal to I101. Rows 102 through 104 contain the other binary decision variables for the synthetic positions, along with constraints that allow us to only select one G, one F, and one U. Rows 106 and 107 contain the lower and upper bounds for the total number of players selected at each of the five real positions.

	G	H	I	J	K	L	M	N	O	P	Q	R	S	T	U	V
99	0	0	0	0	1	0	0	0	0	0	0	0	0			**Salary Cap**
100	**Total**						2	2	1	1	1	277.9	50	<=	50	**Budget**
101	8	=	8													
102	**Guard**						0	1				1	=	1		
103	**Forward**							1	0			1	=	1		
104	**Utility**						1	0	0	0	0	1	=	1		
105																
106	**Lower**						1	1	1	1	1					
107	**Upper**						2	2	2	1	2					

FIGURE 60.3   Solver Model for Daily Fantasy Basketball Contest on February 1, 2019.

TABLE 60.1

Optimal Team for Daily Fantasy Basketball Contest on February 1, 2019

Player	Team	Opponent	Position	Cost	Expected FPs
Mario Hezonja	NYK	BOS	SF	4.2	27.57
Kyle Anderson	MEM	@CHA	PG	4.4	27.99
Paul Millsap	DEN	HOU	PF	5.0	30.19
Wayne Ellington	MIA	OKC	SG	3.6	20.87
Justin Holiday	MEM	@CHA	SG	4.1	22.40
Kenneth Faried	HOU	@DEN	PF	6.9	36.58
Nikola Jokic	DEN	HOU	C	10.8	57.37
Russell Westbrook	OKC	@MIA	PG	11.0	55.94
Total				50.0	277.9

As you can see, we selected eight players total—two PGs, two SGs, one SF, two PFs, and one C, as indicated in cells M100:Q100. The total expected number of FPs is 277.9, as shown in cell R100, and we have exhausted the entire salary cap or budget of $50K, as indicated in cell S100. The players selected are shown in Table 60-1. This is the "best" we can do with this model.

Although we did not explicitly model the requirement to select players from two or more teams and two or more different games, it turns out that we don't have to worry about it in this case. Sometimes that is a way to make finding an optimal answer easier. Had either of those requirements not been met, we could have added additional constraints, but it would likely have made the model larger than what is possible with the free version of Solver that comes with Excel (limits of 200 variables and 100 constraints), requiring an add-in solver engine.

This is indeed a feasible (meets all constraints) and optimal (maximizes expected FPs) solution. However, if you were to enter this lineup in a DFS contest, it would likely not win you any prize money. Why not? Recall that another word for "expected value" is average. In other words, you maximized the average performance of your fantasy team of eight NBA players. But you are competing against hundreds or even many thousands of other participants who are likely using more sophisticated models.

HOW CAN WE IMPROVE OUR PROBABILITY OF WINNING?

For starters, we probably need to change the objective function. Instead of maximizing the expected number of FPs, we probably want to maximize the upper end of our selected athletes' performance. To do this, we need some additional information about the variability in their performance, not just an average or expected performance. This could come in several different ways; it could be in a familiar statistical measure like variance or standard deviation, but it is often shown as a floor and ceiling on their performance. Variability is not always symmetric, and it is desirable to select players who have relatively high floors (low downside) and also high ceilings (upside), as shown in Figure 60-4.

Modeling options (from simple to more complex) could include the following. The first three are shown in modified models in the Chapter60.xlsx file.

- Maximizing the ceiling (straightforward modification of the original model). This is shown in the Model2 worksheet; the

	A	B	C	D	E	F	G
1	**PlayerName**	**Salary($K)**	**Team**	**Position**	**Opp**	**SalaryChange**	**MinPerGame**
2	Mario Hezonja	4.2	NYK	SF	BOS	−200	18.9
3	Kyle Anderson	4.4	MEM	PG	@ CHA	−1200	29.8
4	Kadeem Allen	3.3	NYK	PG	BOS	−200	17.9
5	Paul Millsap	5	DEN	PF	HOU	−400	25.7
6	Bruno Caboclo	3.4	MEM	SF	@ CHA	−500	22.9
7	Wayne Ellington	3.6	MIA	SG	OKC	−700	22
8	Monte Morris	5.1	DEN	PG	HOU	600	24.4
9	Tyler Johnson	4.2	MIA	PG	OKC	−1100	25.6
10	Gary Harris	5.3	DEN	SG	HOU	−200	29.9
11	Justin Holiday	4.1	MEM	SG	@ CHA	−200	32.5

FIGURE 60.4 Additional Data Fields for Daily Fantasy Basketball Contest on February 1, 2019.

players chosen differ dramatically. This is a MaxiMax type formulation. Should the eight players selected all perform well that day, you could do very, very well.

- Maximizing the floor (also a straightforward modification of the original model). This is shown in the Model3 worksheet; once again, the players selected are quite different, as the model seeks to "maximize the minimum" (MaxiMin) number of points scored. This one probably won't score very well, as it is very conservative.

- Maximizing some linear combination of the floor and ceiling (slightly more challenging, but not too difficult). The Model4 worksheet shows an example that includes a 60% weight on the floor and 40% weight on the ceiling, with =(X3 ∗ F3 + Y3 ∗ G3) ∗ I3 in cell T3 for Hezonja's "combined" FPs. Cell X3 contains the weight for the ceiling and cell Y3 contains the weight for the floor. These should sum to 1. Once again, a different solution is obtained. One can adjust the weights as desired.

- Maximizing the probability of a positive payout.

We will briefly discuss this last option, to provide an example of how this could be modeled. Consider an example DFS contest that has 11,000 entries and pays out to the top 2,235 (20%) players. In other words, we have to beat 80% of our competitors to win. If we assume that their scores are normally distributed with an average of 260 and a standard deviation of 30, we can compute =NORMSINV(.80) = 0.84. Then, a positive payout would occur with $260 + (0.84) * 30 = 285$ points. So, to maximize $1 - NORMSDIST(285,260,30,1)$ or other complex function, we need a better solver than the one provided with Excel. We could use something like Analytic Solver or Open Solver (www.opensolver.com) to take on this challenge.

BIBLIOGRAPHY

Adler, J., *Baseball Hacks*, O'Reilly Media, 2006

Agonistes, Dan, "A brief history of run estimation: Runs created," http:// danagonistes.blogspot.com/2004/10/brief-history-of-run-estimation -runs.html

Albright, Samuel Christian, "A statistical analysis of hitting streaks in baseball," *Journal of the American Statistical Association*, 88(424), 1993, 1175–1183

Annis, D. H., "Optimal end-game strategy in basketball," *Journal of Quantitative Analysis in Sports*, 2(2), 2006

Bar-Eli, M., Azar, O. H., Ritov, I., Keidar-Levin, Y., and Schein, G., "Action bias among elite soccer goalkeepers: The case of penalty kicks," *Journal of Economic Psychology*, 28(5), 2007, 606–621

Bats and Stats, "Run Expectancy Matrix," https://batsandstats.com /2016/08/03/one-of-my-favorite-baseball-stats/

Bellman, R., *Dynamic Programming*, Princeton University Press, 1957

Berri, David J., et al., *The Wages of Wins: Taking Measure of the Many Myths in Modern Sport*, Stanford Business Books, 2007

Berry, Scott, Reese, Shane, and Larkey, Patrick, "Bridging different eras in sports," *Journal of the American Statistical Association*, 94, 1999, 661–676

Bialik, C., "Should the outcome of a coin flip mean so much in NFL overtime?" available at http://faculty.haas.berkeley.edu/hender/ot_wsj .pdf

Billie, et al., "Simulating rare baseball events using Monte Carlo methods in Excel and R," http://archives.math.utk.edu/ICTCM/VOL22/S096 /paper.pdf

Birnbaum, P., "Do NFL teams overvalue high draft picks?" available at http://blog.philbirnbaum.com/2006/12/do-nfl-teams-overvalue-high-draft-picks.html

Birnbaum, Phillip, "Are NFL coaches too conservative on fourth down?" Sabermetric Research (2007), http://sabermetricresearch.blogspot.com/2007/01/are-nfl-coaches-too-conservative-on.html

Bocskocsky, Andrew, Ezekowitz, John, and Stein, Carolyn, "Heat check: New evidence on the hot hand in basketball," available at https://papers.ssrn.com/sol3/papers.cfm?abstract_id=2481494

Boswell, Thomas, Palmer, Pete, and Thorn, John, *Total Baseball*, Warner Books, 1989

Braess, D., Nagurney, A., and Wakolbinger, T., "On a paradox of traffic planning", *Transportation Science*, 39(4), November 2005

Brams, S. J., *Mathematics and Democracy: Designing Better Voting and Fair-Division Procedures*, Princeton University Press, 2007

Broadie, Mark, *Every Shot Counts*, Avery, 2014

Burke, B., "Jimmy Garoppolo's edge on the rookie QBs," available at http://www.espn.com/nfl/story/_/id/18741560/jimmy-garoppolo-edge-rookie-qbs

Burke, Brian, "Expected points and expected points added explained" (2014), http://www.advancedfootballanalytics.com/index.php/home/stats/stats-explained/expected-points-and-epa-explained

Cabot, Victor, Sagarin, Jeff, and Winston, Wayne, "A stochastic game model of football," unpublished paper, 1981

Carter, Virgil, and Machol, Robert, "Operations research on football," *Operations Research*, 19(2), 1971, 541–544

Chang, Y. H., Maheswaran, R., Su, J., Kwok, S., Levy, T., Wexler, A., and Squire, K., "Quantifying shot quality in the NBA." In *Proceedings of the 8th Annual MIT Sloan Sports Analytics Conference*, Boston, 2014

Chiappori, P.-A., Levitt, Steven, and Groseclose, Timothy, "Testing mixed-strategy equilibria when players are heterogeneous: The case of penalty kicks in soccer," *American Economic Review*, 92(4), 2002, 1138–1151

Cook, Earnshaw, *Percentage Baseball*, M.I.T. Press, 1966

Davis, N., and Lopez, M., "NHL coaches are pulling goalies earlier than ever," available at https://fivethirtyeight.com/features/nhl-coaches-are-pulling-goalies-earlier-than-ever/

Decroos, T., Bransen, L., Van Haaren, J., and Davis, J. "Actions speak louder than goals: Valuing player actions in soccer." In *Proceedings of the 25th ACM SIGKDD*, 2019, pp. 1851–1861.

Dewan, John, *The Fielding Bible*, Acta Sports, 2006

Donohue, B. "Post-combine thank you and congratulations from Brendan Donohue," available at https://2kleague.nba.com/news/post-combine-thank-you-and-congratulations-from-brendan-donohue/

Dreslough, Clay, "DICE: A new pitching stat," available at https://web.archive.org/web/20070528164743/http://www.sportsmogul.com/content/dice.htmhttps://captaincalculator.com/sports/baseball/defense-independent-component-era-calculator/

Drinen, D., "Approximate value," available at https://www.pro-football-reference.com/blog/index37a8.html

Dubin, J., "Want to confuse an NBA defense? Have a guard set a ball screen," available at https://fivethirtyeight.com/features/want-to-confuse-an-nba-defense-have-a-guard-set-a-ball-screen/

Easley, D. and Kleinberg, J., *Networks, Crowds and Markets: Reasoning about a Highly Connected World*, Cambridge University Press, 2010

Egidi, L. and Ntzoufras I., "Modelling Volleyball data Using a Bayesian Approach." In Math Sport International Conference, 2019

Elam, N. "An examination of the effectiveness of the Elam ending at TBT2018 and TBT2019." *Sports Innovation Journal*, 1, 2020, 120–133

ESPN, "A guide to NFL FPI," available at http://www.espn.com/blog/statsinfo/post/_/id/123048/a-guide-to-nfl-fpi

ESPN, "Warriors draft first woman into NBA 2K league," available at https://www.espn.com/esports/story/_/id/26146486/warriors-draft-first-woman-nba-2k-league

ESPN, "Upset special: Greece stuns US in FIBA semis," available at https://www.espn.com/olympics/wbc2006/news/story?id=2568543

ESPN Analytics, "NFL pass-blocking and pass-rushing rankings", available at https://www.espn.com/nfl/story/_/id/27584726/nfl-pass-blocking-pass-rushing-rankings-2019-pbwr-prwr-leaderboard

Ezekowitz, J., "Intentionally fouling DeAndre Jordan is futile," available at https://fivethirtyeight.com/features/intentionally-fouling-deandre-jordan-is-futile/

Ezekowitz, J., "Up three, time running out, do we foul? The first comprehensive CBB analysis," available at https://harvardsportsanalysis

.wordpress.com/2010/08/24/intentionally-fouling-up-3-points-the-first
-comprehensive-cbb-analysis/

Fairchild, A., Pelechrinis, K., and Kokkodis, M. "Spatial analysis of shots in MLS: A model for expected goals and fractal dimensionality," *Journal of Sports Analytics*, 4(3), 2018, 165–174

Fernandez, J., and Bornn, L. "Wide open spaces: A statistical technique for measuring space creation in professional soccer." In MIT Sloan Sports Analytics Conference, 2018

Fewell, J. H., Armbruster, D, Ingraham, J., Petersen, A, and Waters, J. S., "Basketball teams as strategic networks," *PLOS ONE*, 7(11), 2012, e47445, https://doi.org/10.1371/journal.pone.0047445

Fivethirtyyeight.com, "When we say 70% it really means 70%," https://fivethirtyeight.com/features/when-we-say-70-percent-it-really-means-70-percent/

Franks, A., et al., "Counterpoints: Advanced defensive metrics for NBA basketball." In 9th Annual MIT Sloan Sports Analytics Conference, 2015

Franks, A., Miller, A., Bornn, L., and Goldsberry, K. "Characterizing the spatial structure of defensive skill in professional basketball," *The Annals of Applied Statistics*, 9(1), 2015, 94–121

Gilovich, Thomas, Vallone, Robert, and Tversky, Amos, "The hot hand in basketball, *Cognitive Psychology*, 17, 1985, 295–314

Glickman, M., "Glicko ratings," available at http://www.glicko.net/glicko.html

Gohberg, Israel, Kaashoek, Marinus, and Spitkovsky, Ilya, M. "An overview of matrix factorization theory and operator applications." In *Factorization and Integrable Systems*, Birkhäuser, Basel, 2003, 1–102

Goldman, Steve, *Mind Game*, Workman Publishing, 2005

Gould, Stephen, "The Streak of streaks," *The New York Review of Books* (1988), https://www.nybooks.com/articles/1988/08/18/the-streak-of-streaks/

Gyarmati, Laszlo, Kwak, Haewoon, and Rodriguez, Pablo, "Searching for a unique style in soccer." In ACM SIGKDD workshop on Large Scale Sports Analytics, 2014

Hayes, M., "Analytics driven sixers ride the numbers to NBA playoffs," available at https://www.inquirer.com/philly/sports/sixers/sixers-76ers-philadelphia-analytics-process-numbers-nba-playoffs-miami-heat-brett-brown-bryan-colangelo-20180418.html

Horowitz, Maxim, et al., "nflscrapR: An R package to utilize and analyze data from the NFL API," available at: https://www.rdocumentation.org/packages/nflscrapR/versions/1.8.3

Hughes, G., "Who is responsible for the corner three revolution?" available at https://bleacherreport.com/articles/2146753-whos-responsible-for-the-nbas-corner-three-revolution

James, Bill, *The New Bill James Historical Abstract*, The Free Press, 2001

James, Bill, and Henzler, J., *Win shares*, Stats Inc., 2002

Jones, M. A., "Win, lose, or draw: A Markov Chain analysis of overtime in the National Football League," *College Mathematics Journal*, 35(5), 2004, 330–336

Kahneman, D., and Miller, D. T., "Norm theory: Comparing reality to its alternatives," *Psychological Review*, 93, 1986, 136–153

Kalist, David, and Spur, Stephen, "Baseball errors," *Journal of Quantitative Analysis in Sports*, 2(4), 2006

Kelley, J. L., "A new interpretation of the information rate," *Bell System Technical Journal*, 35(4), 1956, 917–926

Keri, Jonah, *Baseball Between the Numbers,* Basic Books, 2006

Kullowatz, M., "Shots: Confusion in correlations," available at https://www.americansocceranalysis.com/home/2014/12/18/shots-confusion-in-correlations

Lahman, Sean, "Baseball archive database," http://www.seanlahman.com/baseball-archive/statistics/

Lawhorn, Adrian, "'3-D': Late-game defensive strategy with a 3-point lead," available at http://www.82games.com/lawhorn.htm

Le, H., Carr, P., Yue, Y., and Lucey, P., "Data-driven ghosting using deep imitation learning." In 11th Annual MIT Sloan Sports Analytics Conference, 2017

Levitt, Steven, "Why are gambling markets organized so differently than other markets?" *Economic Journal*, 114, 2004, 223–246

Levitt, Steven, and Dubner, Stephen, *Freakonomics*, William Morrow, revised edition 2020

Liew, J., "Do football possession statistics indicate which team will win? Not necessarily," available at http://www.telegraph.co.uk/sport/football/competitions/champions-league/10793482/Do-football-possession-statistics-indicate-which-team-will-win-Not-necessarily.html

Lisk, J., "Sacks are a quarterback stat," available at https://www.thebiglead.com/posts/sacks-are-a-quarterback-stat-01dxqapkgvw9

List of Major League Baseball perfect games, Wikipedia, https://en.wikipedia.org/wiki/List_of_Major_League_Baseball_perfect_games

Lopez, M., "All win probability models are wrong—some are useful," available at https://statsbylopez.com/2017/03/08/all-win-probability-models-are-wrong-some-are-useful/

Lopez, M. J., "Bigger data, better questions, and a return to fourth down behavior: an introduction to a special issue on tracking data in the National football League" Journal of Quantitative Analysis in Sports, vol. 16, no. 2, 2020

Lopez, M., "Rethinking draft curves," available at https://statsbylopez.netlify.app/post/rethinking-draft-curve/

Lowe, Z., "Lights, camera, revolution," available at http://grantland.com/features/the-toronto-raptors-sportvu-cameras-nba-analytical-revolution/

Lucey, P., Bialkowski, A., Monfort, M. Carr, P. and Matthews, I., "'Quality vs quantity': Improved shot prediction in soccer using strategic features from spatiotemporal data." In 9th Annual MIT Sloan Sports Analytics Conference, 2015

Mackay, N., "Predicting goal probabilities for possessions in football," Vrije Universiteit Amsterdam, Technical Report, 2017

Mallepalle, S., Yurko, R., Pelechrinis, K., Ventura, S. L., "Extracting NFL tracking data from images to evaluate quarterbacks and pass defenses," Journal of Quantitative Analysis in Sports, 16(2), 2019, 95–120

Massey, C., and Thaler, R. H., "The loser's curse: Decision making and market efficiency in the National Football League draft," Management Science, 59(7), 2013, 1479–1495

Mauboussin, M., The Success Equation: Untangling Skill and Luck in Business, Sports, and Investing, Harvard Business Review Press, 2012

Mease, David, "A penalized maximum likelihood approach for the ranking of college football teams independent of victory margins," The American Statistician, 57, 2003, 241–248

Microsoft, "Calculate multiple results by using a data table," https://support.office.com/en-us/article/calculate-multiple-results-by-using-a-data-table-e95e2487-6ca6-4413-ad12-77542a5ea50b

Miller, A., Bornn, L., Adams, R., and Goldsberry, K., "Factorized point process intensities: A spatial analysis of professional basketball." In International Conference on Machine Learning, 2014, 235–243

Miller, Joshua, and Sanjurjo, Adam. "Surprised by the hot hand fallacy? A truth in the law of small numbers," *Econometrica*, 86(6), 2018, 2019–2047

Mills, Eldon, and Mills, Harlan, *Player Win Averages*, A. S. Barnes, 1970

Morris, B., "When to go for two, for real," available at http://fivethirty eight.com/features/when-to-go-for-2-for-real/

Narizuka, Takuma, Yamamoto, Ken, and Yamazaki, Yoshihiro, "Statistical properties of position-dependent ball-passing networks in football games," *Physica A: Statistical Mechanics and Its Applications*, 412, 2014, 157–168

NBA Draft Study, https://www.reddit.com/r/nba/comments/36wv9m /trying_to_create_an_nba_draft_trade_value_chart/

NFL Next Gen Stats, available at https://operations.nfl.com/the-game /technology/nfl-next-gen-stats/

NFL Punt Analytics Competition at Kaggle, https://www.kaggle.com /c/NFL-Punt-Analytics-Competition

Norman, M. F., and Yellott, J. I., "Probability matching," *Psychometrika*, 31(1), 1966, 43–60

Oliver, Dean, *Basketball on Paper*, Potomac Books, 2004

Page, Lawrence, et al., "The PageRank citation ranking: Bringing order to the web," Stanford InfoLab, 1999

Palaskas, V., Ntzoufras, I., and Drikos, S., "Bayesian modelling of volleyball sets." In Math Sport International Conference, 2019

Palmer, Pete, and Thorne, John, *The Hidden Game of Baseball*. Doubleday, 1985.

Partnow, S., "NBA offensive styles analysis, part 2: Variety is the spice of life," available at https://theathletic.com/1733785/2020/04/10/nba -offensive-styles-analysis-part-ii-variety-is-the-spice-of-life/

Pelechrinis, K., "iWinRNFL: A simple, interpretable & well-calibrated in-game win probability model for NFL," arXiv preprint arXiv:1704.00197, 2017

Pelechrinis, K., "Decision making in American football: Evidence from 7 Years of NFL data." In ECML/PKDD workshops on Machine Learning and Data Mining for Sports Analytics, 2016

Pelechrinis, K., Papalexakis, E., and Faloutsos, C., "SportsNetRank: Network-based sports ream ranking." In ACM SIGKDD workshop on Large Scale Sports Analytics, 2016

Pelechrinis, K., Yurko, R., and Ventura, S., "Reducing concussions in the NFL: A data-driven approach," *CHANCE*, 32(4), 2019, 46–56

Pelechrinis, Kostas, and Winston, Wayne, "The hot hand in the wild," working paper, 2021, available at https://arxiv.org/abs/2006.14609

Pelton, Kevin, "How real plus minus reveals hidden NBA stars," https://www.espn.com/nba/story/_/id/28309836/how-real-plus-minus-reveal-hidden-nba-stars

Peterson, Alexander, Penner, Orion, and Stanley, Eugene, "Methods for detrending success metrics to account for inflationary and deflationary factors," *Physics of Condensed Matter*, 79(1), 2011, 67–78

Pluto, Terry, *Tall Tales*, Simon and Schuster, 1992

Pope, Devin, Price, J., and Wolfers, J., "Awareness reduces racial bias," *Management Science*, 64(11), 2018, 4988–4995

Poundstone, William, *Fortune's Formula*, Hill and Wang, 2005

Power, P., Ruiz, H., Wei, X., and Lucey, P., "Not all passes are created equal: Objectively measuring the risk and reward of passes in soccer from tracking data." In ACM SIGKDD, 2017

Robinson, B., "Grinding the mocks," available at https://benjamin-robinson.github.io

Rogers, C., "Esports encourage skills development in education," available at https://edtechnology.co.uk/comments/esports-encourges-skills-development-in-education/

Romer, David, "Do firms maximize? Evidence from pro football," *Journal of Political Economy*, 114(21), 2006, 340–365

Rosen, P. A., and Wilson, R. L., "An analysis of the defense first strategy in college football overtime games," *Journal of Quantitative Analysis in Sports*, 3(2), 2007

Rudd, S., "A framework for tactical analysis and individual offensive production assessment in soccer using Markov chains," New England Symposium on Statistics in Sports, 2011

Sagarin, Jeff, "Archived *USA Today* sports ratings," http://www.usatoday.com/sports/sagarin-archive.htm

Sagarin, Jeff, "Current *USA Today* sports ratings," http://www.usatoday.com/sports/sagarin.htm

Schuckers, M., "An alternative to the NFL draft pick value chart based upon player performance," *Journal of Quantitative Analysis in Sports*, 7(2), 2011

Schwarz, Alan, "Study of N.B.A. sees racial bias in calling fouls," *New York Times* (May 2, 2007), http://www.nytimes.com/2007/05/02/sports/basketball/02refs.html?ex=1335844800&en=747ca51bedc1548d&ei=5124

Schwarz, Alan, *The Numbers Game*, Stats Inc, 2002

Seidl, T., Cherukumudi, A., Hartnett, A., Carr, P., and Lucey, P., "Bhostguster: Realtime interactive play sketching with synthesized NBA defenses." In MIT Sloan Sports Analytics Conference, 2018

Shlens, J. "A Tutorial on Principle Component Analysis", arXiv:1404 .1100v1 [cs.LG], 2014

Shockley, R. L., "An Applied Course in Real Options Valuation," Cengage Learning, 2007

Sicilia, A., "On the application of convex hull based spatial metrics in the NBA." In Cascadia Symposium on Statistics in Sports, 2018 (poster)

Sicilia, A., Pelechrinis, K., and Goldsberry, K., "DeepHoops: Evaluating micro-actions in basketball using deep feature representations of spatio-temporal data." In the 25th ACM SIGKDD, 2019

Silver, N., *The Signal and the Noise*, Penguin Books, 2015

Silver, Nate, "Description of RAPTOR," https://fivethirtyeight.com /features/how-our-raptor-metric-works/

Silver, N., and Fischer-Baum, R., "How we calculate NBA Elo ratings," available at https://fivethirtyeight.com/features/how-we-calculate-nba -elo-ratings/

Skinner, B., "The price of anarchy in basketball," *Journal of Quantitative Analysis in Sports*, 6(1), 2010

Sobel, M. E., "Asymptotic confidence intervals for indirect effects in structural equation models," *Sociological Methodology*, 13, 1982, 290–312

Spearman, W., "Beyond expected goals." In MIT Sloan Sports Analytics Conference, 2018

Stephen, E., "Bears fans try—and fail—a Cody Parker field goal challenge in snowy Chicago," available at https://www.sbnation.com/nfl/2019 /1/12/18180142/chicago-bears-fans-field-goal-challenge-fails

Stern, Hal, "On the probability of winning a football game," *The American Statistician*, 45, 1991, 179–183

Stoll, Greg, "Win expectancy finder," https://gregstoll.com/~gregstoll /baseball/stats.html#V.0.1.0.1.0.0

Tango, Tom, Lichtman, Mitchell, and Dolphin, Andrew, *The Book*, Create Independent Publishing Platform, 2014

Tangotiger, "True talent levels for sports leagues," available at http:// www.insidethebook.com/ee/index.php/site/comments/true_talent _levels_for_sports_leagues

Taylor, S., "Learning the NFL draft," available at https://seanjtaylor .github.io/learning-the-draft/

Thomas, Andrew C., "Inter-arrival times of goals in ice hockey," *Journal of Quantitative Analysis in Sports*, 3(3), 2007

Thorn, John, and Palmer, Pete, *The Hidden Game of Baseball: A Revolutionary Approach to Baseball and Its Statistics*, University of Chicago Press, 2015

Tversky, Amos, and Kahneman, Daniel, "Belief in the law of small numbers," *Psychological Bulletin,* 76(2), 1971, 105–110

Vollman, Rob, *Stat Shot: The Ultimate Guide to Hockey Analytics*, ECW Press, 2016

Weisheimer, A., and Palmer, T. N., "On the reliability of seasonal climate forecasts," *Journal of the Royal Society Interface*, 11(96), 2014, 20131162

Wikipedia, "Bowl championship series," https://en.wikipedia.org/wiki /Bowl_Championship_Series#History_leading_to_the_creation_and _dissolution_of_the_BCS

Wolfers, Justin, and Price, Joseph, "Racial Discrimination Among NBA Referees," National Bureau of Economic Research, 2007

Yunes, Tallys, "How to build the best fantasy football team," O. R. by the Beach (2015), https://orbythebeach.wordpress.com/2015/09/28/how -to-build-the-best-fantasy-football-team/

Yurko, R., Matano, F., Richardson, L., Granered, N., Pospisil, T., Pelechrinis, K., and Ventura, S., "Going deep: models for continuous-time within-play valuation of game outcomes in American football with tracking data," *Journal of Quantitative Analysis in Sports*, 2020, doi:10.1515/ jqas-2019–0056

Yurko, R., Ventura, S. and Horowitz, M., "nflWAR: A reproducible method for offensive player evaluation in football," *Journal of Quantitative Analysis in Sports*, 15(3), 2019, 163–183

INDEX

The letters t or f following a page number indicate a table or figure on that page.